Lecture Notes in Mathematics

Edited by A. Dold and B. Eckmann

T0211821

817

Lothar Gerritzen
Marius van der Put

Schottky Groups and Mumford Curves

Springer-Verlag
Berlin Heidelberg New York 1980

Authors

Lothar Gerritzen
Ruhr-Universität Bochum, Institut für Mathematik, Gebäude NA 2/33
Postfach 102148
4630 Bochum 1
Federal Republik of Germany

Marius van der Put
University of Groningen, Department of Mathematics, WSN-gebouw
Paddepoel
Groningen
The Netherlands

AMS Subject Classifications (1980): 10D30, 14G20, 14H30, 14H40, 14Kxx, 30G05, 32Gxx, 32K10

ISBN 3-540-10229-9 Springer-Verlag Berlin Heidelberg New York
ISBN 0-387-10229-9 Springer-Verlag New York Heidelberg Berlin

Library of Congress Cataloging in Publication Data. Gerritzen, Lothar, 1941-
Schottky groups and Mumford curves. (Lecture notes in mathematics; 817)
Bibliography: p. Includes index. 1. Curves, Algebraic. 2. Fields, Algebraic.
3. Discontinuous groups. 4. Automorphic forms. 5. Analytic spaces. I. Put, Marius van der, 1941- joint author. II. Title. III. Series: Lecture notes in mathematics (Berlin); 817.
QA3.L28. no. 817. [QA567]. 510s [512'.33] 80-20755

Printing and binding: Beltz Offsetdruck, Hemsbach/Bergstr.
2141/3140-543210

Introduction

The idea of investigating the p-adic version of classical uniformizations of curves is due to John Tate who showed that an elliptic curve over a p-adic field K whose j-invariant has absolute value greater than 1 can be analytically uniformized. While Tate's original paper has never been published there are good accounts of his work available, see [34].

The generalization of the above result of Tate to curves of higher genus has been given by David Mumford in 1972 in a work called "Analytic construction of degenerating curves over complete local rings".

The main result of the paper states that there is a one-to-one correspondence between

a) conjugacy classes of Schottky groups $\Gamma \subset PGL_2(K)$

b) isomorphism classes of curves C over K which are the generic fibers of normal schemes over the valuation ring $\overset{o}{K}$ of K whose closed fiber is a split degenerate curve.

In these Notes we call the curves that Mumford has associated to p-adic Schottky groups Mumford curves. Manin has called them Schottky-Mumford curves in [26].

When Mumford received the Fields medal in 1974 his discovery was praised by Tate when he described the work of Mumford. He said: "Next I want to mention briefly p-adic uniformization. Motivated by the study of the boundary of moduli varieties for curves, i.e. of how nonsingular curves can degenerate, Mumford was led to introduce p-adic Schottky groups, and to show how one can obtain certain p-adic curves of genus $\geqq 2$ transcendentally as the quotient by such groups of the p-adic projective line minus a Cantor set. The corresponding theory for genus 1 was discovered by the author, but the generalization to higher genus was far from obvious. Besides its significance for moduli, Mumford's construction is of interest in itself as a highly nontrivial example of "rigid" p-adic analysis".

While Mumford worked with formal schemes over the valuation ring of K, several authors, stimulated by Mumford's investigation, developed his construction within the framework of analytic spaces over K.

Manin-Drinfeld [27] and Myers [29] introduced the notion of automorphic forms and made clear that the Jacobian variety of a p-adic Schottky curve can be constructed analytically as an analytic torus given by a period matrix which is symmetric and positive definite. A good account of the analytic theory of Schottky curves has been given by Manin in 1974, see [26].

In recent years a number of new results on Mumford curves have been obtained by different authors as D. Goss, F. Herrlich and the authors of these Notes. It is the purpose of this work to give an introduction into the theory of Mumford curves presenting the major results and describing a variety of explicit examples.

We will employ two different approaches to the topic, one of which relies on p-adic function theory methods and the machinery of automorphic forms. The advantage of this approach lies in the fact that it is quite elementary and we have tried to be as much down to earth as possible.

The second approach works with methods stemming from algebraic, formal and affinoid geometry and exploiting the reduction of spaces. Here it is not always possible to entirely avoid more advanced and less accessible constructions.

Discontinuous subgroups Γ of the group $PGL_2(K)$ of fractional linear transformations of the projective line $\mathbb{P}(K)$ where K is a non-archimedean valued complete algebraically closed field are introduced in Chap. I and used throughout the Notes. The condition "discontinuous" for a subgroup Γ means that the closure of every orbit of Γ is compact and Γ has ordinary points. In the non-elementary cases we associate to Γ a canonical tree T on which Γ acts. Using the tree one shows for a finitely generated Γ the existence of a normal subgroup Γ_0 of Γ of finite index which is a Schottky group. All Schottky groups can be constructed from a fundamental domain consisting of the complement of 2g open disks.

In Chap. II automorphic forms relative to a Schottky group Γ with constant factors of automorphy are constructed and determined as products of the basic forms $\Theta(a, b; z)$. This allows to prove that the field of Γ-invariant meromorphic functions on the domain Ω of ordinary points for Γ is an algebraic function field of one variable whose set of places S coincides with the orbit space Ω/Γ.

In the first section of Chap. III some basic material on affinoid algebras, affinoid and analytic spaces is presented. Especially, reductions of analytic spaces are introduced. In the second part the construction of Ω/Γ (Γ Schottky group, Ω the set of its ordinary points) as analytic space is given. It is shown that Ω/Γ is in fact a non-singular complete curve of genus g. One further obtains that Ω/Γ has a split degenerate reduction.

In Chap. IV domains Ω in $\mathbb{P}(K)$ are characterized among the non-singular one dimensional analytic spaces by the property: "Ω has an analytic reduction $\bar{\Omega}$ consisting of genus zero curves with a tree as intersection graph".

A complete non-singular curve X which has a split degenerate reduction (i.e. the reduction consists of genus zero curves and only nodes as singularities) is shown to have a universal covering $\Omega \to X$. The space Ω has a reduction of the type explained above and it follows that Ω is a domain in $\mathbb{P}(K)$. Moreover the automorphism group Γ (which is the fundamental group of X) of the covering $\Omega \to X$ turns out to be a Schottky group with Ω as set of ordinary points. This amounts to Mumford's theorem: A curve X has a split degenerate reduction if and only if X can be parametrized by a Schottky group.

The main result of Chap. V is the existence of an analytic reduction $\bar{X}$ of a complete non-singular curve X (of genus > 1) satisfying: $\bar{X}$ has only nodes as singularities and $\bar{X}$ has a finite group of automorphisms. Such a reduction is called stable and is uniquely determined by X. This result is very close to Deligne-Mumford's result [5] on the existence of stable algebraic reductions. As a corollary one finds: X is a Mumford curve if and only if X has a finite covering by affinoid subsets of $\mathbb{P}(K)$.

In Chap. VI we present an analytic construction of the Jacobian variety $\mathcal{J}(S)$ for a Mumford curve S together with the canonical mapping ϕ of the curve into its Jacobian and show that $\mathcal{J}(S)$ is an analytic torus $(K^*)^g$ modulo a lattice with a polarization defined by a period matrix. The Riemann theta function $\vartheta(u_1, \dots, u_g)$ on the algebraic torus $(K^*)^g$ associated to a square root of the period matrix for Γ is well-defined. We can prove that the divisor of $\vartheta(c \cdot u(z))$, $c \in (K^*)^g$, is of degree g if $\vartheta(c \cdot u(z))$ does not vanish identically on Ω, where $u : \Omega \to (K^*)^g$ is a lift of the canonical mapping $\phi : S \to \mathcal{J}(S)$. As in the complex Riemann vanishing theorem $\vartheta(u_1, \dots, u_g) = 0$ is the equation for a translate of the hypersurface $\phi(S^{g-1}) \subset \mathcal{J}(S)$.

The starting point of the discussion in Chap. VII on automorphisms is the result that the automorphism group Aut S of a Mumford curve $S = S(\Gamma)$ is canonically isomorphic to the factor group N/Γ where N is the normalizer of Γ in $PGL_2(K)$. We describe various results the most striking of which states that the order of Aut S is less than or equal to $12(g-1)$ if the ground field K has characteristic zero and the characteristic of the residue field is different from 2, 3, 5.

In Chap. VIII we consider the curve T associated to a finitely generated discontinuous group N which does contain transformations of finite order and show how one can describe the divisor class group of degree 0 by automorphic forms with respect to N. The genus of T turns out to be the $\mathbb{Z}$-rank of the commutator factor group of N.

In the first part of Chap. IX we show that the group $H(\mathbb{Z}[\frac{1}{p}])^*$ of invertible Hurwitz quaternions with coefficents in $\mathbb{Z}[\frac{1}{p}]$ is a discrete subgroup of $GL_2(k)$ where k is a finite extension of $\mathbb{Q}_p$. Its image Λ in $PGL_2(k)$ is a discontinuous group. The genera of the Mumford curves parametrized by Λ and the congruence subgroup $\Lambda(2)$ are calculated. The geometry of the curves and their stable reductions is made explicit.

In the second part Whittaker groups are considered. They are subgroups of index 2 of groups generated by elliptic transformations of order 2. They parametrize hyperelliptic curves.

In Chap. X we work with the Laurent series field $k = \mathbb{F}_q((\frac{1}{t}))$ and the discontinuous group $\Gamma(1) = PSL(2, \mathbb{F}_q[t])$ which shares many features with the classical modular group $PSL(2, \mathbb{Z})$. The quotient space with respect to $\Gamma(1)$ is the affine line and can be completed by adjoining a parabolic point. The algebra of modular forms for $\Gamma(1)$ is determined.

We like to thank Dr. F. Herrlich for contributing ideas in the preparation of Chap VII and for his help in proof reading. We are grateful to Prof. Dr. S. Bosch who read part of the manuscript and suggested improvements. Also we like to express our gratitude to

Mrs. Marianne Puhlvers for the excellent job of typing the manuscript and her patience with the authors.

Table of Contents

Chapter I Discontinuous groups

Introduction: The field k is supposed to be complete with respect to a non-archimedean valuation. By K we denote a complete and algebraically closed field containing k. We work with the projective line over k as analytic variety. However, in this chapter almost no function theory is needed and it suffices to consider $\mathbb{P}^1(k)$ and $\mathbb{P}^1(K)$ as topological spaces. The projective linear group PGL(2, k) acts in the ususal way on $\mathbb{P} = \mathbb{P}^1(K)$. A subgroup Γ of PGL(2, k) is called discontinuous if the closure of every orbit of Γ in $\mathbb{P}$ is compact and Γ has ordinairy points.

Let Γ be discontinuous and let $\mathcal{L}$ be its set of limit points. As in the complex case, $\mathcal{L}$ is compact, nowhere dense. Further $\mathcal{L}$ is perfect if $\mathcal{L}$ contains more than two points.

Unlike the complex case $\mathbb{P} - \mathcal{L}$ is always connected. Another feature which differs from the classical case is that a parabolic element of infinite order does not generate a discontinuous group. (§1). In §2 one associates to $\mathcal{L}$ (and more generally to a compact set X in $\mathbb{P}$) an infinite tree T. The group Γ acts on this tree and for a finitely generated Γ the quotient T/Γ is a finite graph. This tree is in fact the same tree introduced by D. Mumford [28]. Using the action of Γ on T one shows the following structure theorem (§3): If Γ is finitely generated then Γ has a normal subgroup Γ_0 of finite index such that Γ_0 is a finitely generated free group.

A finitely generated free, discontinuous group is called a Schottky group. Again, using the action on the tree, one shows in §4 that every Schottky group has a nice fundamental domain F:

$F = \mathbb{P}$ - (2g open disks). Let us, for convenience, assume that $\infty \in F$ and call the open disks $B_1, B_2, \ldots, C_1, \ldots, C_g$. Then the disks are in "good position", which means that the corresponding closed disks are disjoint. Moreover Γ has free generators $\gamma_1, \ldots, \gamma_g$ satisfying γ_i maps $\mathbb{P} - B_i$ onto C_i^+, and $\mathbb{P} - B_i^+$ onto C_i.

§1 Groups acting on $\mathbb{P}^1$

(1.1) In what follows k denotes a field which is complete with respect to a non-archimedean valuation $|\;|$. This means that there is given a map $|\;| : k \to \mathbb{R}$ with the properties:

1) $|x| \geq 0$ and $|x| = 0$ if and only if $x = 0$.

2) $|xy| = |x|\,|y|$.

3) $|x + y| \leq \max(|x|, |y|)$.

4) there is an $x \in k$ with $|x| \neq 0, 1$.

5) k is complete with respect to the metric $d(x, y) = |x - y|$.

The most interesting examples are possibly $\mathbb{Q}_p$, the field of p-adic numbers and $\mathbb{F}_p((t))$, the field of Laurent-series in t with coefficients in the finite field $\mathbb{F}_p$.

The field of p-adic numbers is the completion of $\mathbb{Q}$ with respect to the valuation $|\;|_p$ (or metric $d(x, y) = |x - y|_p$) defined by $|p^m \frac{t}{n}| = p^{-m}$ if $m \in \mathbb{Z}$ and $(p, t) = (p, n) = 1$.

The valuation on $\mathbb{F}_p((t))$, which consists of the expressions $\sum_{n >> -\infty} a_n t^n$ and is the quotient field of the formular power series ring $\mathbb{F}_p[\![t]\!]$, is defined by $|\Sigma a_n t^n| = \max\{p^{-n} | a_n \neq 0\}$.

In this section we collect some of the properties of (non-archimedean) valued fields. The valuation ring k^o of k is given by $\{\lambda \in k \,|\, |\lambda| \leq 1\}$. Its unique maximal ideal k^{oo} equals $\{\lambda \in k \,|\, |\lambda| < 1\}$ and $\bar{k} = k^o/k^{oo}$ is called the residue field of k. The value group $|k^*| = \{|\lambda| \,|\, \lambda \in k, \lambda \neq 0\}$ of k is a subgroup of $\mathbb{R}_{>0}$. We say that the valuation of k is discrete if $|k^*| \cong \mathbb{Z}$; in this case k^o is a Noetherian ring. The valuation is called dense if $|k^*|$ is a dense subgroup of $\mathbb{R}_{>0}$;

in this case k^o is not Noetherian. By $\sqrt{|k^*|}$ is meant $\{a \in \mathbb{R}_{>0} |$ for some $n \geq 1$, $a^n \in |k^*|\}$.

For every field extension ℓ of k there is a valuation on ℓ which extends the valuation on k. This extension is unique if ℓ is an algebraic field-extension of k. In particular, the algebraic closure of k has a unique valuation and the completion of this field is again algebraically closed. We denote by $K \supset k$ an algebraically closed field which is complete with respect to a valuation extending the valuation of k. As we have seen, K exists.

Now and then we will work with maximally complete fields k. That is, k has the property that every sequence $B_1 \supset B_2 \supset B_3 \supset \ldots$ of disks (open or closed) in k has a non-empty intersection. This property is equivalent to: for every valued field extension $\ell \supset k$ one has $|\ell^*| \supsetneq |k^*|$ or $\bar{\ell} \supsetneq \bar{k}$.

A field k is called a <u>local field</u> if k is locally compact. This is equivalent to $|k^*| \cong \mathbb{Z}$ and $\bar{k}$ is finite. Every local field is a finite extension of either $\mathbb{Q}_p$ or $F_p((t))$ (and conversely).

Finally we recall that a complete field k has the <u>Hensel-property</u>, i.e.: if $F \in k^o[t]$ is a monic polynomial and if its image $\bar{F} \in \bar{k}[t]$ is a product of two monic polynomials f_1, f_2 with g.c.d. 1, then $F = F_1F_2$ where F_i are monic polynomials with $\bar{F}_i = f_i$ $(i = 1,2)$.

(1.2) The <u>projective line</u> over k is denoted by $\mathbb{P}^1(k)$. As usual, each point p of $\mathbb{P}^1(k)$ represent a line $L \subset k^2$ through $(0,0)$. If $L = \{\lambda(x_0, x_1) | \lambda \in k\}$ then we will write $p = [x_0, x_1]$. The field k is identified with $\mathbb{P}^1(k) - \{[0, 1]\}$ by means of the map $\lambda \to [1,\lambda]$ and the point $[0, 1]$ will be denoted by ∞. This identification leads to writing $z \in k \cup \{\infty\}$ for the elements of $\mathbb{P}^1(k)$.

Let $\sim$ denote the equivalence relation on $k^2 - \{(0, 0)\}$ given by $(x, y) \sim (x', y')$ if $(x, y) = \lambda(x', y')$ for some $\lambda \in k^* (= k - \{0\})$. Then $\mathbb{P}^1(k) = k^2 - \{(0, 0)\}/\sim$ and $\mathbb{P}^1(k)$ inherents a topology from k. Further $\mathbb{P}^1(k)$ is compact if and only if k is locally compact.

We abbreviate in the sequel $\mathbb{P}^1(K)$ by $\mathbb{P}(K)$ or $\mathbb{P}$.

For $\begin{pmatrix} a & b \\ c & d \end{pmatrix} \in GL(2, k)$ = invertible 2×2-matrices over k, we consider the "fractional linear" automorphism ϕ of $\mathbb{P}^1(k)$ given by $z \mapsto \frac{az + b}{cz + d}$, or in the homogeneous coordinates "$[x_0, x_1]$" ϕ is given by $[x_0, x_1] \mapsto [cx_1 + dx_0, ax_1 + bx_0]$. The group of automorphisms of $\mathbb{P}^1(k)$ thus obtained is $PGL(2, k) = GL(2, k)/\{\begin{pmatrix} \lambda & 0 \\ 0 & \lambda \end{pmatrix} | \lambda \in k^*\}$. In more than one aspect (namely algebraically and analytically) those are the only automorphisms of $\mathbb{P}^1(k)$. ALso GL(2, k) and PGL(2, k) inherit in an obvious way a topology from k. They are never compact. But if k is locally compact then GL(2, k) and PGL(2, k) have interesting maximal compact subgroups. We will return to this in §2.

(1.3) Let Γ be a subgroup of PGL(2, k). An element $p \in \mathbb{P}$ is called a limit point of Γ if there exists $q \in \mathbb{P}$ and an infinite sequence $\{\gamma_n | n \geq 1\} \subset \Gamma$ (i.e. $\gamma_n \neq \gamma_m$ if $n \neq m$) with $\lim \gamma_n(q) = p$. If Γ is not discrete in PGL(2, k) then there exists a sequence $\{\gamma_n\}$ in Γ with $\lim \gamma_n = \gamma \in PGL(2, k)$. So $\lim \gamma_n(\gamma^{-1}(p)) = p$ for all $p \in \mathbb{P}$ and every point of $\mathbb{P}$ is a limit point for Γ. Let $\mathcal{L}$ denote the set of all limit points of Γ.

We will call Γ a discontinuous group if

(a) $\mathcal{L} \neq \mathbb{P}$.

(b) $\overline{\Gamma p}$ (= the closure of the orbit of p) is compact for all $p \in \mathbb{P}$.

Condition (b) is superfluous if k is locally compact. In the sequel we will use the following terminology: an element $\gamma \in PGL(2, k)$ represented by a matrix $A = \begin{pmatrix} a & b \\ c & d \end{pmatrix} \in GL(2, k)$ is called elliptic, parabolic or hyperbolic according to the following three cases:

- the eigenvalues of A are different but have the same absolute value -
- the eigenvalues are equal - or
- the eigenvalues have different absolute value.

In general the eigenvalues of A are not in k but in some finite extension of k which carries a unique valuation extending the valuation on k. Obviously the choice of k is unimportant in the definition above. Further if γ is elliptic or parabolic then $\frac{(a+d)^2}{ad-bc}$ has absolute value ≤ 1. If γ is hyperbolic then $\left|\frac{(a+d)^2}{ad-bc}\right| > 1$.

Let $GL(2, k^o)$ denote the 2×2-matrices over k^o with determinant invertible in k^o. Further $PGL(2, k^o)$ will denote the image of $GL(2, k^o)$ in $PGL(2, k)$.

(1.4) Lemma: Let $\gamma \in PGL(2, k)$.

(1) γ is elliptic or parabolic if and only if a conjugate of γ^2 lies in the subgroup $PGL(2, k^o)$ of $PGL(2, k)$.

(2) γ is hyperbolic if and only if γ is conjugated to an element of $PGL(2, k^o)$ represented by a matrix $\begin{pmatrix} q & 0 \\ 0 & 1 \end{pmatrix}$, $q \in k$, $0 < |q| < 1$.

Proof:

(1) If γ^2 (or γ^n for some $n \geq 1$) is conjugated to an element in $PGL(2, k^o)$, then clearly $\left|\frac{(a+d)^2}{ad-bc}\right| \leq 1$ for a representation $\begin{pmatrix} a & b \\ c & d \end{pmatrix}$ of γ. So γ is elliptic or parabolic.

Conversely, suppose that γ is parabolic of elliptic. If γ represented by $B \in GL(2, k)$ and γ^2 by B^2 then there is a $\lambda \in k^*$ such that $\begin{pmatrix} a & b \\ c & d \end{pmatrix} = A = B^2$ satisfies $|ad - bc| = 1$. Then $|a + d| \leq 1$.

Put $N = k^o \oplus k^o \subset k \oplus k$ and $M = N + A(N)$. Then M is a finitely generated k^o-module, invariant under A since $A^2 - (a + d) A + (ad - bc) = 0$. Let $\{e_1, \ldots, e_n\}$ be a minimal set of generators of M as k^o-module. If there exists a non-trivial relation $\lambda_1 e_1 + \ldots + \lambda_n e_n = 0$, $\lambda \in k$, then we may assume $\max |\lambda_i| = 1$ and also that one of λ_i equals 1. This contradicts the minimality of n. So $n = 2$ and $\{e_1, e_2\}$ is a free base of the k^o-module M. Let $C : k^2 \to k^2$ be the k-linear map given by $C(1, 0) = e_1$, $C(0, 1) = e_2$. Then $A \in CGL(2, k^o)C^{-1}$ and γ^2 lies in a conjugate of the subgroup $PGL(2, k^o)$ of $PGL(2, k)$.

(2) Clearly $\begin{pmatrix} q & o \\ o & 1 \end{pmatrix}$, $0 < |q| < 1$, and all its conjugates are hyperbolic. On the other hand, if γ is hyperbolic then γ can be represented by $A = \begin{pmatrix} a & b \\ c & d \end{pmatrix}$ with $a + d = 1$ and $|ad - bc| < 1$. The characteristic polynomial P of A is $X^2 - X + (ad + bc) \in k^o[X]$. The reduction $\bar{P}$ in $\bar{k}[X]$ has two roots $\bar{0}, \bar{1} \in \bar{k}$. Since k has the Hensel property, there are roots λ_o, λ_1 of P in k^o with $\bar{\lambda}_o = \bar{0}$, $\bar{\lambda}_1 = \bar{1}$. So the eigenvalues of A has the required form $\begin{pmatrix} q & o \\ o & 1 \end{pmatrix}$, $q \in k$, $0 < |q| < 1$.

(1.5) Examples: Let Γ be generated by one element $\gamma \in PGL(2, k)$.

(1) If γ is hyperbolic then Γ is discontinuous. The set $\mathcal{L}$ of limit points of Γ consists of the two points of $\mathbb{P}^1(k)$ corresponding to the two eigenvectors of γ. These points are the two fixed points of γ.

(2) If γ is elliptic or parabolic and Γ is discontinuous then γ has finite order. In $PGL(2, K)$ the element γ is either conjugated to $\delta_1 : z \mapsto \lambda z$ and $|\lambda| = 1$, or to $\delta_2 : z \mapsto z + b$. The group generated by δ_1 is discontinuous if and only if $\{\lambda^n | n \in \mathbb{Z}\}$ is compact and 1 is not a limit point of this set. This means that γ should be a root of unity. So δ_1 and γ have finite order. The group generated by δ_2 is discontinuous if and only if $\overline{\{nb | n \in \mathbb{Z}\}}$ is compact and does not have 0 as limit point. That means that $nb = 0$ for some $n \neq 0$. So γ_2 and γ have finite order. We note that γ has two fixed points if γ is elliptic and has one fixed point if γ is parabolic. We have also shown the following result:

(3) If γ is parabolic ($\neq$ id) and if Γ is discontinuous then k has characteristic $p \neq 0$ and γ has order p.

(1.6) A subgroup Γ of $PGL(2, k)$ is called a Schottky group if

(a) Γ is fintely generated.

(b) Γ has no elements ($\neq 1$) of finite order.

(c) Γ is discontinuous.

According to (1.5) condition (b) can be replaced by: every $\gamma \in \Gamma$, $\gamma \neq 1$, is hyperbolic. We start now the investigation of $\mathcal{L}$ = the limit points of a discontinuous group Γ. We may (and will) assume that $\infty \notin \mathcal{L}$.

Proposition: Γ is a discontinuous group and $\infty \notin \mathcal{L}$. Then

(1.6.1) Represent any $\gamma \in \Gamma$ by $\begin{pmatrix} a & b \\ c & d \end{pmatrix} \in GL(2, k)$ (a, b, c, d depending on γ). For any $\delta > 0$ the set $\{\gamma \in \Gamma \mid |c|^2 \leq \delta |ad - bc|\}$ is finite. Moreover Γ is finite or countable.

(1.6.2) For $a \in \mathbb{P}$, $\mathcal{L}(a) \subseteq \overline{\Gamma a}$ denotes the set of $b \in \mathbb{P}$ for which there exists an infinite sequence $\{\gamma_n\} \subset \Gamma$ with $\lim \gamma_n(a) = b$.

Given three different points $a_1, a_2, a_3 \in \mathbb{P}$. Then there exists an i with $\mathcal{L}(a_i) = \mathcal{L}$.

(1.6.3) $\mathcal{L} = \mathcal{L}(\infty) = \overline{\Gamma(\infty)} - \Gamma(\infty)$ is compact. $\mathcal{L}$ has no interior. $\mathcal{L}$ is perfect if $\mathcal{L}$ contains more than two elements.

(1.6.4) Suppose that k is a local field. Then any discrete subgroup Λ of $PGL(2, k)$ is discontinuous and has a set of limit points $\mathcal{L} \subset \mathbb{P}(k)$.

Proof: Any infinite sequence in Γ (or Λ in case (1.6.4)) has a subsequence $\gamma_n = \begin{pmatrix} a_n & b_n \\ c_n & d_n \end{pmatrix}$ such that $\frac{a_n}{c_n}, \frac{b_n}{d_n}, \frac{d_n}{c_n}$ (equal to $\gamma_n(\infty)$, $\gamma_n(0)$, $-\gamma_n^{-1}(\infty)$) are convergent. In case of the group Λ we can change the coordinates such that all limits are $\neq \infty$.

$$\text{Then } \lim \begin{pmatrix} \frac{a_n}{c_n} & \frac{b_n}{c_n} \\ 1 & \frac{d_n}{c_n} \end{pmatrix} = \begin{pmatrix} a & b \\ 1 & d \end{pmatrix}.$$ From the discreteness of Γ (or Λ) it follows that $ad = b$.

For $q \in \mathbb{P}$ we find $\lim \gamma_n(q) = a$ unless $q = -d$ and the sequence $\frac{d_n}{c_n}$ is constant.

(1) If the sequence γ_n satisfies $|c_n|^2 \leq \delta|a_n d_n - b_n c_n|$ then we obtain the contradiction $ad - b = (\frac{a_n}{c_n}\frac{d_n}{c_n} - \frac{b_n}{c_n}) \neq 0$. This proves the first statement of (1.6.1). The second one follows at once.

(2) It follows from our considerations above that $\mathcal{L} = \mathcal{L}(a_1) \cup \mathcal{L}(a_2)$ if $a_1 \neq a_2$. Moreover if $a_1 \notin \mathcal{L}$ then for any infinite sequence γ_n as above we have $\lim \gamma_n(a_1) = a$ since a_1 is unequal to $-d \in \mathcal{L}$.

It follows that $\mathcal{L} = \mathcal{L}(a_1)$. Let now a_1, a_2, a_3 be three different points in $\mathcal{L}$. Then $a_3 \in \mathcal{L}(a_1) \cup \mathcal{L}(a_2)$, say $a_3 \in \mathcal{L}(a_1)$.
Then $\mathcal{L} = \mathcal{L}(a_3) \cup \mathcal{L}(a_1) \subseteq \mathcal{L}(a_1) \cup \mathcal{L}(a_1) = \mathcal{L}(a_1) \subseteq \mathcal{L}$.

(3) Suppose that $\mathcal{L}$ contains at least 3 points. Then $\mathcal{L} = \mathcal{L}(a)$ for some $a \in \mathcal{L}$. Then clearly $\mathcal{L}(a) = \overline{\Gamma a}$. Hence $\mathcal{L}$ is compact and perfect.

Further $\mathcal{L}(\infty) = \overline{\Gamma(\infty)} - \Gamma(\infty) = \mathcal{L}$ since ∞ is the fixed point of only finitely many elements of Γ. So $\mathcal{L}$ has no interior points.

(4) Let $q \in \mathbb{P}$ and let a sequence in $\Lambda(q)$ be given. We have to show the existence of a convergent subsequence with limit in $\mathbb{P}^1(k)$. We take a subsequence $\gamma_n(q)$, with $\{\gamma_n\}$ as above. Then $\lim \gamma_n(q) = a \in \mathbb{P}^1(k)$ or $q = -d$ and $\lim \gamma_n(q) = \infty \in \mathbb{P}^1(k)$. So we have shown that $\overline{\Lambda(q)}$ is compact and $\mathcal{L} \subseteq \mathbb{P}^1(k)$.

(1.7) Examples:

(1) Suppose that the discontinuous group Γ contains no hyperbolic elements, then either (a) Γ is finite.

or (b) the characteristic of k is $p \neq 0$; Γ_0 the subset of parabolic elements is an infinite normal subgroup of Γ isomorphic to a discrete subgroup of k; Γ/Γ_0 is a finite group of roots of unity in k^*. Further $\mathcal{L}$ consists of one point.

<u>Proof:</u> We may suppose that k is algebraically closed. Further we may assume $0 \in \mathcal{L}$ and $\infty \notin \mathcal{L}$. Choose $\gamma \in \Gamma$, an infinite sequence $\{\gamma_n\} \subset \Gamma$ with $\lim \gamma_n(\infty) = 0$, and matrices $\begin{pmatrix} a & b \\ c & d \end{pmatrix}$, $\begin{pmatrix} a_n & b_n \\ c_n & d_n \end{pmatrix}$ with determinant 1 representing γ, and γ_n.

Then $\lim |c_n| = \infty$; $\lim \frac{a_n}{c_n} = 0$. Since $|a_n + d_n| \leq 1$ also $\lim \frac{d_n}{c_n} = 0$.

Since $a_n d_n - b_n c_n = 1$ also $\lim \frac{b_n}{c_n} = 0$.

Further $b = \mathrm{Tr}\, \{\begin{pmatrix} a & b \\ c & d \end{pmatrix} \begin{pmatrix} 0 & 0 \\ 1 & 0 \end{pmatrix}\} = \lim \frac{1}{c_n} \mathrm{Tr}\, \{\begin{pmatrix} a & b \\ c & d \end{pmatrix} \begin{pmatrix} a_n & b_n \\ c_n & d_n \end{pmatrix}\} = 0.$

The last equality holds since $\begin{pmatrix} a & b \\ c & d \end{pmatrix} \begin{pmatrix} a_n & b_n \\ c_n & d_n \end{pmatrix}$ has determinant 1, represents an element of Γ and its trace has absolute value ≤ 1.

So the elements $\gamma \in \Gamma$ are represented by matrices of the form $\begin{pmatrix} a & 0 \\ c & a^{-1} \end{pmatrix}$. Let $\phi : \Gamma \to k^*$ be given by $\phi(\gamma) = a^{-2}$. Then $\Gamma_0 = \ker \phi$ is the subset of all parabolic elements of Γ and Γ/Γ_0 is isomorphic to a group of unity in k^*. Represent any $\gamma \in \Gamma_0$ by $\begin{pmatrix} 1 & 0 \\ c & 1 \end{pmatrix}$.

Then $\{c \in k \mid \begin{pmatrix} 1 & 0 \\ c & 1 \end{pmatrix} \in \Gamma_0\}$ is a discrete additive subgroup of k. The group Γ/Γ_0 acts by conjugation on $\Gamma_0 \subseteq k$. This action is in fact multiplication by $\phi(\gamma)$ and it easily follows that Γ/Γ_0 is finite. So Γ_0 is isomorphic to an infinite discrete subgroup of k and k must have characteristic $p \neq 0$. Further clearly $\mathcal{L} = \{0\}$.

(2) <u>Suppose that the discontinuous group Γ has two limitpoints. Then Γ has a normal subgroup of finite index generated by one hyperbolic element.</u>

<u>Proof:</u> Let $\mathcal{L} = \{0, \infty\}$ and let Γ_0 be the normal subgroup of index 1 or 2 of Γ consisting of the $\gamma \in \Gamma$ for which $\gamma(0) = 0$, $\gamma(\infty) = \infty$.

Then any $\gamma \in \Gamma_0$ has the form $\gamma(z) = az$ with $a \in k^*$. Further $\{a \in k^* \mid (z \mapsto az) \in \Gamma_0\}$ is a discrete subgroup of k^*. Such a subgroup is generated by two elements q, ε, where $0 < |q| < 1$ and ε is a nth -

root of unity ($n \geq 1$). From this the statement follows.

(3) Remark: Example (1.7.1) is dicussed in Ch. X. The example (1.7.2) is the simplest case of the type of discontinuous groups that we find interesting, namely finitely generated discontinuous groups. We will show that later on (§3):

A Schottky group is a free group on g hyperbolic generators.
A finitely generated discontinuous group has a normal subgroup of finite index which is a Schottky group.

Example (1.7.2) is the simple case $g = 1$. For $g > 1$ we have according to (1.6.3) and (1.7.2) that $\mathcal{L}$ is an infinite perfect set.

§2 The tree of a compact subset of $\mathbb{P}$.

(2.1) Graphs and trees: In the sequel we will meet some graphs and trees. We will use the following notations. A graph consists of a set of vertices V and for every pair $a, b \in V$ a set of edges $E(a, b)$. We will always assume that $E(a, b) = E(b, a)$. For an edge $e \in E(a, b)$ the vertices a, b are called the endpoints of e. The set $E(a, b)$ can be non-empty. We will only be interested in locally finite graphs, that is: 1) every $E(a, b)$ is finite and 2) $\{b \in V | E(a, b) \neq \emptyset\}$ is finite for all a. For such a graph we make as usual pictures like

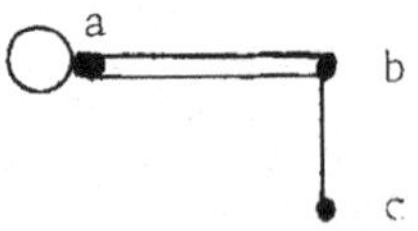

This picture means that $V = \{a, b, c\}$; $E(a, a)$ has one element; $E(a, b)$ has two elements, $E(b,c)$ has one element; the other $E(.,.)$ are empty. Two vertices a, b in G are said to be connected if there is a "path" $a = a_1, a_2, a_3, \ldots, a_n = b$ such that $E(a_i, a_{i+1}) \neq \emptyset$ for $i = 1, \ldots, n - 1$.

A loop in G is a subgraph of G of the form

etc.

An endvertex of G is a vertex which is the endpoint of exactly one edge. A graph G is said to be a tree if G is connected (i.e. every pair of vertices is connected) and G has no loops. Any two points in a tree are connected by a unique path (without trivial, repetitions). The distance between vertices a, b in a tree is d(a, b) = - 1 + the number of vertices in the unique path from a to b.

A halfline in a tree is a subtree of the form  ;

a line in a tree is a subtree of the form  .

(2.2) For a subset X of $\mathbb{P}^1(k)$ we write $X^{(3)}$ for $X \times X \times X - \Delta$ where $\Delta = \{(x_1, x_2, x_3) \in X \times X \times X \mid x_1 = x_2 \text{ or } x_1 = x_3 \text{ or } x_2 = x_3\}$.
For $a = (a_0, a_1, a_\infty) \in X^{(3)}$ we denote by γ_a the unique automorphism of $\mathbb{P}^1(k)$ with $\gamma_a(a_i) = i$ for $i = 0, 1, \infty$.

The <u>standard reduction</u> R: $\mathbb{P}^1(\bar{k})$ is the map given by $[x_0, x_1] \mapsto [\bar{x}_0, \bar{x}_1]$ where $\max(|x_0|, |x_1|) = 1$ is assumed and $\bar{x}_0$, $\bar{x}_1$ denote the residue classes in $\bar{k}$ of x_0 and x_1. Of course R is continuous (where $\mathbb{P}^1(\bar{k})$ has the discrete topology) and surjective. In particular R(X) is finite if X is compact.

For $a \in X^{(3)}$ we consider the reduction $R_a : \mathbb{P}^1(k) \overset{\gamma_a}{\to} \mathbb{P}^1(k) \overset{R}{\to} \mathbb{P}^1(\bar{k})$.
This map is again surjective and continuous. Moreover
$R_a(a_0) = \bar{0}$, $R_a(a_1) = \bar{1}$ and $R_a(a_\infty) = \bar{\infty}$.

For $a, b \in X^{(3)}$ we consider another "reduction" (explanation of this term will be given in III) namely

$R_{a,b} : \mathbb{P}^1(k) \overset{\gamma_a \times \gamma_b}{\to} \mathbb{P}^1(k) \times \mathbb{P}^1(k) \overset{R \times R}{\to} \mathbb{P}^1(\bar{k}) \times \mathbb{P}^1(\bar{k})$.

We will study this map in detail. Let $\gamma_a\gamma_b^{-1} : \mathbb{P}^1(k) \to \mathbb{P}^1(k)$ be given by $\gamma_a\gamma_b^{-1}([y_o, y_1]) = [\gamma y_1 + \delta y_o, \alpha y_1 + \beta y_o]$. On $\mathbb{P}^1(k) \times \mathbb{P}^1(k)$ we use the coordinates $([x_0, x_1], [y_o, y_1])$. One easily sees that the image of $\gamma_a x \gamma_b$ is the subset on $\mathbb{P}'(k) \times \mathbb{P}'(k)$ given by one (quadratic bihomogeneous) equation $F = 0$ where

$$F = -x_0(\alpha y_1 + \beta y_o) + x_1(\gamma y_1 + \delta y_o) \text{ and } \max(|\alpha|, |\beta|, |\gamma|, |\delta|) = 1.$$

The reduction $\bar{F} = -x_0(\bar{\alpha} y_1 + \bar{\beta} y_o) + x_1(\bar{\gamma} y_1 + \bar{\delta} y_o)$ of F is a polynomial over $\bar{k}$. Its zeroset $Z(\bar{k})$ is a subset of $\mathbb{P}^1(\bar{k}) \times \mathbb{P}^1(\bar{k})$ satisfying: im $(R_{a,b}) \subseteq Z(\bar{k})$. We consider two cases.

<u>Case</u> (a) : det $\begin{pmatrix} \bar{\alpha} & \bar{\beta} \\ \bar{\gamma} & \bar{\delta} \end{pmatrix} \neq 0$ (equivalently $\begin{pmatrix} \alpha & \beta \\ \gamma & \delta \end{pmatrix} \in GL(2, k)$ or $\gamma_a\gamma_b^{-1} \in PGL(2, k^o)$).

Let A denote the automorphism of $\mathbb{P}^1(\bar{k})$ given by the matirx $\begin{pmatrix} \bar{\alpha} & \bar{\beta} \\ \bar{\gamma} & \bar{\delta} \end{pmatrix}$.

Then we have a commutative diagramm

$$\begin{array}{ccccc} \mathbb{P}^1(k) & \xrightarrow{\gamma_b} & \mathbb{P}^1(k) & \xrightarrow{\gamma_a\gamma_b^{-1} \times id} & \mathbb{P}^1(k) \times \mathbb{P}^1(k) \\ & & \downarrow R & & \downarrow R \times R \\ & & \mathbb{P}^1(\bar{k}) & \xrightarrow{A \times id} & \mathbb{P}^1(\bar{k}) \times \mathbb{P}^1(\bar{k}). \end{array}$$

This implies that im $(R_{a,b}) = $ im $(A \times id) = Z(\bar{k}) \cong \mathbb{P}^1(\bar{k})$. The points $a, b \in X^{(3)}$ will be called <u>equivalent</u> in this situatuion. So a, and b are equivalent if and only if $\gamma_a\gamma_b^{-1} \in PGL(2, k^o)$. This is independent of the choice of X.

<u>Case</u> (b) : det $\begin{pmatrix} \bar{\alpha} & \bar{\beta} \\ \bar{\gamma} & \bar{\delta} \end{pmatrix} = 0$. Then $\bar{F}$ is reducible and in fact $\bar{F} = (a_0 x_0 + a_1 x_1)(b_o y_o + b_1 y_1)$. So $Z(\bar{k}) = (p \times \mathbb{P}^1(\bar{k})) \cup (\mathbb{P}^1(\bar{k}) \times q) \subset \subset \mathbb{P}^1(\bar{k}) \times \mathbb{P}^1(\bar{k})$.

In other words $Z(\bar{k})$ consists of two lines intersecting in the point (p,q).

In order to calculate im $(R_{a,b})$ we make a linear change of coordinates such that $F = x_0 y_o + u_1 x_0 y_1 + u_2 x_1 y_o + u_3 x_1 y_1$ with all $|u_i| < 1$

$$\bar{F} = x_0 y_o.$$

We have to see which solutions of $\bar{F}$ in $\mathbb{P}^1(\bar{k}) \times \mathbb{P}^1(\bar{k})$ lift to a solution of F in $\mathbb{P}^1(k) \times \mathbb{P}^1(k)$. If $\bar{F}([x_0, x_1], [y_o, y_1]) = 0$ and $x_0 \neq 0$ or $y_o \neq 0$ then clearly this lifts to a solution of F in $\mathbb{P}^1(k) \times \mathbb{P}^1(k)$.

Now the case $x_0 = y_o = 0$. For a lifting $([x_0, x_1], [y_o, y_1])$ of this point we assume $x_1 = y_1 = 1$. So we have to solve the equation $x_0y_o + u_1x_0 + u_2y_o + u_3 = 0$ or $(x_0 + u_2)(y_o + u_1) = u_1u_2 - u_3$ with $x_0, y_o \in k$ and $\max(|x_0|, |y_o|) < 1$.

There is just one case for which the equation has no solution in k, namely if there is no element $\lambda \in k$ with $|u_1u_2 - u_3| < |\lambda| < 1$. That means that the valuation of k is discrete and $(u_1u_2 - u_3)$ generates the maximal ideal of k^o (i.e. $(u_1u_2 - u_3)$ is a prime divisor). Any field extension k' of k with a bigger value group contains a solution of the equation above. In particular $k' = k(\sqrt{u_1u_2 - u_3})$ contains a solution.

(2.3) we now suppose that X is compact. Points a, $b \in X^{(3)}$, a not equivalent to b are called connected if the finite set $R_{a,b}(X)$ does not contain the intersection of the two lines (see (2.1) case (b)).

Further a, b are said to unconnected if the intersection of the two lines does belong to $R_{a,b}(X)$. We picture the image of $R_{a,b}$ and $R_{a,b}(X)$ by lines and dots:

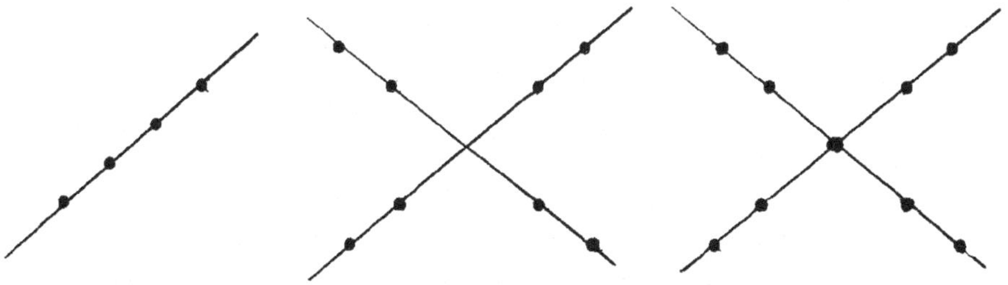

a equivalent with b a connected with b a unconnected with b.

Every $a \in X^{(3)}$ provides a finite partition of X into open comapct subsets, namely $\{R_a^{-1}(p) \mid p \in R_a(X)\}$. Then a and b are equivalent if and only if they give identical partitions of X. Points a and b are connected if and only if their partitions can be written in the form $\{X_1, \ldots, X_s\}$ and $\{X_1', \ldots, X_t'\}$ such that $X_1 = X_2' \cup \ldots \cup X_t'$ and $X_1' = X_2 \cup \ldots \cup X_s$.

The tree of X, denoted by T(X) is defined as follows:
the vertices of T(X) are the equivalence classes [a] of the elements $a \in X^{(3)}$; {[a], [b]} is an edge of T(X) if a and b are connected.

If two points a, $b \in X^{(3)}$ are close together then $\gamma_a \gamma_b^{-1} \in PGL(2, k^o)$ as is easily seen. So the equivalence relation on $X^{(3)}$ is open. Since $X^{(3)}$ is a countable union of compact sets it follows that T(X) is countable or finite. Further T(X) is finite if and only if X is finite. In order to show that T(X) is not only a graph but in fact a locally finite tree we need the following lemma.

(2.4) Lemma: Let $[a] \in T(X)$ and let $R_a(X) = \{p_1, \ldots, p_s\}$ $(s \geq 3)$. If $R_a^{-1}(p_i)$ consists of more than one point then there exists precione $[b_i] \in T(X)$ such that $\{[a], [b_i]\}$ is an edge and $R_{b_i} R_a^{-1}(\{p_1, \ldots, p_{i-1}, p_{i-1}, \ldots, p_s\})$ is one point. In particular the number of edges through [a] is equal to the number of i, $1 \leq i \leq s$, such that $R_a^{-1}(p_i)$ contains more than one point.

Proof: First we remark the following: if R_a and $R_{a'}$ separate both the triple $(x_0, x_1, x_\infty) \in \mathbb{P}^1(k)^{(3)}$ then $a \sim a'$. This follows immediately from the picture on the last page.

Now the lemma. For convenience we put i = 1. Suppose that $c \in X^{(3)}$ is connected with a (above the point p_1). Let X_i denote $R_a^{-1}(p_i)$ and choose $b_o \in X_1$, $b_\infty \in X_2$. Then for some $b_1 \in X_1$ we have $c \sim (b_o, b_1, b_\infty) = b$. For convenience we take $b_o = 0$, $b_\infty = \infty$ and we make the identification $k \cup \{\infty\} \simeq \mathbb{P}^1(k)$. Then X_1 is a compact subset of k.

If $|b_1| < |d|$ for some $d \in X_1$ then $R_b(d) = R_b(\infty)$ which contradicts the assumption that b and a are connected. Hence $|b_1| = \max \{|x_1| \; |x_1 \in X_1\} = \rho$. For any such choice of b_1, X_1 is divided by b into maximal sets of diameter $< \rho$ and $X_2 \cup \ldots \cup X_s$ is mapped into $\infty \in \mathbb{P}^1(\bar{k})$. So we have found a unique element of T(X), connected with a that divides X_1.

The lemma implies that any chain without repetitions in T(X) can be explained in partitions of X as follows:

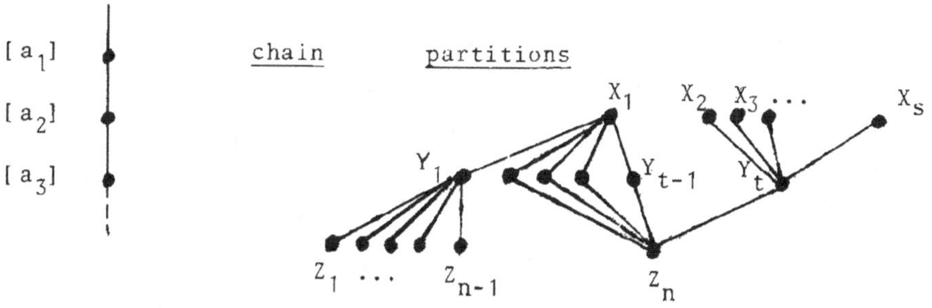

From this it follows that T(X) contains no cycles. We are also interested in halfline's in T(X). For a halfline in T(X) we have a series of partitions of X, denoted by $\{X_1^n, \ldots, X_{s(n)}^n\}$ $(n \geq 1)$ and $X_1^n \supsetneq X_1^{n+1}$ for all $n \geq 1$.

Since X is compact $\bigcap_{n\geq 1} X_1^n$ contains at least one point, say 0, and suppose $\infty \in X_2^1$. Let $\rho_n = \max \{|x| \, | x \in X_1^n$, where again we have used an identitication $k \cup \{\infty\} \simeq \mathbb{P}^1(k)$.

Clearly $X_1^n = \{x \in X_1 | \, |x| \leq \rho_n\}$ and $\lim \rho_n = 0$ since X is compact. It follows that $\bigcap_{n\geq 1} X_1^n$ consists of precisely one point, which is a limit point of X.

Conversely, starting with a limit point z of X and $[a] \in T(X)$ there is a unique halfline $\{[a_n] \, | n \geq 1\}$ defined by $[a_1] = [a]$, $[a_{n+1}]$ is the unique element of T(X) connected with $[a_n]$ above $R_{a_n}(z)$.

So we have a bijection between the halfline's starting at $[a] \in T(X)$ and the limitpoints of X. We use the following terminology: $[a] \in T(X)$ and $z \in X$, then $\{[a], [b]\}$ is the edge in the *direction* of x if $[b]$ corresponds to the point $R_a(z)$.

Next we want to show that T(X) is connected. Given $[a], [b] \in T(X)$. If $\{[a], [b]\}$ is not an edge then there exists $z \in X$ such that $R_{a,b}(z)$ lies on the intersection of the two lines of the reduction. Define $[a_1]$, by $\{[a], [a_1]\}$ is the edge in the direction z. By induction we find a chain $[a], [a_1], [a_2], \ldots$. According to results on the halflines above the chain must be finite.

Summarising we have found:

(2.5) *Proposition:* T(X) *is a locally finite tree and has at most countably many vertices*. *The halflines of* T(X) *(disregarding their endpoints) are in bijective correspondence with the limitpoints of* X.

(2.5.1) *Remarks and examples:* If the residue field $\bar{k}$ is infinite then any locally finite tree T on a finite or countable set of vertices can be realised as a T(X). If $\bar{k}$ is finite then $T \cong T(X)$ if and only if each vertex of T is connected with at most (1 + the number of elements of $\bar{k}$) other vertices.

(2.5.2) k is any complete non-archimedean valued field;
$X = \{q^n \mid n \in \mathbb{Z}\} \cup \{0, \infty\}$, where $q \in k$, $0 < |q| < 1$. Any element in $X^{(3)}$ is equivalent with a unique $(0, q^n, \infty)$. Further $(0, q^n, \infty)$ and $(0, q^n, \infty)$ are connected if and only if $|n - m| = 1$. So T(X) is the line

The two halflines of T(X) correspond with the two limitpoints $0, \infty$ of X.

(2.5.3) $k = \mathbb{Q}_p$, the p-adic numbers and $X = \mathbb{P}^1(\mathbb{Q}_p)$. Then $PGL(2, \mathbb{Q}_p)$ acts transitively on $X^{(3)}$. Then element $[(0, 1, \infty)] \in T(X)$ is connected with precisely $p + 1$ other element of $T(X)$, namely $[0, p^{-1}, \infty)]$, $[(0, p, \infty)]$, $[(1, 1+p, \infty)], \ldots, [(p-1, p-1+p, \infty)]$. The same holds for any vertex of $T(X)$.

(2.6.) <u>Other interpretations of the tree T(X).</u>

(2.6.1) <u>Mumford's tree:</u> The connection with the work of D. Mumford [28] can be given as follows. One considers finitely generated, rank 2, k^o-submodules of $k \oplus k$. Such a module is necessarily free on two generators. Two modules M and M' are said to be equivalent if $M = \lambda M'$ for some $\lambda \in k^*$. The equivalence class of M is denoted by $[M]$. We assume now that k is a local field. Then we define a tree Δ by:

1) the vertices of Δ an the elements $[M]$.

2) $[M_1]$ and $[M_2]$ are connected by an edge if there are $M_1' \sim M_1$ and $M_2' \sim M_2$ such that $M_1' \supset M_2'$ and $M_1'/M_2' \simeq \bar{k}$, the residue field of k. One easily finds that Δ is indeed a tree and that every vertex is the endpoint of exactly $\# \mathbb{P}^1(\bar{k})$ edges.

Let $\pi \in k^o$ generate the maximal ideal and let $[M]$ be a vertex. Then $[M]$ is connected with the $\# \mathbb{P}^1(\bar{k})$ vertices $[N]$ given by $\pi M \subsetneq N \subsetneq M$. The group $PGL(2, k)$ acts in a natural way on the tree.

Any triple $a = (a_0, a_1, a_\infty)$ of different points in $\mathbb{P}^1(k)$ defines a vertex $[M(a)]$ as follows:

Choose $A_o, A_1, A_\infty \in k \oplus k$, representing a_0, a_1, a_∞, and let $\lambda A_o + \lambda_1 A_1 + \lambda_\infty A_\infty = 0$ be a non-trivial relation with $\lambda_o, \lambda_1, \lambda_\infty \in k$. Then $M(a)$ denotes the k^o-module generated by $\lambda_o A_o, \lambda_1 A_1, \lambda_\infty A_\infty$. One easily verifies that $[M(a)]$ does not depend on the choices that are made. Moreover one verfies that $[M(a)] = [M(b)]$ if and only if a and b are equivalent in the sense of (2.2).

Now let X be a compact subset of $\mathbb{P}^1(k)$. The tree T(X) is isomorphic to the subtree of Δ generated by the vertices $\{[M(a)] \mid a \in X^{(3)}\}$. In Mumford's work one takes for X the set $\mathcal{L}_0$ of the fixed points of the hyperbolic elements in a Schottky group. In our set up we will take for X the set of limit points $\mathcal{L}$ of the group. But since $\mathcal{L}$ is the closure of $\mathcal{L}_0$, the trees $T(\mathcal{L}_0)$ and $T(\mathcal{L})$ coincide. So we have shown that our tree for a Schottky group coincides with the one constructed by Mumford.

If the field k is not locally compact, the family {[M]} no longer forms a locally finite tree. however our subfamily defined by a compact subset X gives rise to a locally finite tree T(X).

(2.6.2) Maximal compact subgroups of GL(2, k).

Again, in order to simplify, we assume that the field k is locally compact. The group GL(2, k) acts on $k \oplus k$ in the usual way. For every finitely generated, rank 2, k^o-module M we define $GL(M) = \{\gamma \in GL(2, k) \mid \gamma M = M\}$. All the GL(M) are conjugated to $GL(k^o \oplus k^o) = GL(2, k^o)$ and the GL(M) are precisely the maximal compact subgroups of GL(2, k). Further GL(M) = GL(M') if and only if [M] = [M']. This leads to an identifications of the tree of $\mathbb{P}^1(k)$ with the Tits-building of GL(2, k).

Since PGL(2, k) acts transitively on the tree we obtain also an isomorphism of the tree with $PGL(2, k)/PGL(2, k^o)$.

(2.6.3) Closed disks. (For convenience we work over the field K). A closed disk in $\mathbb{P}$ is a subset given by an inequality $|z - a| \geq \rho$ or $|z - a| \geq \rho$ (with $p \in |K^*|$). This notion is invariant under the action of the group PGL(2, K). For a triple $a = (a_0, a_1, a_\infty)$ we have defined a reduction $R_a : \mathbb{P} \to \mathbb{P}(\bar{K})$. For any $p \in \mathbb{P}(\bar{K})$ the set $R_a^{-1}(\mathbb{P}(R) - \{p\})$ if a closed disk. And in fact any closed disk can be obtained in this way.

So a closed disk defines a reduction. Two closed disks are said to be equivalent if they define equivalent reductions. One easily sees that disks D_1, D_2 are equivalent if and only if there is a $\gamma \in PGL(2, K)$ $\gamma(D_1) = \{z \in K \mid |z| \leq 1\}$ and $\gamma(D_2) = \{z \in \mathbb{P} \mid |z| \geq 1\}$ or $\gamma(D_2) = \gamma(D_1)$. If one takes a fixed choice for the point at infinity, then there is a bijection between reduction $R_a : \mathbb{P} \to \mathbb{P}(\bar{K})$ and closed disks, not containing ∞. This bijection is given by $(R_a) \mapsto R_a^{-1}(\mathbb{P}(\bar{K}) - R_a(\infty))$.

Suppose that ∞ does not lie in the compact set X. The vertices [a] of T(X) correspond then to closed disks B[a] in K. The vertices [a] and [b] are connected by an edge if and only if

1) $B[a] \subsetneq B[b]$ or $B[b] \subsetneq B[a]$

2) there are no closed disks B[c] properly between B[a] and B[b].

(2.6.4) In Chap. III and Chap. IV we will construct a "residue space over $\bar{k}$" corresponding to the open subset $\Omega = \mathbb{P} - \tilde{X}$, where $\tilde{X}$ is the set of limit points of X. The residue space over $\bar{k}$ consists of components ($\simeq$ to a Zariski-open subset of $\mathbb{P}^1(\bar{K})$); the intersectionsgraph of those components turns out to be isomorphic to the tree T(X).

§3 Structure theorem for discontinuous groups

(3.1) Theorem:

(1) Let Γ be a finitely generated discontinuous group. Then Γ has a normal subgroup Γ_0 of finite index, which is a Schottky group.

(2) Any Schottky group Γ_0 is a free group (non-abelian if the number of generators is > 1).

Proof: Using (1.7) we will disregard in the sequel the case where Γ has ≤ 2 limitpoints. Let $X \subset \mathbb{P}^1(k)$ be a compact Γ-invariant set. Then clearly $X \supset \mathcal{L}$ = the set of limit points of Γ by (1.6). The group Γ also acts in an obvious way on $X^{(3)}$ and preserves equivalence and connectedness of points in $X^{(3)}$. So Γ acts on the tree $T(X)$. The theorem will follow from a study of the action of Γ on $T(X)$.

(3.2) *Lemma: Let* Γ *be a discontinuous group*, X *an invariant compact set.*

(3.2.1) *The stabilizer of a vertex or an edge in* $T(X)$ *is a finite subgroup of* Γ.

(3.2.2) *Assume that* $\mathcal{L}$ *is the set of limit points of* X. *Then* $T(X)/\Gamma$ *is finite if and only if* Γ *is finitely generated*.

Proof: (1) Let $G \subset \Gamma$ be the stabilizer of $[a] \in T(X)$. For any $g \in G$ there exists a unique $\bar{g} \in PGL(2, \bar{k})$ such that the diagram is commutative.

$$\begin{array}{ccc} \mathbb{P}^1(k) & \xrightarrow{\quad g \quad} & \mathbb{P}^1(k) \\ \Big\downarrow R_a & & \Big\downarrow R_a \\ \mathbb{P}^1(\bar{k}) & \xrightarrow{\quad \bar{g} \quad} & \mathbb{P}^1(\bar{k}) \end{array}$$

The resulting homomorphism $G \to PGL(2, \bar{k})$ has a finite image G, since any $\bar{g} \in \bar{G}$ is determined by its permutation on the finite set $R_a(X)$.

The kernel H of this homomorphism consists of the $h \in \Gamma$ which leave every $R_a^{-1}(p)$, $p \in \mathbb{P}^1(\bar{k})$, invariant. Take $p \notin R_a(X)$, if possible, then H must be finite since $\mathcal{L} \cap R_a^{-1}(p) = \emptyset$. If $R_a(X)$ happens to be $\mathbb{P}^1(\bar{k})$ then we have to use a more complicated argument. If H were infinite then H has a limit point in each $R_a^{-1}(p)$.

So H has more than 2 limitpoints. On the other hand H lies in a conjugate of the subgroup $PGL(2, k^o)$ of $PGL(2, k)$. So H contains no hyperbolic elements. This contradicts (1.7).

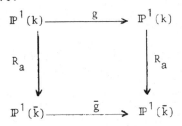

(2) Suppose that Γ is finitely generated. Let Δ be a finite subset of Γ such that $\Delta = \Delta^{-1}$ and $1 \in \Delta$, and Δ generates Γ. Choose $p \in T(X)$ and a finite subtree V of T(X) containing $\Delta(p)$. As one easily verifies $W = \bigcup_{\gamma\in\Gamma} \gamma V$ is again a subtree of T(X). For convience we introduce the following terminology: let q be a vertex of a tree T (T infinite and locally finite). Then $T - \{q\} = T_1 \cup \ldots \cup T_s$ which is a disjoint union of subtrees of T. The <u>finite side</u> of q is $\cup \{T_i | T_i \text{ finite}\}$. The point q is called <u>one-sided</u> if there is just one i for which T_i is infinite. In our situation above we may assume that V contains for every $v \in V$ also the finite side of V. Clearly W must then also have this property. We now want to show W = T(X) and consequently $T(X)/\Gamma$ is finite since V is finite.

We have assumed that $\mathcal{L}$ contains more than 2 points. As a consequence any one-sided $q \in T(X)$ lies in the finite side of some more-sided element of T(X). So it suffices to show that any more-sided (i.e. not one-sided) vertex q of T(X) belongs to W.

Since q is more-sided there exists a halfline L (without repetitions) starting at p through q. This halfline corresponds with a limit point z of the set X. By assumption $z \in \mathcal{L}$. Let z_o be an ordinary point for the group Γ (i.e. $z_o \notin \mathcal{L}$) and take a sequence of elements $\gamma_1, \gamma_2, \gamma_3, \ldots$ in Γ with $\lim \gamma_n(z_o) = z$. ($\{\gamma_n\}$ exists according to (1.6)). Then all the segments $[p, \gamma_1(p)]$, $[\gamma_1(p), \gamma_2(p)]$, $[\gamma_2(p), \gamma_3(p)]$, ... belong to W. ($[q_1, q_2]$ = all points of T(X) on the unique path without repetitions from q_1 to q_2). The path without repetitions obtained from those segments is the unique line L starting from p in the direction z. Hence $q \in W$.

On the other hand, suppose that $T(X)/\Gamma$ is finite. Choose a point $p \in T(X)$ and let $V_m = \{v \in T(X) | d(p, v) \leq m\}$. Here d denotes the obvious distance function on T(X) given by $d(q_1, q_2) + 1$ = the number of elements of the segment $[q_1, q_2]$. Each V_m is a finite subtree of T(X) and $\cup V_m = T(X)$.

So V_m maps surjectively to $T(X)/\Gamma$ if $m \geq n$ (for some n).

Put $\Delta = \{\gamma \in \Gamma | \gamma V_n \cap V_{n+1} \neq \emptyset\}$. Then Δ is finite and generates a subgroups Γ' of Γ. We will show by induction on $m \geq 1$ that there exists for any $q \in V_{n+m}$ a $\gamma \in \Gamma'$ with $\gamma(q) \in V_n$.

Let $m = 1$ then $\gamma(q) \in V_n$ holds for some $\gamma \in \Gamma$; but by definition $\gamma^{-1} \in \Delta \subset \Gamma'$. Now let $q \in V_{n+m}$, then q is connected with a $q' \in V_{n+m-1}$ for some $\gamma_1 \in \Gamma'$. Then $\gamma_1(q) \in V_{n+1}$ and for some $\gamma_2 \in \Gamma'$ we have $\gamma_2\gamma_1(q) \in V_n$.

It follows that $\bigcup_{\gamma \in \Gamma'} \gamma V_n = T(X)$. Let $\gamma \in \Gamma$, there exists $\gamma_1 \in \Gamma'$ such that $\gamma_1\gamma(p) \in V_n$. By definition $\gamma_1\gamma \in \Gamma'$ and so $\gamma \in \Gamma'$. Hence $\Gamma = \Gamma'$ and Γ finitely generated.

(3.3) Lemma:

(3.3.1) If $\gamma \in \Gamma$ has finite order then γ fixes a vertex or an edge of $T(X)$.

(3.3.2) Let Γ be finitely generated. Then the elements of finite order of Γ belong to finitely many conjugacy classes.

Proof: (1) As in the proof of (3.2.2) we use the distance function d on $T = T(X)$ given by $1 + d(q_1, q_2)$ = the number of vertices in the segment $[q_1, q_2]$.

Let $q \in T$ with $q \neq \gamma q$ be given. Consider the segments $V_o = [q, \gamma q]$, $V_1 = [\gamma q, \gamma^2 q], \ldots, V_{n-1} = [\gamma^{n-1} q, q]$ where n = the order of γ. The path $V_o, V_1, \ldots, V_{n-1}$ in T must be trivial since T in a tree. Hence There is some $i \leq n - 1$ such that the intersection $V_{i-1} \cap V_i$ contains more than one point. Then also $V_o \cap V_1 = (\gamma^{i-1})^{-1}(V_{i-1} \cap V_i)$ contains more then one point. If $V_o \cap V_1 = \{q, \gamma(q)\}$ then using again that T has no loops one finds $d(q, \gamma(q)) = 1$ and $\gamma^2(q) = q$. This means that γ fixes the edge $\{q, \gamma(q)\}$ of T.

If $V_o \cap V_1 \neq \{q, \gamma(q)\}$ then there exists a $a \in [q, \gamma(q)]$ with $p \neq q, \gamma(q)$ and $\gamma(p) \in [q, \gamma(q)]$, $\gamma(p) \neq q, \gamma(q)$. This means that $d(p, \gamma(p)) < d(q, \gamma(q))$. So by induction on the number $d(q, \gamma(q))$ it follows that γ fixes a vertex or an edge of T.

(2) Let $\{t_1, \ldots, t_s\}$ be a set of representatives for the edges of T/Γ. Let $S \subset \Gamma$ be the set elements that stabilize some t_i or α_j. Then S is finite. Let $\gamma \in \Gamma$ be of finite order. Then γ fixes a vertex or an edge h of T. For a suitable $\gamma_1 \in \Gamma$. $\gamma_1(h) \in \{t_1, \ldots, t_s, \alpha_1, \ldots, \alpha_t\}$ and $\gamma_1\gamma\gamma_1^{-1}$ fixes an element of $t_1, \ldots, t_s, \alpha_1, \ldots, \alpha_t$. So $\gamma_1\gamma\gamma_1^{-1} \in S$.

(3.4) Proof: of (3.1.1)
Since Γ is finitely generated there is a subring $R \subset k$, finitely generated over the prime ring of k (i.e. over Z or $\mathbb{Z}/p\mathbb{Z}$) such that $\Gamma \subset PGL(2, R) \subset PGL(2, k)$.

Let $S \in \Gamma$ be a finite set of elements of finite order such that every $\gamma \in \Gamma$ of finite order is conjugated to some elements of S. For a maximal ideal $\underline{m}$ of R the field $R/\underline{m}$ is a finite field and we consider the map $\phi : PGL(2, R) \to PGL(2, R/\underline{m})$. Since S is finite we can find $\underline{m}$ such that $s \in S$, $s \neq 1$, then $\phi(s) \neq 1$.

Let $\Gamma_0 = \{\gamma \in \Gamma | \phi(\gamma) = 1\}$. Then Γ_0 is a normal subgroup of Γ with a finite index since $PGL(2, R/\underline{m})$ is a finite group. If $t \in \Gamma_0$ has finite order, then $s = \gamma t \gamma^{-1} \in \gamma_o$ for suitable $\gamma \in \Gamma$. Since $\Gamma_0 \cap S = \{1\}$ by construction it follows that Γ_0 contains no elements ($\neq 1$) of finite order.

Finally $T(X)/\Gamma_0$ is finite since Γ_0 has finite index in Γ. Hence by (3.2.2) (in this part no condition on X is used) the group Γ_0 is finitely generated. Hence Γ_0 is a Schottky group.

(3.5) Proof: of (3.1.2).
A Schottky group Γ_0 acts freely on the tree T(X). The quotientgraph $T(X)/\Gamma_0$ is finite (for a good choice of X). So T(X) is the universal

covering of the graph $T(X)/\Gamma_0$ and Γ_0 is isomorphic to the fundamental group of the graph $T(X)/\Gamma_0$. The fundamental group of a finite graph is a free finitely generated group (this follows easily from van Kampen's theorem on fundamental group).

(3.6) Remarks: The reasoning of (3.4) is very close to the proof of a lemma of Selberg ([36], lemma 8 on p. 154). It is also possible to avoid the use of 2×2-matrices at this point. One can replace (3.4) by a proof using only the trees and (3.2) (see [37], Prop 11, on p. 160).

(3.7) Examples (1)

Suppose that $\mathcal{L}$ has two points. By (1.7.2) we know that the essential part of Γ is a normal subgroup Γ_0 generated by one element γ, represented by $\begin{pmatrix} q & o \\ o & 1 \end{pmatrix}$ with $0 < |q| < 1$. Take $X = \{q^n | n \in \mathbb{Z}\} \cup \{0,\infty\}$. As in (2.5.2) $T(X)$ is a line ...  ...

and γ acts as a translation over 1. The quotientgraph is 

and has fundamental group $\mathbb{Z}$.

(2) Suppose that $\mathcal{L}$ has one point.

By (1.7.1) the essential part of Γ is $\Gamma_0 = \{z \mapsto z + a | a \in A\}$ where A is a discrete additive subgroup of k and k has characteristic $p \neq 0$. Let $X = A \cup \{\infty\}$. This is a compact Γ_0-invariant set.

Let $0 < \rho_1 < \rho_2 < \rho_3 < \ldots$ denote the set of absolute values of A. Then $\lim p_n = \infty$ and each subgroup $A_n = \{a \in A | |a| \leq \rho_n\}$ of A is finite. Let a_n denote an element of A with $|a_n| = \rho_n$. Then each triple in $X^{(3)}$ is equivalent to $(b, a_n + b, \infty)$ for some $b \in A$. Further $(a, a_n + b, \infty)$ if and only if $b - b' \in A_n$. So we can indentify $T(X)$ with the disjoint union of the group A/A_n $(n \geq 1)$.

Let $\sigma_n : A/A_n \to A/A_{n+1}$ denote the obvious grouphomomorphism. Then $p, q \in T(X)$ with $p \neq q$ form an edge if and only if for some m, $\sigma_m(p) = q$ or $\sigma_m(q) = p$.

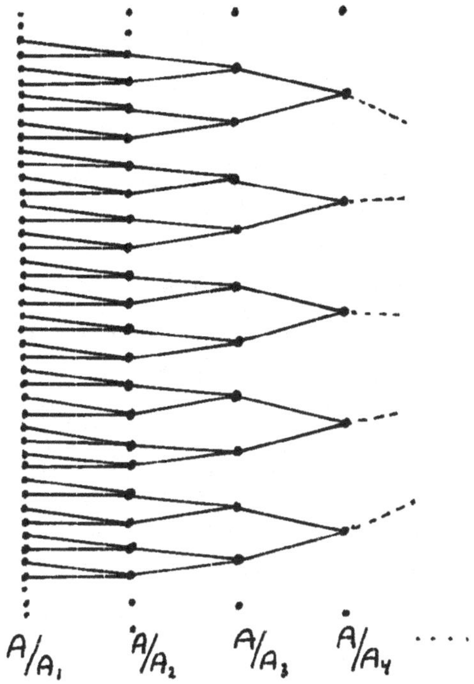

The action of Γ_0 on this tree is the following: let $\gamma \in \Gamma_0$ be given by $z \mapsto z + a$. Then γ acting on $T(X)$ leaves every A/A_n invariant and acts on A/A_n as $(b + A_n) \mapsto (a + b) + A_n$. It follows that the graph of $T(X)/\Gamma_0$ is the halfline:

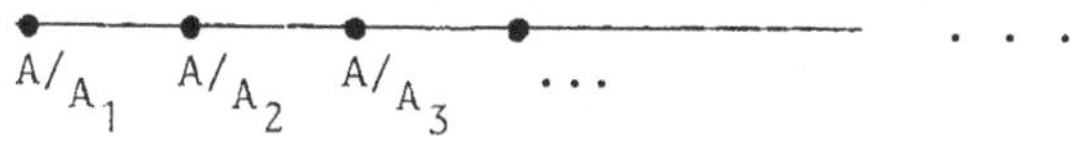

§4 Fundamental domains for a Schottky group

(4.1) Construction of Schottky groups:

Sometimes the field k is too small for our purpose. For instance, if the field k has a discrete valuation then the sets $\{z \in k \mid |z| < \rho_1\}$ and $\{z \in k \mid |z| < \rho_2\}$ can be equal for $\rho_1 \neq \rho_2$.

The best way to overcome this problem is to use affinoid algebras and k-analytic spaces. In chapter III an introduction to k-analytic spaces will be given. However for the modest needs of this section a simple device works equally well. Namely k is embedded in a field K which is algebraically closed and complete. Further $\mathbb{P}^1(k)$ is seen as the set of "k-rational points" of $\mathbb{P}^1(K)$. As usual $K \cup \{\infty\}$ is identified with $\mathbb{P}^1(K)$ and the elements of this set are denoted by z.

By an <u>open disk</u> we mean $B(a, \rho^-) = \{z \mid |z - a| < \rho\}$ with $a \in K$, $\rho \in |K^*|$. By a <u>closed disk</u> we mean $B(a, \rho^+) = \{z \mid |z - a| \leq \rho\}$ with $a \in K$, $\rho \in |K^*|$. Both $B(a, \rho^-)$ and $B(a, \rho^+)$ are in a topological sense open and closed. Since the valuation of K is not discrete, a $B(a, \rho^-)$ can never be equal to a $B(b, \alpha^+)$ (with $a, b \in K$; $\rho, \alpha \in |K^*|$). So the definition "open disk" and "closed disk" make sense.

<u>Lemma</u> (4.1.2)

<u>Let</u> $B = B(a, \rho_1^-)$, $C = B(b, \rho_2^-)$, <u>be two open disks with</u> $a, b \in k$. <u>Suppose that the corresponding closed disks</u> B^+, C^+ <u>are disjoint</u>. <u>Then the following properties are equivalent:</u>

(i) <u>There exists</u> $\gamma \in PGL(2, k)$ <u>with</u> $\gamma(\mathbb{P}^1(k) - B) = C^+$ <u>and</u> $\gamma(\mathbb{P}^1(k) - B^+) = C$.

(ii) $\rho_1\rho_2 \in |k^*|$.

<u>Proof:</u> Using the k-rational transformation $\delta : z \mapsto \frac{z - a}{z - b}$, the sets B, B^+, C, C^+ are transformed into

$B_1 = \{z \mid |z| < \lambda_1\}$

$B_1^+ = \{z \mid |z| \leq \lambda_1\}$

$C_1 = \{z \mid |z| > \lambda_2^{-1}\}$ where $\lambda_i = \frac{\rho_i}{|a - b|}$ for $i = 1,2$.

$C_1^+ = \{z \mid |z| \geq \lambda_2^{-1}\}$

Let (ii) be given then for some $q \in K^*$ we have $|q|^{-1} = \lambda_1\lambda_2$. Define $\gamma \in PGL(2, k)$ by $\gamma(z) = qz$. This clearly has the properties $\gamma(\mathbb{P}^1(K) - B_1) = C_1^+$ and $\gamma(\mathbb{P}^1(K) - B_1^+) =$ and C_1. Then $\delta\gamma\delta^{-1}$ has the properties required in (i).

Now assume that (i) is satisfies then $\gamma_1 = \delta\gamma\delta^{-1} \in PGL(2, k)$ has similar properties for B_1, B_1^+ etc. It follows that γ_1 can be represented by $\gamma_1(z) = v\,\frac{z - n}{z - p}$ with $v \in k$ and $n \in k$, $p \in \mathbb{P}^1(k)$, $|n| < \lambda_1 < |p|$. (If $p = \infty$ this expression means $v(z - n)$)). Let $z \in K$ satisfy $|z| = \lambda_1$ then we must have $|\gamma_1(z)| = \lambda_2$. This implies $\lambda_1\lambda_2 \in |k^*|$ and also $\rho_1\rho_2 \in |k^*|$.

(4.1.3) After this lemma we can start with a geometric constructing that leads to Schottky groups. Let be given 2g open disks $B_1, C_1, \ldots, B_g, C_g$ with centers in k and such that:

(a) (radius of B_i) $\cdot$ (radius of C_i) $\in |k^*|$ for $i = 1, \ldots, g$.

(b) the corresponding closed disks $B_1^+, C_1^+, \ldots, C_g^+$ are disjoint.

Make a choice of $\gamma_1, \ldots, \gamma_g \in PGL(2, k)$ such that $\gamma_i(\mathbb{P}^1(K) - B_i) = C_i^+$ and $\gamma_i(\mathbb{P}^1(K) - B_i^+) = C_i$ for $i = 1, \ldots, g$.

Let Γ be the subgroup of $PGL(2, k)$ generated by $\gamma_1, \ldots, \gamma_g$.

Let F and F^o denote the sets: $\mathbb{P}^1(K) - (\cup B_i \cup \cup C_i)$ and $\mathbb{P}^1(K) - (\cup B_i^+ \cup \cup C_i^+)$.

Proposition:

a) Γ is a Schottky group with $\{\gamma_1, \dots, \gamma_g\}$ as free generators

b) $\bigcup_{\gamma \in \Gamma} \gamma F = \mathbb{P}^1(K) - \mathcal{L}$, where $\mathcal{L}$ is the set of limit points of Γ.

c) $\gamma F \cap F \neq \emptyset$ if and only if $\gamma \in \{1, \gamma_1, \dots, \gamma_g, \gamma_1^{-1}, \dots, \gamma_g^{-1}\}$.

d) $\gamma F^o \cap F = \emptyset$ if $\gamma \neq 1$.

Remark: The complex analogue of this proposition is in fact Schottky's original definition of those groups.

Definition: Let Γ be a Schottky group. A set F is called a good fundamental domain for Γ if

a) $F = \mathbb{P}^1(K) - (\bigcup_{i=1}^{g} B_i \cup \bigcup_{i=1}^{g} C_i)$ where $B_1, C_1, \dots, B_g, C_g$ are $2g$ open disks with centers in k.

b) the closed disks $B_1^+, C_1^+, \dots, C_g^+$ are disjoint.

c) Γ is generated by elements $\gamma_1, \dots, \gamma_g$ that satisfy $\gamma_i(\mathbb{P}^1(K) - B_i) = C_i^+$ and $\gamma_i(\mathbb{P}^1(K) - B_1^+) = C_i (i = 1, \dots, g)$.

Proof of the proposition:

A reduces word w in $\delta_1, \dots, \delta_g$ is an expression $w = \delta_s \delta_{s-1}, \dots, \delta_1$ with all $\delta_i \in \{\gamma_1, \dots, \gamma_g, \gamma_1^{-1}, \dots, \gamma_g^{-1}\}$ and such that no succession of γ_j and γ_{j-1} occurs. With induction on s one verifies:
"$w = \gamma_s \gamma_{s-1}, \dots, \gamma_1$ is reduced and $\gamma_s = \gamma_i^{-1}$ or γ_i then $w(F^o) \subseteq B_i$ or C_i."

At one it follows that Γ is a free group on $\gamma_1, \dots, \gamma_g$ and that $\gamma F^o \cap F = \emptyset$ if $\gamma \neq 1$.

Further let $s \geq 2$, then again by induction one easily sees that $w(F) \cap F = \emptyset$. So we have shown the statements c) and d).

From c) it follows that $F \cap \mathcal{L} = \emptyset$ and consequently $\mathcal{L} \subset \mathbb{P}^1(K) - \bigcup_{\gamma} \gamma F$.

In order to show equality and to show that Γ is discontinuous, we study the set $\mathbb{P}^1(K) - \underset{\gamma}{\cup}\gamma F$.

Every $\gamma \in \Gamma$ can uniquely be written as a reduced word; the length of this expression will be denoted by $\ell(\gamma)$. For every reduced word $w = \delta_s, \ldots, \delta_1$ we define an open disk $B(w)$ as follows:
$B(w) = w(\mathbb{P} - B_i^+)$ if $\delta_1 = \gamma_i$ and $B(w) = w(\mathbb{P} - C_i^+)$ if $\delta_1 = \gamma_i^+$.
One easily verifies the following statements:

1) $B(w) \subset B(w')$ if and only if $w = w't$ and $\ell(w) = \ell(w') + \ell(t)$.

2) $w(\infty) \in B(w)$.

3) $F = \mathbb{P} - \underset{\ell(w)=1}{\cup} B(w)$ and $\mathbb{P} - \underset{\ell(\gamma)<n}{\cup} \gamma F = \underset{\ell(w)=n}{\cup} B(w)$.

We want to show that radii $r(w)$ of the open disks $B(w)$ are going to zero if $\ell(w) \to \infty$.

In general, if $B(m, r^-)$ is an open disk and $\gamma \in PGL(2, k)$ has the property $\infty \notin \gamma(B(m, r^-))$, then $\gamma(B(m, r^-)) = B(\gamma(m), |\frac{d\gamma}{dz}(m)|\, r^-)$.
If $B(w) \subsetneqq B(w')$ then we can write $w' = \delta_s \ldots \delta_1$ with $s = \ell(w')$ and $w = \delta_s \ldots \delta_1 t$ with $\ell(w) = s + \ell(t)$.

$B(w) = \delta_s \ldots \delta_2 B(\delta_1 t) \subsetneqq (w') = \delta_s \ldots \delta_2 B(\delta_1)$ and it follows that $\frac{r(w)}{r(w_1)} = \frac{r(\delta_1 t).}{r(\delta_1)}$. Let $\rho < 1$ denote the maximum of $\frac{r(\delta_1 t)}{r(\delta_1)}$ taken over all possiblities for δ_1 and t. Then $r(w) \leq \rho r(w')$. By induction on finds $r(w) \leq \rho^{\ell(w)}c$ for some constant c. So the radii are going to zero and $\mathbb{P} - \cup\gamma F$ is a compact set containing $\mathcal{L}$. Any point $z \in \mathbb{P} - \cup\gamma F$ is the intersection of a sequence of open disks $B(w_1) \supset B(w_2) \supset B(w_3) \supset \ldots$ where $\ell(w_n) = n$. It follows that $z = \lim w_n(\infty)$. So we have shown that $\mathcal{L} = \mathbb{P} - \underset{\gamma\in\Gamma}{\cup}\gamma F$.

Finally we have to show that Γ is discontinuous. That means that we have to show "$\overline{\Gamma p}$ is compact for all $p \in \mathbb{P}$".

The set $\Gamma p \cap F$ is finite (in fact has ≤ 2 elements). Hence $\Gamma p \cap \bigcup_{\ell(\gamma)\leq n} \gamma F$ is finite for every n. So $\overline{\Gamma p} \subseteq \mathcal{L} \cup \Gamma p$ and is compact.

(4.1.4) Remarks: The construction of (4.1.3) gives in fact every Schottky group in PGL(2, k). That means every Schottky group has a good fundamental domain. A direct proof of this statement can be found in [8]. The method used there is "Ford's method of isometric circles" (see [25] p 114 and on). We will explain this method applied to the non-archimedean case.

Let Γ be a Schottky group in PGL(2, k) such that ∞ is not a limit point. Let $\lambda : \Gamma \to K^*$ be a grouphomomorphism. We define functions w_γ on $\mathbb{P}$ as follows: $w_\gamma = 1$ if $\gamma = 1$ and

$$w_\gamma(z) = \lambda(\gamma) \frac{ad - bc}{(cd + d)^2} = \lambda(\gamma) \frac{d}{dz} \gamma(z) \text{ where } \gamma \text{ is represented by } \begin{pmatrix} a & b \\ c & d \end{pmatrix}.$$

The chain rule for differentiation yields the (1-cocycle-) relation:

$w_{\alpha\beta}(z) = w_\alpha(\beta z) w_\beta(z)$. For $\gamma \in \Gamma$, $\gamma \neq 1$, we consider the disks $B_\gamma = \{z \in k \mid |w_\gamma(z)| > 1\}$ and $B_\gamma^+ = \{z \in k \mid |w_\gamma(z)| \geq 1\}$. The statement in [8] is the following:

<u>There is a grouphomomorphism</u> λ <u>and a generating subset</u> $\{\gamma_1, \ldots, \gamma_g\}$ <u>of</u> Γ <u>such that</u> $F = \mathbb{P}^1(K) - (\bigcup_{i=1}^{g} B_{\gamma_i} \cup \bigcup_{i=1}^{g} B_{\gamma_i^{-1}})$ <u>is a good fundamental for</u> Γ.

We note that for any choice of λ and $\gamma \neq 1$ we have $\gamma(\mathbb{P}^1(K) - B_\gamma) = B^+_{\gamma^{-1}}$ and $(\mathbb{P}^1(K) - B^+_\gamma) = B_{\gamma^{-1}}$.

We will give another proof of: <u>every Schottky group has a good fundamental domain</u>. Our proof will be based on the action of Γ on the tree T(X).

(4.2.) Before we start the actual construction of a fundamental domain we generalize the construction of a reduction $R_{a,b}$ given in (2.2). Let X be a finite subset of $\mathbb{P}^1(k)$ and $S = X^{(3)}/\sim = T(X) =$ the tree of X. We consider $R_S : \mathbb{P}^1(k) \xrightarrow{\gamma_{a_1} \times \ldots \times \gamma_{a_n}} \mathbb{P}^1(k)^n \xrightarrow{R \times \ldots \times R} \mathbb{P}^1(\bar{k})^n$, where $S = \{[a_1], \ldots, [a_n]\}$ and R denotes the canonical reduction of $\mathbb{P}^1(k)$. On $\mathbb{P}^1(k)^n$ and $\mathbb{P}^1(\bar{k})^n$ we use the coordinates $(\ulcorner x_1, y_1 \urcorner, \ldots, \ulcorner x_n, y_n \urcorner)$. The image of $\gamma_{a_1} \times \ldots \times \gamma_{a_n}$ is the subset of $\mathbb{P}^1(k)^n$ given by a set of equations $\{F_{ij} | 1 \leq i < j \leq n\}$. The equation F_{ij} is derived from $\gamma_{aj}\gamma_{a_i}^{-1}(\ulcorner x_i, y_i \urcorner) = \ulcorner x_j, y_j \urcorner$. As in (2.2), F_{ij} is quadratic bihomogeneous and normalized such that all its coefficients are in k^o and $\overline{F_{ij}} \neq 0$. In fact $\overline{F_{ij}}$ is the product of a linear term in $\ulcorner x_i, y_i \urcorner$ and one in $\ulcorner x_j, y_j \urcorner$. Let $Z \subset \mathbb{P}^1(\bar{k})^n$ be the subset given by all equations $\overline{F_{ij}} = 0$. We want to show the following:

Proposition: 1) $Z = L_1 \cup \ldots \cup L_n$ where the line $L_i \subset \mathbb{P}^1(\bar{k})$ is given by $L_i = p_1^i \times \ldots \times p_{i-1}^i \times \mathbb{P}^1(\bar{k}) \times p_{i+1}^i \times \ldots \times p_n^i$.

2) The intersection graph of the components of Z is equal to T(X).

3) $\text{im}\, R_S \subseteq Z$ and $Z - \text{im}\, R_S \subset \bigcup_{i \neq j} L_i \cap L_j$.

4) if $\text{im}\, R_S \neq Z$ then the valuation of k is discrete and for any field extension k' of k with a larger value group, one has "$\text{im}\, R_S = Z$".

5) $R_S : X \to Z$ is injective and every point of $R_S(X)$ lies on just one component of Z.

Proof: We will not give all details, but sketch the proof. We use induction on the number of elements of X. If X has three elements then $\# S = 1$ and the statement is trivial. Let $X' = X \cup \{p\}$ and let S' and S correspond to X' and X. There are three cases:

1) $R_S(p) \notin \bigcup_{i \neq j} (L_i \cup L_j) \cup R_S(X)$.

2) $R_S(p) = R_S(q)$ for some $q \in X$.

3) $R_S(p) \in \bigcup_{i \neq j} (L_i \cap L_j)$.

In the first case S = S' and there is nothing to be done. In the second case, S' has one new element, namely $[(p, q, r)]$ where r is some element of X unequal to q. An easy calculation shows that $R_{S'} : \mathbb{P} \to Z'$

is obtained by adding a new line to Z, on which $R_{S'}(p)$ and $R_{S'}(q)$ are different points. This new line intersects Z at the point $R_S(q)$. We have pictured the situation:

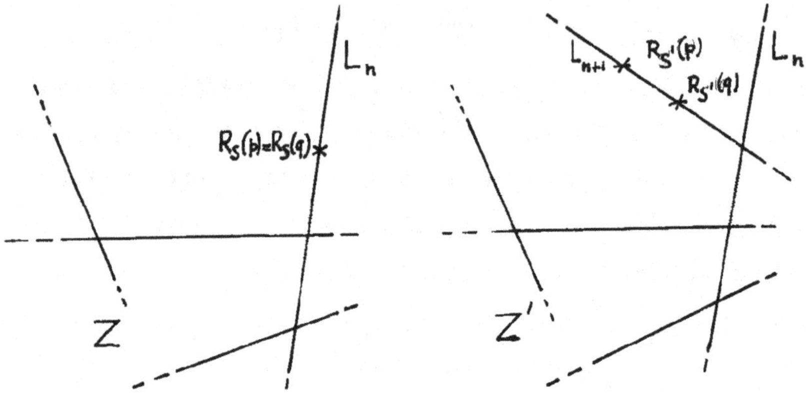

In the third case there is again one new element in S' namely $[(p, q_1, q_2)]$ for well chosen q_1, q_2. The change from Z to Z' can be

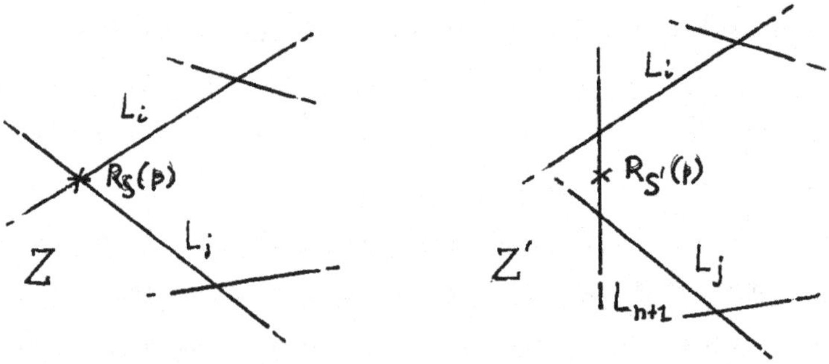

This sketches the proof of 1), 2) and 5).

The properties 3) and 4) follow as in (2.2) where the case n = 2 is treated.

4.3) Theorem: For any Schottky group there exists a good fundamental domain.

<u>Proof:</u> A graph is called <u>combinatorical</u> if $E(v_1, v_2)$ contains at most one element and $E(v, v) = \emptyset$ for all $v \in V$. So a combinatorical graph is given by a set of "vertices" V and a set of "edges" $E \subset V \times V$ satisfying $(v, v) \notin E$ for all $v \in V$ and if $(v_1, v_2) \in E$ then $(v_2, v_1) \in E$.

Let the Schottky group Γ act on some tree $T(X)$ (as in §3, we take X infinite compact and the limit set of X equal to $\mathcal{L}$). <u>The quotient-graph</u> $G = T(X)/\Gamma$ is defined by:

a) the vertices of G are $\dfrac{\text{the vertices of } T(X)}{\text{action of } \Gamma}$

b) the edges between $g_1, g_2 \in G$ are

$$\frac{\text{the edges of } T(X) \text{ with endpoints mapped to } g_1, g_2}{\text{action of } \Gamma}$$

In general $T(X)/\Gamma$ will not be a combinatorical graph. Let d again denote the distance function on the tree $T(X)$. Then $T(X)/\Gamma$ is a combinatorical graph if $d = (p, \gamma(p)) \geq 3$ holds for all $p \in T(X)$ and $\gamma \neq 1$, $\gamma \in \Gamma$. If $T(X)$ does not have this property then we can enlarge X to a suitable X' such that $T(X')$ has this property. This is done as follows: let $\{p_1, q_1\}$ be an edge of $T(X)$; choose a point $x_1 \in \mathbb{P}^1(K)$ (not necessarily in $\mathbb{P}^1(k)$) such that $T_{p,q}(x)$ lies on the two lines of im $R_{p,q}$. We make such choices for a finite number of edges $\{p_i, q_i\}$, $(i = 1, \dots, n)$ representing the edges of $T(X)/\Gamma$. Enlarge X to $X_1 = X \cup \bar{\Gamma}x_1 \cup \dots \cup \bar{\Gamma}x_n$. Then the tree $T(X_1)$ is obtained from $T(X)$ by subdividing every edge. If one takes for every edge $\{p_i, q_i\}$ two different points $x_i, y_i \in \mathbb{P}^1(K)$ mapping to the intersection of the two lines of im $R_{p,q}$ then $X_2 = X \cup {}^-x_1 \cup {}^-y_1 \cup \dots \cup {}^-y_n$ is compact and $T(X_2)$ is obtaines from $T(X)$ by replacing every edge 
by  .

So $T(X_2)/\Gamma$ is a combinatorical graph. From now on we suppose that $T(X)/\Gamma = G$ is already a combinatorical graph. Take a maximal number of edges $M_1, \ldots, M_s$ out of G such that the remainder is still connected. Then $s = g$ and the remainder is a tree. We choose points $z_1, \ldots, z_g$ in $\mathbb{P}(K)$ (not neccessarily in $\mathbb{P}(k)$) such that $X' = X \cup \bar{\Gamma} z_1 \cup \ldots \cup \bar{\Gamma} z_g$ satisfies: $T(X')$ is obtained from $T(X)$ by subdivision of all edges that project to $M_1, M_2, \ldots,$ or M_g. Clearly $T(X')/\Gamma = G'$ is the graph obtained from G by subdividing the edges $M_1, \ldots, M_g$. Since $T(X')$ is the universal covering of G' one can lift any subtree of G' uniquely to $T(X')$ if the lift of one point of that subtree is prescribed. This implies that there is a subtree T of $T(X')$ with a finite set of vertices S such that:

a) S has 2g endpoints, they are mapped in pairs to the g points of $G' - G$.

b) $T/\sim \; = G'$ where $\sim$ is the identification stated in a).

c) if $s \in S$ is not an endpoint and if $\{q, s\}$ is an edge in $T(X')$ then $q \in T$.

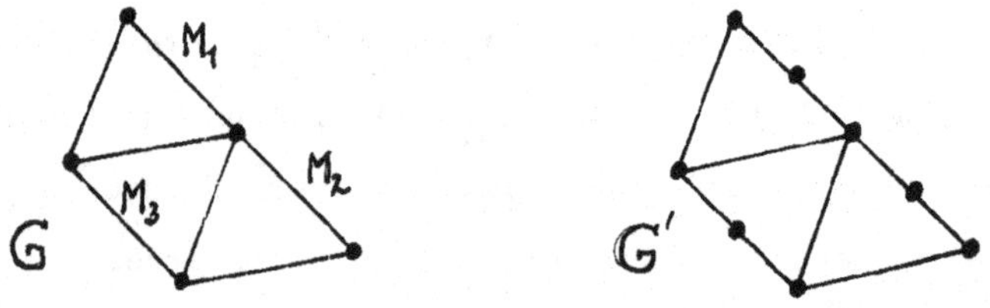

Let $\gamma_1, \ldots, \gamma_g \in \Gamma$ be elements such that the 2g endpoints $[b_1], \ldots, [b_g], [c_1], \ldots, [c_g]$ of S satisfy $\gamma_i[b_i] = [c_i]$ for $i = 1, \ldots, g$. An argument similar to the one used in (3.2) shows that $\gamma_1, \ldots, \gamma_g$ are free generators of Γ. For an endpoint $t \in \{[b_1], \ldots, [c_g]\}$ of S the set $R_t(X')$ consist of three points say t_1, t_2, t_3.

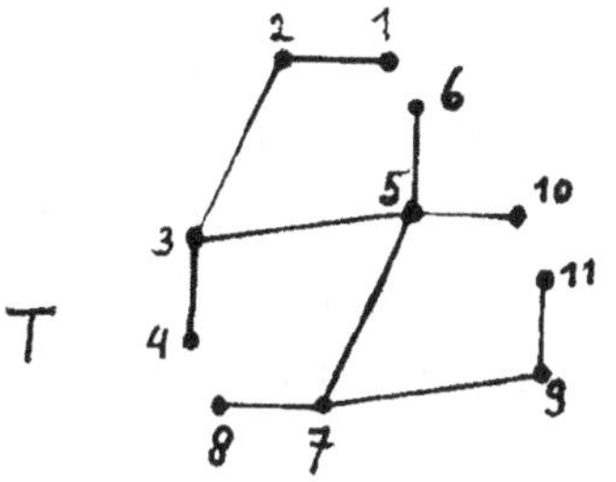

Here t_1, t_3 are the points for which $X' \cap R_t^{-1}(t_i)$ is infinite. Further t_3 corresponds to the one edge t has in T and $X' \cap R_t^{-1}(t_2)$ consists of just one point belonging to the set $\Gamma z_1 \cup \Gamma z_2 \cup \ldots \dot{\cup} \Gamma z_g$.

We consider now the reduction $R_S : \mathbb{P} \to Z$. Using (4.2) we find that $Z \cong T$ and that $R_S(X)$ consists of 4g points, 2 on each line of Z corresponding to an endpoint of T namely t_1 and t_2. We can make a choice for the point at infinity such that $R_S(\infty)$ does not lie on the 2g lines of Z corresponding to $\{[b_1], \ldots, [c_g]\}$. For any point $p \in Z$ lying on just one component, the set $R_S^{-1}(\{p\})$ is an open disk. Put $B_i = R_S^{-1}(\{[b_i]_1\})$ and $C_i = R_S^{-1}(\{[c_i]_1\})$. The B_i and C_i are open disks, such that the corresponding 2g closed disks B_i^+, C_i^+ are disjoint. From $\gamma_i[b_i] = [c_i]$ one easily deduces that in fact $\gamma_i(\mathbb{P} - B_i^+) = C_i$ and $\gamma_i(\mathbb{P} - B_i) = C_i^+$. So we have found a fundamental domain for Γ.

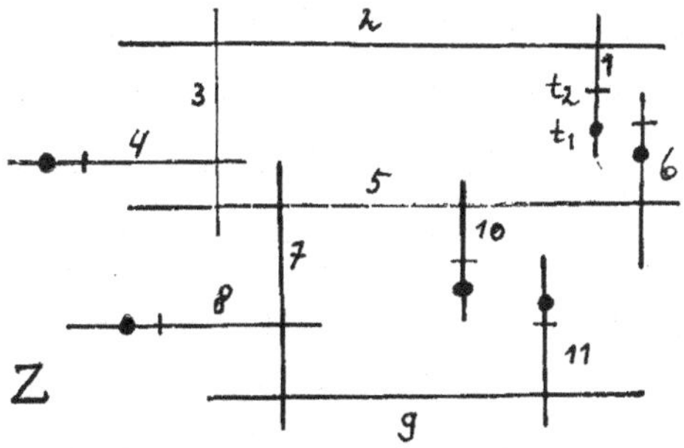

(4.4) Concluding Remarks:

(1) The centers of the 2g open disks that we have constructed are in k. If the valuation of k is dense then by (2.1) case (b) or (4.2.4) one sees that all the new points can actually be taken in $\mathbb{P}^1(k)$. So in this case the radii can also be taken in $|k^*|$. If the value group of k is discrete, say $|k^*| = \{|\pi|^n | n \in \mathbb{Z}\}$, then our construction shows that the radii can be taken in the group $\{|\pi|^{n/2} | n \in \mathbb{Z}\}$. In general this is the best that one can obtain.

(2) We can also find a homomorphism $\lambda : \Gamma \to K^*$ (and in fact with values in k^*) such that the $B_1, \ldots, B_g, C_1, \ldots, C_g$ are equal to the open disks $B_{\gamma_1}, \ldots, B_{\gamma_g}, B_{\gamma_1^{-1}}, \ldots, B_{\gamma_g^{-1}}$ of (4.1.4). Namely, the values $\lambda(\gamma_i) = \lambda_i$ can be taken such that $B_{\gamma_i} = B_i$ for $i = 1, \ldots, g$. Using $1 = w_\gamma(\gamma^{-1}z)\, w_{\gamma^{-1}}(z)$ one finds that $B_{\gamma_1^{-1}}, \ldots, B_{\gamma_g^{-1}}$ are just $C_1, \ldots, C_g$.

(3) We have seen that the fundamental domain gives some insight into Schottky groups and also provides a construction of all Schottky groups. But the fundamental domain is also useful for the construction of the quotient Ω/Γ where $\Omega = \mathbb{P}^1(K) - \mathcal{L}$. Using (4.1.3) one sees that at least set-theoretically $\Omega/\Gamma = F/\sim$ where $\sim$ is the equivalence relation on F derived from a pairwise indentification of the 2g "boundary components" $B_1^+ - B_1, C_1^+ - C_1, \ldots, C_g^+ - C_g$ of F. This identification is given by then "biholomorphic" mappings $\gamma_i : B_i^+ - B_i \tilde{\to} C_i^+ - C_i$ $(i = 1, \ldots, g)$. We will make this precise in Chapter III when the correct notion of k-analytic spaces is available.

(4) Let Γ be a Schottky group on $g > 1$ generators. We have associated to Γ a compact set $\mathcal{L}$, a tree $T(\mathcal{L})$ and a graph $T(\mathcal{L})/\Gamma$. Especially the finite graph $T(\mathcal{L})/\Gamma$ is interesting. In Ch. III (2.12.3) and Ch. IV it is shown that this graph is in fact the intersection graph of the stable reduction of Ω/Γ.

Chapter II Mumford curves via automorphic forms

A non-singular projective curve S over a ground field K is completely determined by its field K(S) of rational functions on S. The point set S is identified with the set of all places of the algebraic function field K(S) over K.

In this chapter we describe the p-adic automorphic forms relative to a p-adic Schottky group Γ and deduce the result that the field of Γ-invariant meromorphic functions is indeed an algebraic function field of one variable over K.

In §1 we give the definition of analytic and meromorphic functions on domains of the projective line $\mathbb{P}(K)$ over K and derive those properties of meromorphic functions that are needed in the sequel. It is an elementary fact that any analytic function defined on an affinoid domain of $\mathbb{P}(K)$ has just a finite number of zeroes.

In §2 we will prove through explicitly writing down infinite products that there exists automorphic forms with constant factors of automorphy on the domain Ω of ordinary points of a Schottky group Γ.

The construction of the p-adic functions $\Theta(a,b;z)$, see (2.3), was first given by Manin-Drinfeld [27], §2 and by Myers [29], Chap. III, §2, but already in Schottky's original paper [35] „Über eine specielle Function, welche bei einer bestimmten linearen Transformation ihres Arguments unverändert bleibt" published in 1887 where Schottky introduced what is now known as complex Schottky groups he worked with very similar infinite products. However in the complex case there is the additional difficulty that maybe those products do not always converge. We are not aware of any article reconsidering or solving this convergence problem of Schottky.

In §2 we also collect some formulas on the automorphic forms $\Theta(a,b;z)$ and their factors of automorphy.

The main result of §3 states that any automorphic form with constant factors of automorphy is up to a constant a finite product of functions $\Theta(a,b;z)$. This result had first been given in [9] but the proof has been somewhat simplified by the introduction of a method that allows the determining of the number of zeroes of an analytic function on a domain F by its behaviour at the boundary ∂F of F.

This result is similar to the method in the complex case of integrating a differential form around ∂F and determining its valuc through sums of residues. It is also crucial in Chap. VI, §3, when the number of zeroes of the thetafunction $\vartheta(u(z))$ on the fundamental domain F is computed.

In §4 we present in a most elementary way some results on analytic mappings of polydomains that are used in the course of §5. The only theorem that is really necessary is the wellknown local invertability theorem for which we indicate an elementary proof.

In §5 we shown that the field of Γ-invariant meromorphic functions on Ω is indeed an algebraic function field of one variable whose set of places is exactly the orbit space Ω/Γ. The main point in the argumentation is to show the existence of onc non-constant Γ-invariant function and this is derived from the fact the canonical mapping

$$\phi : \Omega^r \to (K^*)^g$$

which describes the automorphy factors of a certain parametrized family of automorphic forms has non-discrete fibers for $r > g$. This is completely analog to one of the standard proofs in the respective theorems of the complex case. For later use we also compute the functional determinant of the mapping $\phi : \Omega^g \to (K^*)^g$.

§1 Analytic and meromorphic functions

The ground field K is throughout this chapter assumed to be algebraically closed, complete and non-trivial with respect to a non-archimedean valuation. We fix a rational function z on the projective line $\mathbb{P}^1(K) = \mathbb{P}$ with a simple zero and a simple pole and identify $\mathbb{P}$ with $K \cup \{\infty\}$ through $x \to z(x)$.

(1.1) A K-valued function $f(z_1,\ldots, z_n)$ on the unit polydisk

$$E^n := \{(z_1,\ldots, z_n) \in K^n : |z_i| \leq 1\}$$

of the n-dimensional affine space K^n is called analytic if there is a power series expansion

$$f(z_1,\ldots, z_n) = \sum_{(\nu_1,\ldots, \nu_n) \in \mathbb{N}^n} a_{\nu_1 \ldots \nu_n} z_1^{\nu_1} \ldots z_n^{\nu_n}$$

which is convergent for all points of E^n.

It is a simple exercise to see that the above power series converges on E^n if and only if the sequence $(a_{\nu_1 \ldots \nu_n})$ of coefficients tends to zero which means that for any real $\varepsilon > 0$ for almost all multi-indices $(\nu_1,\ldots, \nu_n)$ the member $a_{\nu_1 \ldots \nu_n}$ has absolute value less than ε

The analytic functions on E^n clearly constitute a K-algebra which we will denote by T_n and which is sometimes called Tate-algebra and also of strictly convergent power series in n variables.

For any analytic function f(z) on E^n we define the norm of uniform convergence on E^n through

$$\|f\|_{E^n} := \sup_{z \in E^n} |f(z)|.$$

As for $(z_1,\ldots, z_n) \in E^n$ we clearly have

$$|a_{\nu_1 \ldots \nu_n} z_1^{\nu_1} \ldots z_n^{\nu_n}| \leq |a_{\nu_1 \ldots \nu_n}|,$$

we get that

$$\|f\|_{E^n} \leq \sup_{\nu_1 \ldots \nu_n \in \mathbb{N}^n} |a_{\nu_1 \ldots \nu_n}|.$$

Proposition 1: $\|f\|_{E^n} = \sup_{\nu_1 \ldots \nu_n \in \mathbb{N}^n} |a_{\nu_1 \ldots \nu_n}|.$

Proof: We may assume that

$$\sup |a_{\nu_1 \ldots \nu_n}| = 1.$$

If $\bar{K}$ is the residue field of K and if we denote by $\bar{w}$ the residue class of $w \in \overset{o}{K}$ = valuation ring of K, then

$$\overline{f(z)} \in \bar{K}$$

for all $z \in E^n$, and we have completed the proof if we show that $\overline{f(z)}$ is not identically trivial in z. But

$$\overline{f(z)} = \sum_{\nu_1 \ldots \nu_n} \bar{a}_{\nu_1 \ldots \nu_n} \cdot \bar{z}_1^{\nu_1} \ldots \bar{z}_n^{\nu_n}$$

which shows that $\overline{f(z)}$ is a polynomial in $\overline{z_1}, \ldots, \overline{z_n}$. As the residue field $\bar{K}$ is also algebraically closed and thus not finite, a non-trivial polynomial is never a zero function on $\bar{K}^n$.

(1.2) Now we are going to derive some elementary properties of analytic functions on affinoid domains of $\mathbb{P}(K)$.

Any domain

$$D := \{z \in K : |z - m| < \rho\}$$

with $m \in K$, $\rho \in |K|$ = value group of K, or

$$D := \{z \in \mathbb{P}(K) : |z - m| > \rho\}$$

with $\rho \in |K|$ is called an open disk on $\mathbb{P}(K)$.
The complements of open disks are called affinoid disks.

Any finite intersection of affinoid disks is called a connected affiniod domain F of $\mathbb{P}(K)$.

A K-valued function f(z) on F is called analytic, if it can be uniformly approximated on F by rational functions in z which have no pole on F. The algebra A(F) of all analytic functions on F is clearly a Banach algebra with recept to the norm of uniform convergence $\| \|_F$ on F.

Assume now that

$F = \{z \in K : |z| \leq \rho_o, \ |z - m_i| \geq \rho_i \quad \text{for } i = 1,2,\ldots, n\}$

where all $\rho_i \in |K|$ and where $m_i \in K$ with $|m_i| \leq \rho_o$ and such that

$$|m_i - m_j| \leq \rho_i$$

for all i,j.

This means that F is obtained by cutting out n open disks from an affinoid disk in K.

If $c_i \in K$ with $|c_i| = \rho_i$ and if

$$w_o(z) = \frac{z}{c_o}$$

$$w_i(z) = \frac{c_i}{z-m_i} ,$$

the norm $\|w_i(z)\|_F$ is 1 for all i.

<u>Proposition:</u> Any analytic function f(z) on F has up to additive constants a unique decomposition

$$f(z) = f_o(z) + f_1(z) + \ldots + f_n(z)$$

where any $f_i(z)$ is strictly convergent power series in $w_i(z)$, i.e.

$$f_i(z) = \sum_{\nu=o}^{\infty} a_{i\nu} w_i(z)^{\nu}$$

where the sequence $(a_{i\nu})_{\nu\geq o}$ tends to zero.

Moreover: $\|f\|_F = \max\limits_{i=o}^{n} \|f_i\|_F \quad (*)$

if $f_i(\infty) = 0$ for $i \geq 1$.

Proof: We assume first that a function f(z) has a decomposition as stated above and show that (*) holds. We may assume that

$$f_i(z)$$

have a zero at ∞ for $i \geq 1$ which is equivalent that $a_{io} = 0$ for $i \geq 1$. As the functions $w_i(z)$ can be transformed into one another by fractional linear transformations it is enough to show that

$$\| f_o \|_F \leq \| f \|_F$$

under the assumption that

$$\| f_o \|_F \geq \| f_i \|_F$$

for all i.

But if we assume $\| f_o \|_F = 1$, then $\overline{f_o(z)}$ - residue class of $f_o(z) \in K$ in the residue field $\bar{K}$ is a polynomial in $\overline{w_o}$. The case where all $\rho_i = \rho_o$ is the most difficult one.

But in this case

$$w_i(z) = \frac{a_i}{b_i - \overline{w_o}}$$

with a_i, $b_i \in \bar{K}$, and these function are linearly independent over $\bar{K}$ which gives (*).

If some ρ_i are smaller that ρ_o we enlarge them and then derive (*).

Now the above decomposition is clearly possible for rational functions without poles in F because of the partial fraction decomposition. Because of (*) the same is true for uniform limits of rational functions.

<u>Proposition:</u> An analytic function $f(z) \not\equiv 0$ of F has just a finite number of zeroes.

Proof: We conduct induction on n. The case n = 1 is an immediate consequence of the Weierstraß preparation theorem for one variable, see Chap. III, (1.1) or the article [16] of Grauert-Remmert.

If $f(z)$ and $f_i(z)$ are as in the above proposition, put

$$\tilde{f}(z_i) = \sum_{\nu=o}^{\infty} a_{i\nu} z_i^{\nu}.$$

Then $\tilde{f}(z_o,\ldots, z_n) = \tilde{f}_o(z_o) + \ldots + \tilde{f}_n(z_n)$ is regular with respect to any variable z_i for which $\|f_i\|_F = \|f\|_F$. It is a consequence of the Weierstraß preparation theorem that there is an analytic function $e(z_1,\ldots, z_n)$ on E^n which is a unit in the algebra of analytic functions on E^n such that

$$e(z) \cdot \tilde{f}(z)$$

is a polynomial with respect to some z_i.
Substituting w_i for z_i and multiplying with a power of w_i^{-1}, we can reduce to the case for n - 1.

(1.3) In the next paragraph we need the notion of analytic and meromorphic functions on a Stein domain of $\mathbb{P}(K)$.

Any domain Ω of $\mathbb{P}(K)$ for which there is a sequence $(\Omega_n)_{n\geq 1}$ of connected affinoid domains Ω_n such that

$$\Omega_n \subset \Omega_{n+1}$$

for all n and

$$\Omega = \bigcup_{n=1}^{\infty} \Omega_n$$

is called a Stein domain of $\mathbb{P}(K)$.

A K-valued function $f(z)$ on Ω is called analytic, if the restriction of $f(z)$ on every Ω_n is analytic on Ω_n. A function $f(z)$ on Ω with values in $K \cup \{\infty\}$ is called meromorphic on Ω_n, i.e. $f(z)|\Omega_n$ is the quotient $\frac{g(z)}{h(z)}$ of two analytic functions $g(z)$, $h(z)$ where the denominator is non-trivial.

For a more systematic account on analytic functions and spaces see Chap. III, and especially III, (1.18.3).

§2 Construction of automorphic forms

(2.1) Let $\Gamma \subset SL_2(K)$ be a Schottky group for which the point ∞ is not a limit point of Γ. It has been proved in Chap. I, (4.3) that there exists for Γ a good fundamental domain F. We use the notation of Chap. I (4.1.3), but we will write B_i' or B_{i+g} instead of C_i.

Thus the complement of F in $\mathbb{P}$ is the disjoint union of open disks $B_1, B_2, \ldots, B_{2g}$ and there are transformations $\gamma_i \in \Gamma$ for $1 \leq i \leq g$ such that

$$\gamma_i(\mathbb{P} - B_i) = B_{i+g}^{+} = \text{closure of } B_{i+g}$$
$$\gamma_i(\mathbb{P} - B_i^{+}) = B_{i+g} = B_i'.$$

Moreover it is assumed that the configuration of the closed disks $B_1^+, B_2^+, \ldots, B_{2g}^+$ is also pairwise disjoint.

Now $\{\gamma_1, \gamma_2, \ldots, \gamma_g\}$ is a basis of the group Γ and any $\alpha \in \Gamma$ has a unique representations

$$\alpha = \delta_1\delta_2 \ldots \delta_r$$

with $\delta_i \in \{\gamma_1, \gamma_2, \ldots, \gamma_g, \gamma_1^{-1}, \gamma_2^{-1}, \ldots, \gamma_g^{-1}\}$ and such that $\delta_i\delta_{i+1} \neq \text{id}$ for all i. The integer r is called the lenght $\ell(\alpha)$ of α with respect to the basis $\{\gamma_1, \ldots, \gamma_g\}$.

Let $\Omega_n := \bigcup_{\ell(\gamma)\leq n} \gamma(F)$ be the union of the translates $\gamma(F)$ for the transformations of lenght $\leq n$. It has been proved in Chap. I, (4.1.3) that the complement of Ω_n is the disjoint union of $2g(2g-1)^n$ open disks whose maximal radius σ_n is going to zero if n goes to infinity.

The union $\bigcup_{\gamma\in\Gamma} \gamma(F)$ of all the translates of F is denoted by Ω.

Thus $\bigcup_{n=0}^{\infty} \Omega_n = \Omega$.

(2.1.2) Proposition:

Take two points a, b $\in \Omega$ and $\varepsilon > 0$. Then for almost all $\gamma \in \Gamma$ we get

$$|\gamma(a) - \gamma(b)| \leq \varepsilon.$$

Proof: Let a, b $\in \Omega_n$ and let n be large enough such that $\sigma_n \leq \varepsilon$.

If a transformation $\alpha \in \Gamma$ is of lenght $> 2n$, then

$$\alpha(\Omega_n) \cap \Omega_n = \emptyset.$$

Because otherwise there would exist $\beta_1, \beta_2 \in \Gamma$ of lenght $\leq n$ and a point $z_o \in F$ such that

$$\beta_1(z_o) = \alpha \cdot \beta_2(z_o).$$

Now the lenght $\ell(\alpha_1\alpha_2)$ of a product is obviously $\geq \ell(\alpha_1) - \ell(\alpha_2)$ and therefore

$$\ell(\alpha\beta_2) > 2n - n > n.$$

Now $\beta_1 = \alpha\beta_2$ because $\beta_1^{-1}\alpha\beta_2(z_o) = z_o$ and the fixed points of all the transformations $\gamma \neq$ id are outside of Ω. But this is a contradiction $\ell(\beta_1) \leq n$.

Now $\alpha(\Omega_n)$ is a disk with a finite number of holes and therefore $\alpha(\Omega_n)$ must sit on one of the maximal disks of $\mathbb{P} - \Omega_n$ whose radius is $\leq \sigma_n \leq \varepsilon$. Thus $\alpha(\Omega_n)$ has diameter $\leq \varepsilon$ and we get the desired result:

$$|\alpha(a) - \alpha(b)| \leq \varepsilon.$$

(2.2) Let (a_i), (b_i) be two sequences of points in $\mathbb{P}$ for which $\lim_{i\to\infty} |a_i - b_i| = 0$. Assume for any n: for almost all indices i the points a_i, b_i are not in Ω_n.

If one of the points a, b is ∞, we define the expression $\frac{z-a}{z-b}$ as follows:

$$\frac{z-a}{z-b} = 1 \text{ if } a = b = \infty$$

$$\frac{z-a}{z-\infty} = z - a \text{ if } a \neq \infty$$

$$\frac{z-\infty}{z-b} = \frac{1}{z-b} \quad \text{if } b \neq \infty .$$

<u>Lemma</u>: The infinite product

$$\prod_{i=1}^{\infty} \frac{z - a_i}{z - b_i}$$

is a welldefined meromorphic function on Ω which has no zeroes outside $\{a_i\}$ and no poles outside $\{b_i\}$.

<u>Proof:</u> Fix an integer n. Then there is an integer N such that $a_i, b_i \notin \Omega_n$ for $i \geq N$.

Put

$$f_k(z) = \prod_{i=N}^{k} \frac{z - a_i}{z - b_i}.$$

Now $f_k(z)$ is an analytic function on Ω_n and

$$f_{k-1}(z) - f_k(z) = \left(\frac{z - a_{k+1}}{z - b_{k+1}} - 1\right) f_k(z)$$

$$= \frac{b_{k+1} - a_{k+1}}{z - b_{k+1}} \cdot f_k(z).$$

For any point $z \in \Omega_n$ have $|z - b_{k+1}| \geq \sigma_n$ as $b_{k+1} \notin \Omega_n$.

Thus

$$\| f_{k+1} - f_k \|_{\Omega_n} \leq \sigma_n^{-1} |b_{k+1} - a_{k+1}| \cdot \| f_k \|_{\Omega_n}.$$

From this we deduce that

$$\lim_{k\to\infty} \| f_{k+1} - f_k \|_{\Omega_n} = 0$$

and this proves that the sequence (f_k) convergence uniformly on Ω_n towards an analytic function $f^*(z)$. The infinite product

$$\prod_{i=1}^{\infty} \frac{z - a_i}{z - b_i}$$

represents the meromorphic function

$$f^*(z) \prod_{i=1}^{N-1} \frac{z - a_i}{z - b_i}.$$

As this convergences is independent of n we get a meromorphic function on Ω.

Clearly $f^*(z)$ has no pole on Ω_n and therefore $\prod_{i=1}^{\infty} \frac{z - a_i}{z - b_i}$ can have no poles outside of $\{b_i\}$. As

$\prod_{i=1}^{\infty} \frac{z - a_i}{z - b_i} \cdot \prod_{i=1}^{\infty} \frac{z - b_i}{z - a_i} \equiv 1$ we see that $\prod_{i=1}^{\infty} \frac{z - a_i}{z - b_i}$ has no zeros outside $\{a_i\}$ as $\prod_{i=1}^{\infty} \frac{z - b_i}{z - a_i}$ has no poles outside $\{a_i\}$.

(2.3) Let a, b be two points of Ω. Then

$$\gamma(a),\ \gamma(b) \notin \Omega_n$$
$$|\gamma(a) - \gamma(b)| \leq \varepsilon$$

for almost all $\gamma \in \Gamma$ and for each n and ε.

The infinite product

$$\Theta(a, b; z) := \prod_{\gamma \in \Gamma} \frac{z - \gamma(a)}{z - \gamma(b)}$$

is therefore a meromorphic function on Ω.

The function $\Theta(a, b; z)$ is remarkable because it is an automorphic form with constant factors of automorphy:

(2.3.1) $\Theta(a, b; z) = c(\alpha) \cdot \Theta(a, b; \alpha(z))$, $c(\alpha) \in K^*$, for any $\gamma \in \Gamma$.

Proof: There is a constant $\eta_\alpha \in K^*$ such that

$$\frac{\alpha(z) - a}{\alpha(z) - b} = \eta_\alpha \cdot \frac{z - \alpha^{-1}(a)}{z - \alpha^{-1}(b)}.$$

This is true because the right-hand side as well as the left-hand side is a rational function with a simple zero at the point $\alpha^{-1}(a)$ and a simple pole at the point $\alpha^{-1}(b)$ and no other zeroes or poles. So up to some multiplicative constant they must be equal.

This constant η_α can be computed by letting the variable z assume some point z_0.

If $\alpha^{-1}(b) \neq \infty$, $\alpha^{-1}(a) \neq \infty$, the function $\frac{z - \alpha^{-1}(a)}{z - \alpha^{-1}(b)}$ has the value 1 at the point ∞ and thus $\eta_\alpha = \frac{\alpha(\infty) - a}{\alpha(\infty) - b}$ and this quotient is well-defined and $\neq 0$ as $\alpha(\infty) \neq a$, $\alpha(\infty) \neq b$.

Let Γ' consist of all γ such that $\alpha^{-1}\gamma(a) \neq \infty$ and $\alpha^{-1}\gamma(b) \neq \infty$. Then $\Gamma - \Gamma'$ has not more that two elements because no transformation $\neq$ id has a fixed point in Ω.

$$\prod_{\gamma\in\Gamma} \frac{\alpha z - \gamma a}{\alpha z - \gamma b} = \prod_{\gamma\in\Gamma'} \frac{\alpha(\infty) - \gamma a}{\alpha(\infty) - \gamma b} \cdot \frac{z - \alpha^{-1}\gamma a}{z - \alpha^{-1}\gamma b} \cdot \prod_{\gamma\in\Gamma-\Gamma'} \frac{\alpha z - \gamma a}{\alpha z - \gamma b}.$$

As $\Gamma - \Gamma'$ is finite, we clearly get a constant $c(\alpha)$ such that

$$c(\alpha) \cdot \Theta(a, b; z)) = \Theta(a, b; z).$$

If $a, b \notin \Gamma\infty$, then we obtain

$$\Theta(a, b; \alpha(z)) = \Theta(a, b; \alpha(\infty)) \cdot \Theta(a, b; z).$$

We get $c(\alpha\beta) = c(\alpha) \cdot c(\beta)$ and $c(\alpha^{-1}) = c(\alpha)^{-1}$.

(2.3.2) If $\Gamma a \neq \Gamma b$, then $\Theta(a, b; z)$ has simple zeroes at the points of Γa and simple poles at the points of Γb and no other zeroes or poles. If $\Gamma a = \Gamma b$, then $\Theta(a, b; z)$ has no zeroes and no poles on all of Ω. Because if $b = \alpha(a)$ then

$$\Theta(a, \alpha(a); z) = \frac{z - a}{z - \alpha(a)} \cdot \frac{z - \alpha^{-1}(a)}{z - \alpha^{-1}\alpha(a)} \cdot \prod_{\gamma\neq\alpha^{-1}, \text{ id}} \frac{z - \gamma(a)}{z - \gamma\alpha(a)} =$$

$$= \frac{z - \alpha^{-1}(a)}{z - \alpha(a)} \prod_{\gamma\neq \text{id}, \alpha^{-1}} \frac{z - \alpha(a)}{z - \gamma\alpha(a)}$$

and this infinite product has no zero or pole at a.

(2.3.3) $\Theta(a, b; z) = \Theta(\alpha(a), \alpha(b); z)$

because

$$\prod_{\gamma\in\Gamma} \frac{z - \gamma a}{z - \gamma b} = \prod_{\gamma\in\Gamma} \frac{z - \gamma\alpha a}{z - \gamma\alpha b}$$

as $\Gamma = \Gamma\alpha$.

(2.3.4) $\Theta(a, \alpha(a); z) = \Theta(b, \alpha(b); z)$ for all $a, b \in \Omega$.

Proof:
$$\prod_{\gamma\in\Gamma} \frac{z-\gamma a}{z-\gamma\alpha a} \cdot \Big(\prod_{\gamma\in\Gamma} \frac{z-\gamma b}{z-\gamma\alpha b}\Big)^{-1} =$$

$$= \prod_{\gamma\in\Gamma} \frac{z-\gamma a}{z-\gamma\alpha a} \cdot \frac{z-\gamma\alpha b}{z-\gamma b} =$$

$$= \prod_{\gamma\in\Gamma} \frac{z-\gamma a}{z-\gamma b} \cdot \prod_{\gamma\in\Gamma} \frac{z-\gamma\alpha b}{z-\gamma\alpha a} =$$

$$= \Theta(a, b; z) \cdot \Theta(b, \alpha; z) \equiv 1.$$

(2.3.5) Define $u_\alpha(z) = \Theta(a, \alpha(a); z)$. This is an analytic function without zeroes on Ω.

$$u_{\alpha\beta}(z) = u_\alpha(z) \cdot u_\beta(z)$$

Proof:
$$u_{\alpha\beta}(z) = \Theta(a, \alpha\beta a; z) =$$
$$= \Theta(a, \beta a; z) \cdot \Theta(\beta a, \alpha\beta a; z) =$$
$$= u_\beta(z) \cdot \Theta(b, \alpha b; z) =$$
$$= u_\beta(z) \quad u_\alpha(z).$$

This shows that $u_\alpha(z) = u_\beta(z)$ if $\alpha \equiv \beta \mod [\Gamma,\Gamma]$ where $[\Gamma,\Gamma]$ denotes the commutator subgroup of Γ.

Denote $u_i(z) = u_{\gamma_i}(z)$. Any $\alpha \in \Gamma$ is congruent mod $[\Gamma,\Gamma]$ to some $\beta = \gamma_1^{n_1} \dots \gamma_g^{n_g}$, where $n_i \in \mathbb{Z}$. Then of course

$$u_\alpha(z) = u_1(z)^{n_1} \dots u_g(z)^{n_g}.$$

Especially $u_\alpha(z) \equiv 1$ if $\alpha \in [\Gamma,\Gamma]$. We will show in §2, that $u_\alpha(z)$ is not a constant if $\alpha \notin [\Gamma,\Gamma]$.

(2.3.6) The automorphy factor $c(\alpha)$ of the form $\Theta(a, b; z)$ does depend on a and b as well.

Let

$$\Theta(a, b; z) = c(a, b; \alpha) \cdot \Theta(a, b; \alpha(z)).$$

Then

$$c(a,\ b;\ \alpha) = \frac{u_\alpha(a)}{u_\alpha(b)}$$

and thus the automorphy factors depend analytically on the parameters a, b.

Proof: $c(a,\ b;\ \alpha) = \dfrac{\Theta(a,\ b;\ z)}{\Theta(a,\ b;\ \alpha(z)} =$

$$= \prod_{\gamma\in\Gamma} \frac{\left(\frac{\gamma a - z}{\gamma b - z}\right)}{\left(\frac{\gamma a - \alpha z}{\gamma b - \alpha z}\right)} = \prod_{\gamma\in\Gamma} \frac{\left(\frac{\gamma a - z}{\gamma a - \alpha z}\right)}{\left(\frac{\gamma b - z}{\gamma b - \alpha z}\right)}.$$

We choose for z some point $z_o \notin \Gamma\infty$.

As a function of a we get

$$\frac{\gamma a - z_o}{\gamma a - \alpha z_o} = \frac{\gamma\infty - z_o}{\gamma\infty - \alpha z_o} \cdot \frac{a - \gamma^{-1} z_o}{a - \gamma^{-1}\alpha z_o}.$$

As functions of a and b, we get

$$\frac{\left(\frac{\gamma a - z_o}{\gamma a - \gamma z_o}\right)}{\left(\frac{\gamma b - z_o}{\gamma b - \alpha z_o}\right)} = \frac{\left(\frac{a - \gamma^{-1} z_o}{a - \gamma^{-1}\alpha z_o}\right)}{\left(\frac{b - \gamma^{-1} z_o}{b - \gamma^{-1}\alpha z_o}\right)}.$$

By using this formula we obtain

$$c(a,\ b;\ \alpha) = \frac{\Theta(z_o,\ \alpha z_o;\ a)}{\Theta(z_o,\ \alpha z_o;\ b)} = \frac{u_\alpha(a)}{u_\alpha(b)}.$$

§ 3 Determination of automorphic forms

(3.1) The transformation γ_i has two fixed points m_i, m_i'; one in B_i and the other in B_i'. Let $m_i \in B_i$ and $m_i' \in B_i'$. Then

$$B_i = \{z \in K : |z - m_i| < r_i\}$$
$$B_i' = \{|z \in K : |z - m_i'| < r_i'\}.$$

Let $w(z) = \frac{z - m_i}{z - m_i'}$. Then $w(\gamma_i(z))$ has also a simple zero at m_i and a simple pole at m_i' and thus

$$w(\gamma_i(z)) = \frac{\gamma_i(z) - m_i}{\gamma_i(z) - m_i'} = q_i \quad \frac{z - m_i}{z - m_i'}$$

with $q_i \in K^*$.

If $z \in \partial B_i = \{z \in K : |z - m_i| = r_i\}$, then $\gamma_i(z) \in \partial B_i'$ and

$$|z - m_i'| = |m_i - m_i'|$$

$$|\gamma_i z - m_i| = |m_i - m_i'|.$$

Thus $\quad w(\gamma_i z) \quad = \quad \frac{|m_i - m_i'|}{r_i'} \quad = |q_i| \cdot \frac{r_i}{|m_i - m_i'|}$

and

$$|q_i| = \frac{|m_i - m_i'|^2}{r_i \cdot r_i'} > 1.$$

Let now ρ be a real number, $\rho > 1$ but $\rho r_i < |m_i - m_i'|$.
If $|z - m_i| = \rho \cdot r_i$, then $z \notin B_i^+ \cup B_i'$. $\gamma_i z \in B_i'$ and

$$|\gamma_i z - m_i'| = \frac{1}{|q_i|} \quad \frac{|z - m_i'|}{|z - m_i|} \quad \cdot |\gamma_i z - m_i|$$

$$= \frac{r_i \quad r_i'}{|m_i - m_i'|^2} \quad \frac{|m_i \cdot m_i'|}{r_i} \cdot |m_i - m_i'|$$

$$= \frac{r_i'}{\rho}.$$

Let $R_i = \{z \in K : \frac{r_i}{\rho} \leq |z - m_i| \leq \rho \cdot r_i\}$ and

$\quad R_i' = \{z \in K : \frac{r_i'}{\rho} \leq |z - m_i'| \leq \rho \cdot r_i'\}$.

Then for $\rho > 1$ close to 1 the transformation γ_i maps R_i onto R_i'.

We choose ρ small enough > 1 such that R_i and R_i' do not meet B_j or B_j' for $j \neq i$ and for all i.

Then R_i, R_i' are contained in Ω, in fact

$$R_i \subset F \cup \gamma_i^{-1} F$$
$$R_i' \subset F \cup \gamma_i F.$$

Sometimes it is convenient to write

$$R_{i+g} = R_i'$$
$$m_{i+g} = m_i'$$
$$r_{i+g} = r_i'.$$

Any analytic function f(z) on the ring domain R_i can be developped into a convergent Laurent series

$$f(z) = \sum_{\nu=-\infty}^{+\infty} f_\nu (z - m_i)^\nu$$

and if f(z) has no zero in R_i, we find a uniquely determined n such that

$$|f(z)| = |f_n| \cdot |z - m_i|^n$$

for all points $z \in R_i$.

We call n the order of f with respect to the disk B_i and write $n = \operatorname{ord}_{B_i} f$.

<u>Proposition:</u> Let f(z) be an analytic function on $\tilde{F} = F \cup (\bigcup_{i=1}^{2g} R_i)$ which has no zero on $\bigcup_{i=1}^{2g} R_i$.

Then the number of zeroes of f (counting multiplicities) is equal to

$$- \sum_{i=1}^{2g} \operatorname{ord}_{B_i} f.$$

<u>Proof:</u> 1) We show first that the result is true if $f(z) = \frac{z-a}{z-b}$. As f(z) has no pole in $\tilde{F}$ the point b is outside of $\tilde{F}$, say $b \in B_1$, $|b - m_i| \le \frac{r_i}{\rho}$ as f(z) has no zero on R_i.

If i = 1, then we get for all $z \in R_k$:

$$\left|\frac{z-a}{z-b}\right| = 1.$$

Thus ord $_{B_k} f(z) = 0$ for all k. As f(z) has no zero in $\tilde{F}$ the result has been proved in this case.

If $i > 1$, then we get

$$\left|\frac{z - a}{z - b}\right| = \frac{|m_1 - a|}{|z - b|} \quad \text{on } R_1.$$

Thus ord $_{B_1} f(z) = -1$.

Also

$$\left|\frac{z - a}{z - b}\right| = \frac{|z - m_i|}{|m_i - m_1|} \quad \text{on } R_i$$

and thus ord $_{B_i} f(z) = +1$.

If $j \neq i$, 1 we get

$$\left|\frac{z - a}{z - b}\right| = \frac{|m_j - a|}{|m_j - b|}$$

and thus ord $_{B_j} f(z) = 0$.

Thus also in this case the statement is correct.

2) Let $f_1(z) \cdot f_2(z) = f(z)$. Now

$$\text{ord}_{B_i} f_1(z) \cdot f_2(z) = \text{ord}_{B_i} f_1(z) + \text{ord}_{B_i} f_2(z)$$

and therefore

$$\sum_{i=1}^{2g} \text{ord}_{B_i} f(z) = \sum_{i=1}^{2g} \text{ord}_{B_i} f_1(z) + \sum_{i=1}^{2g} \text{ord}_{B_i} f_2(z).$$

The number of zeroes of f(z) is the sum of the number of zeroes of $f_1(z)$ and the number of zeroes of $f_2(z)$.

Using this technique we see that the result is true for any rational function.

3) If f(z) is arbitrary we find a rational function h(z) and an analytic function e(z) without zeroes on $\tilde{F}$ such that

$$f(z) = h(z) \cdot e(z).$$

Now $|\frac{1}{e(z)}|$ is bounded on $\tilde{F}$; let $|\frac{1}{e(z)}| \leq M$ and $0 < \varepsilon < \frac{1}{M}$. There exists a rational function p(z) such that

$$\sup_{z \in \tilde{F}} |e(z) - p(z)| \leq \varepsilon.$$

Then $|c(z)| = |p(z)|$ for all $z \in \tilde{F}$ and therefore $\mathrm{ord}_{B_i} e(z) = \mathrm{ord}_{B_i} p(z)$. As p(z) has no zeroes on $\tilde{F}$ we have

$$\sum_{i=1}^{2g} \mathrm{ord}_{B_i} e(z) = 0.$$

This completes the proof.

<u>Proposition:</u> Let f(z) be an analytic function on $\tilde{F}$ without zeroes on $R = \bigcup_{i=1}^{2g} R_i$. If $\mathrm{ord}_{B_i} f(z) = 0$ for all i, then $|f(z)| \equiv \text{const}$ on $\tilde{F}$.

<u>Proof:</u> We have seen in the proof of the last proposition that there is a rational function p(z) such that $|p(z)| = |f(z)|$ on F. Take the rational function p(z) with the least number of poles on Ω such that $|p(z)| = |f(z)|$ on $\tilde{F}$.

Let $p(z) = a_0 \cdot \frac{(z - a_1)\ldots(z - a_r)}{(z - b_1)\ldots(z - b_r)}$. If $a_1, b_1 \in B_i$, then $|\frac{z - a_1}{z - b_1}| \equiv \text{const}$ on F and $|f(z)| \equiv \text{const}\ |\frac{(z - a_2)\ldots(z - a_r)}{(z - b_2)\ldots(z - b_r)}|$.

This cannot happen because r is minimal. If $a_1 \in B_k$, then no $b_i \in B_k$ and $\mathrm{ord}_{B_k} f(z)$ is the number of indices j such that $a_j \in B_k$. Thus $r = 0$ and $|f(z)| \equiv \text{const}$.

<u>Remark:</u> Let us briefly indicate that the propositions of this sections are really analogues of thc classical residue theroem.

Let $D = \{z \in K : |z - m| < r\}$ be an open disk of K and let f(z) be an analytic function on the boundary

$$\partial D = \{z \in K : |z - m| = r\}$$

of D relative to m.

Then $f(z)$ has Laurent series expansion

$$f(z) = \sum_{n=-\infty}^{\infty} a_n (z - m)^n$$

We call the coefficient a_{-1} the residue of $f(z)$ with respect to D and denote it by $\text{res}_D f(z)$.

This definition does not depend on the center m of D, because if m' is another point of D we obtain an expansion

$$f(z) = \sum_{n=-\infty}^{\infty} b_n (z - m)^n$$

but $b_{-1} = a_{-1}$.

This follows from the simple computation:

$$\frac{1}{z - m'} = \frac{1}{(z - m) + (m - m')} = \frac{1}{z - m} \cdot \frac{1}{1 - \frac{m - m'}{z - m}}$$

$$= \frac{1}{z - m} \sum_{i=o}^{\infty} (m - m')^i \frac{1}{(z - m)^i}$$

$$= \sum_{i=1}^{\infty} (m - m')^{i-1} \frac{1}{(z - m)^1} = \frac{1}{z - m} + \frac{(m - m')}{(z - m)^2} + \dots$$

Of course, $\text{ord}_D f(z) = \text{res}_D \frac{f'(z)}{f'(z)}$, where $f'(z)$ is the derivative of $f(z)$ with respect to z, if $\text{ord}_D f(z)$ is defined as above and char $K = 0$. The first proposition of this section can be generalized to res and is thus seen to be formally the same as the residue theorem. We do not want to go into more details because we do not need this concept in the sequel; but see Chap. III, (1.18.4).

(3.2) Let E be some closed disk in $K \subset \mathbb{P}$, say

$E = \{z \in K : |z - b| \leq \tau\}$. Put

$$E_n = E \cap \Omega_n.$$

Then E_n is an affinoid domain, say

$E_n = \{z \in K : |z - b| \leq \tau, |z - b_1| \geq \tau_1, \dots, |z - b_r| \geq \tau_r\}$. For a given $\varepsilon > 0$ there is a number n_o such that $\tau_1, \dots, \tau_r \leq \varepsilon$ if $n \geq n_o$.

Now let $f(z)$ be an analytic function on Ω which is bounded on $E \cap \Omega$. Let M be a constant such that

$$|f(z)| \leq M \text{ on } E \cap \Omega.$$

Now we can develop $f(z)$ on E_n as follows:

$$f(z) = f_o(z) + f_1(z) + \ldots + f_r(z)$$

and $f_o(z)$ is analytic on E

$$f_i(z) \text{ analytic on } \mathbb{P} - \{z : |z - b_i| < \tau_i\}$$

and $f_i(\infty) = 0$.

As $f_o(z)$ is bounded on E we may assume that $f_o(z) \equiv 0$ by considering the difference $f(z) - f_o(z)$ otherwise.

Now

$$f_i(z) = \sum_{\nu=1}^{\infty} a_{i\nu} \frac{1}{(z - b_i)^\nu}$$

and

$$\| f_i(z) \|_{E_n} = \max_{\nu=1} |a_i| \cdot \frac{1}{\tau_i^\nu}.$$

As $\| f(z) \|_{E_n} = \max_{i=1}^{r} \| f_i(z) \|_{E_n}$, we obtain: $\max_{\nu=1}^{\infty} |a_i| \cdot \frac{1}{\tau_i^\nu} \leq M$.

Let $\partial E = \{z \in K : |z - b| = \tau\}$.

We assume that ∂E is contained in E_n.

Then $|b_i - z| = \tau$ for all $z \in \partial E$.

Then $\sup_{z \in \partial E} |f_i(z)| = \max_{\nu=1}^{\infty} |a_{i\nu}| \cdot \frac{1}{\tau^\nu} \leq \max_{=1} |a_{i\nu}| \cdot \frac{1}{\tau_i^\nu} \cdot (\frac{\tau_i}{\tau})^\nu \leq M \cdot \frac{\tau_i}{\tau}$.

Thus

$$\sup_{z \in \partial E} |f(z)| \leq M \max_{i=1}^{r} \frac{\tau_i}{\tau}.$$

The radii τ_i depend on n and their maximum tends to zero as n goes to infinity.

Thus $\|f\|_{\partial E} = 0$ and $f \equiv 0$ on ∂E and thus $f \equiv 0$ on Ω. This proves the following.

Proposition: Any bounded analytic function on Ω is a constant.

(3.3) A meromorphic function $f(z)$ on Ω is called an automorphic form with constant automorphy factors, if for any $\alpha \in \Gamma$ there is a constant $c(\alpha) \in K^*$ such that

$$f(z) = c(\alpha) \cdot f(\alpha z).$$

Proposition: Let $f(z) \not\equiv 0$ be an automorphic form. If $f(z)$ has no poles on Ω, then $f(z)$ has no zeroes on Ω.

Proof: 1) Assume first that $f(z)$ has no zero on $\bigcup_{i=1}^{2g} R_i$. Then

$$\operatorname{ord}_{B_i} f(\gamma_i z) = -\operatorname{ord}_{B_i'} f(z).$$

This can be seen as follows: if $\operatorname{ord}_{B_i'} f(z) = 0$ then $|f(z)| \equiv$ const on R_i'. But then also $|f(\gamma_i z)| \equiv$ const on R_i as the domain R_i is mapped onto R_i' by γ_i.

If $f(z) = \left(\frac{z - m_i}{z - m_i'}\right)^k$, then $f(\gamma_i z) = \left(q_i \frac{z - m_i}{z - m_i'}\right)^k$ and thus $\operatorname{ord}_{B_i} f(\gamma_i z) = \operatorname{ord}_{B_i} f(z) = k$. Now $\operatorname{ord}_{B_i'} f(z) = -k$.

By combining these two special cases we get the general case.

If $f(z)$ is an automorphic form we obtain

$$\operatorname{ord}_{B_i} f(z) + \operatorname{ord}_{B_i'} f(z) = 0.$$

Using Proposition (3.1.1) we conclude that $f(z)$ has no zeroes on F. But then $f(z)$ has no zero on Ω.

2) If $f(z)$ has zeroes $a_1, a_2, \ldots, a_r$ on $\bigcup_{i=1}^{2g} R_i$ respecting multiplicities we consider

$$f^*(z) = f(z) \cdot \prod_{i=1}^{r} \Theta(b, a_i; z)$$

where b is some point of Ω outside of $\bigcup_{i=1}^{2g} R_i$. Then $f^*(z)$ is an analytic automorphic form on Ω without zeroes on $\bigcup_{i=1}^{2g} R_i$. Therefore $f^*(z)$ has no zeroes. But b is a zero of $f^*(z)$ if $r \geq 1$. Thus $r = 0$ and we are done.

(3.2) The functions $\Theta(a, b; z)$ defined in (2.3) are automorphic forms and any finite product $\Theta(a_1, b_1; z) \ldots \Theta(a_r, b_r; z)$ is also an automorphic form.

<u>Theorem:</u> Let $f(z)$ be an automorphic form on Ω. Then

$$f(z) = a_0 \cdot \Theta(a_1, b_1; z) \ldots \Theta(a_r, b_r; z)$$

with $a_0 \in K$, $a_i, b_i \in \Omega$ for $i \geq 1$.

<u>Proof:</u> If a is a zero (resp. pole) of $f(z)$ of order k, then αa is also a zero (resp. pole) of $f(z)$ of order k for any $\alpha \in \Gamma$.

If $f(z) \neq 0$, then $f(z)$ has only a finite number of zeroes (resp. poles) in F. As $\Theta(a, b; z)$ has a simple pole at b and a simple zero at a if $\Gamma a \neq \Gamma b$, we find by the standard method a decomposition

$$f(z) = f_1(z) \cdot \Theta(a_1, b_1; z) \ldots \Theta(a_r, b_r; z)$$

where $f_1(z)$ is an automorphic form without zeroes or without poles. By Proposition (2.3.1) $f_1(z)$ has then neither zeroes nor poles.

Let $\text{ord}_{B_i} f_i(z) = n_i$. The automorphic forms

$$u_i(z) = \Theta(a, \gamma_i a; z) = \prod_{\gamma \in \Gamma} \frac{z - \gamma a}{z - \gamma\gamma_i a}$$

have neither poles nor zeroes on Ω and are independent of a. Let $a \in F$, $a \notin R$. Then with the exception of $\gamma = \text{id}$ and $\gamma = \gamma_i^{-1}$ we have

$$\gamma a, \gamma\gamma_i a \in B_k$$

for the same k.

Therefore

$$|u_i(z)| = \left|\frac{z - a}{z - \gamma_i a} \cdot \frac{z - \gamma_i^{-1} a}{z - a}\right| \cdot \text{ const on } \tilde{F}$$

$$= \left|\frac{z - \gamma_i^{-1} a}{z - \gamma_i a}\right| \cdot \text{ const.}$$

Now $\gamma_i a \in B_i'$ and $\gamma_i^{-1} \in B_i$ and thus

$$\text{ord}_{B_k} u_i(z) = \begin{cases} +1 : & i = k \\ 0 : & i \neq k \end{cases}$$

$$\text{ord}_{B_k'} u_i(z) = \begin{cases} -1 : & i = k \\ 0 : & i \neq k. \end{cases}$$

Let $d(z) = u_1(z)^{n_1} \ldots u_g(z)^{n_g}$. Then

$$\text{ord}_{B_i} \frac{f_1(z)}{e(z)} = 0$$

for $1 \leq i \leq 2g$. Using Proposition (2.1.2) we find that $\left|\frac{f_1(z)}{e(z)}\right| = $ const on F and thus on Ω as $\frac{f_1(z)}{e(z)}$ is automorphic.

Using (3.2) we find that $\frac{f_1(z)}{e(z)} = a_0 \in K$.

The proof is complete.

(3.2) Let $\alpha = \gamma_1^{n_1} \ldots \gamma_g^{n_g} \in \Gamma$ and $u_\alpha(z) = \Theta(a, \alpha a; z)$. Then $u_\alpha(z) = u_1(z)^{n_1} \ldots u_g(z)^{n_1}$ and

$$\text{ord}_{B_i} u_\alpha(z) = n_i.$$

Thus $u_\alpha(z) \not\equiv$ const if one of the $n_i \neq 0$. This shows

<u>Proposition:</u> $u_\alpha(z)$ is not a constant if $\alpha \notin [\Gamma, \Gamma]$.

§4 <u>Analytic mappings</u>

(4.1) Let $\Omega_1, \ldots, \Omega_r$ be Stein domains of $\mathbb{P}(K)$ and consider the product space

$$\Omega := \Omega_1 \times \ldots \times \Omega_r$$

as subspace of the r-fold product of $\mathbb{P}(K)$.

A K-valued function $f(z)$ on Ω is called analytic, if there exists a series

$$\sum_{\nu=1}^{\infty} p_\nu(z)$$

whose partial sums converge uniformly on every domain $F = F_1 \times \ldots \times F_r$ where F_i are affinoid domains of Ω_i, toward the function $f(z)$ and for which the members $p_\nu(z)$ have the form:

$$p_\nu(z_1, \ldots z_r) = f_{\nu 1}(z_1), \ldots, f_{\nu r}(z_r)$$

with analytic function $f_{\nu i}(z_i)$ on Ω_i. In case Ω is polydisk in K^n we recover the definition given in §1.

(4.2) Let $\Omega'_1, \ldots \Omega'_s$ be also Stein domains of $\mathbb{P}(K)$ and consider also the s-fold product space $\Omega' = \Omega'_1 \times \ldots \times \Omega'_s$.

Then a mapping

$$\phi : \Omega \to \Omega'$$

is called analytic if for any analytic function $f(z)$ on Ω' the composite

$$f(\phi(z))$$

is also analytic on Ω.

If $\phi(z) = (g_1(z), \ldots, g_s(z))$, then clearly ϕ is analytic if and only if all $g_i(z)$ are analytic functions on Ω.

It is a standard result of analytic geometry that any non-empty fiber of an analytic mapping $\phi : \Omega \to \Omega'$ is an analytic subset of Ω who has everywhere at least the dimension $r - s$.

At this point we do not want to define the notions of analytic subsets and their dimension and neither prove the above result which can easily be derived from Krull's dimension theorem. The reader can consult on these questions the book of Abhyankar [1] especially §27, or the book of Nagata [30] .

(4.3) We will apply this result in the course of the next paragraph, but we could easily avoid it and only make use of the well-known and basic local invertability theorem for analytic mappings that we are going to mention now. Assume that $r = s$ and that the analytic mapping $\Phi : \Omega \to \Omega'$ is given through

$$\Phi(z_1,\ldots, z_r) = (g_1(z),\ldots, g_r(z)).$$

Then the functional determinant $\Delta(z)$ of $\Phi(z)$ is defined to be the determinant of the $r \times r$ matrix

$$\begin{pmatrix} \frac{\partial g_1}{\partial z_1}(z),\ldots, & \frac{\partial g_1}{\partial z_r}(z) \\ \vdots & \\ \frac{\partial g_r}{\partial z_1}(z),\ldots, & \frac{\partial g_r}{\partial z_r}(z) \end{pmatrix} .$$

Proposition: If z_o is a point of Ω for which $\Delta(z_o) \neq 0$, then there is a polydisk P around z_o and an affinoid subdomain X of Ω' such that Φ induces a bianalytic mapping from P onto X.

Indication of a proof: As the statement is local we may assume that $z_o = 0$ and $\phi(z_o) = 0$ and that $\Omega = \Omega'$ is the unit polydisk in K^r. After a linear change of the variables we may assume that

$$g_i(z_1 ,\ldots, z_n) = z_i + h_i(z)$$

and that $\frac{\partial h_i}{\partial z_j}(0) = 0$ for any i, j.

If we look at the power series expansion of the functions $h_i(z)$ relative to the variables $z_1,\ldots, z_r$, we see that there is a constant $\rho > 0$ such that

$$\|h_i\|_{E^r_\rho} < 1$$

for all i where

$$E^r_\rho = \{(z_1,\ldots, z_r) \in E^r : |z_i| \leq \rho\}.$$

After a change of the variables ($\frac{z_i}{c}$ instead of z_i with $c \in K$ and $|c| = \rho$) we may assume that $\rho = 1$.

Then $\|z_i - g_i(z)\|_{E^r} \leq \delta < 1$ for some $\delta < 1$. If now $f(z_1,\dots, z_r)$ is any analytic function on E^r, then so is

$$\hat{f}(z) := f(g_1(z),\dots, g_r(z))$$

and a simple computation shows that

$$\|\hat{f}(z) - f(z)\| \leq \delta \cdot \|f(z)\|.$$

So we get

$$f(z) = \hat{f}(z) + f_1(z)$$

with $\|f_1\| \leq \delta \cdot \|f\|$. By iterating this procedure with function $f_k(z) = \hat{f}_k(z) + f_{k+1}(z)$ and $f_o = f_1 + \hat{f}_o = \hat{f}$ we get

$$f(z) = \hat{f}(z) + \hat{f}_1(z) + \hat{f}_2(z) + \dots$$

where the series on the right side converges uniformly on E^r as $\|f_k\| \leq \delta^k\|f\|$. As the right side is a strictly convergent series in $g_1(z),\dots, g_r(z)$ we see that ϕ is bianalytic from E^r onto E^r.

This proof also makes it clear that the domain X can be chosen to be

$$X = \{z : |w_i(z) - q_i| \leq 1\}$$

where $w_1(z),\dots, w_r(z)$ is a set of linearly independent linear function in $z_1,\dots, z_r$ and $q = (q_1,\dots, q_r) = \phi(p)$.

§5 The field of Γ-invariant meromorphic functions

(5.1) Denote by C the g-fold product of the multiplicative group K^* of K. The mapping

$$\phi : \Omega \to C$$

defined by $\phi(z) = (u_1(z),\dots, u_g(z))$ is analytic and $\phi(\infty) = (1,\dots, 1$ as $u_i(\infty) = 1$.

Denote by Ω^r the r-fold product of Ω and by

$$\phi_r = \phi : \Omega^r \to C$$

the mapping given by

$$\phi_r(z_1,\ldots, z_r) = \phi(z_1) \cdot \phi(z_2) \ldots \phi(z_r).$$

Then ϕ_r is also analytic. Any fibre $\phi^{-1}(c)$, $c \in C$, is an analytic subset of Ω^r whose dimension is at least $r - g$ if it is not empty, because $\dim C = g$, $\dim \Omega^r = r$. Let σ be a permutatuion of the set of integers $\{1, 2,\ldots, g\}$. Then

$$\phi(z_1,\ldots, z_g) = \phi(z_{\sigma(1)},\ldots, z_{\sigma(g)}).$$

Let $\sigma \cdot (z_1,\ldots, z_g) = (z_{\sigma(1)},\ldots, z_{\sigma(g)})$ and let S_g be the group of permutations of $\{1, 2,\ldots, g\}$.

If now $r \geq g + 1$ and $(z_1,\ldots, z_r)$ a point in Ω^r then there exists a point $(w_1,\ldots, w_r) \in \Omega^r$ such that

$$\phi(z_1,\ldots, z_r) = \phi(w_1,\ldots, w_r)$$
$$(w_1,\ldots, w_r) \notin S_g \cdot (z_1,\ldots, z_r).$$

Consider now the automorphic forms

$$f_1(z) = \Theta(z_1, \infty; z) \cdot \ldots \cdot \Theta(z_r, \infty; z)$$
$$f_2(z) = \Theta(w_1, \infty; z) \cdot \ldots \cdot \Theta(w_r, \infty; z).$$

As $\Theta(a, \infty; z) = u_\alpha(a) \cdot \Theta(a\ \infty; \alpha z)$, we get

$$f_1(z) = u_i(z_1) \cdot \Theta(z_1, \infty; \gamma_i z) \cdot \ldots \cdot u_i(z_r) \cdot \Theta(z_r, \infty; \gamma_i z)$$
$$f_1(z) = u_i(z_1) \cdot \ldots \cdot u_i(z_r) \cdot f_1(\gamma_i z)$$
$$f_2(z) = u_i(w_1) \cdot \ldots \cdot u_i(w_r) \cdot f_2(\gamma_i z).$$

Therefore $f_1(z)$ and $f_2(z)$ are automorphic forms with the same factors of automorphy.

The quotient

$$h(z) = \frac{f_1(z)}{f_2(z)}$$

is therfore a Γ-invariant meromorphic function on Ω.

As the fibre through $(z_1, \ldots, z_r)$ is at least one-dimensional we find a point $w = (w_1, \ldots, w_r)$ in it such that $w_1 \notin \Gamma z_1 \cup \ldots \cup \Gamma z_r$. Then $h(z)$ is not a constant as it has a pole at the point w_1.

Denote by S the orbit set Ω/Γ consisting of all orbits Γa and by $K(S)$ the field of all Γ-invariant meromorphic functions on Ω.

Theorem: $K(S)$ is a function field of one variable over K.

Proof: We have seen that $K(S)$ does contain non-constant functions. If $h(z)$ is such a function we find a representation

$$h(z) = a_0 \cdot \Theta(a_1, b_1; z) \ldots \Theta(a_r, b_r\, z)$$

as $h(z)$ is automorphic.

Let $s = a \cdot \Gamma$ be a point in S. Then define

$$\operatorname{ord}_s h(z) = \operatorname{ord}_a h(z).$$

This characterization shows that $\operatorname{ord}_s h(z) = 0$ for almost all $s \in S$ and that

$$\sum_{s \in S} \operatorname{ord}_s h(z) = 0.$$

Also any $h(z) \in K$ has at least one pole because a function without pole is bounded on Ω.

The proof that $K(S)$ is a function field of one variable is now quite formal, see e.q. [41].

(5.2) We take now a closer look at the mapping

$$\phi = \phi_g : \Omega^g \to G$$

by calculating the functional determinant $\Delta(z_1, \ldots, z_g)$ of ϕ.

We write for the moment

$$\phi(z_1, \ldots, z_g) = \phi(z) = (\psi_1(z), \ldots, \psi_g(z))$$

$\psi_i(z) = u_i(z_1)u_i(z_2) \ldots u_i(z_g)$.

$\psi_{ij}(z) := \frac{\partial \psi_i}{\partial z_j} = u_i(z_1) \ldots u_i'(z_j)u_i(z_{j+1}) \ldots u_i(z_g)$ where $u'(z)$ denotes the derivative with respect to z.

Then

$$\psi_{ij}(z) = \psi_i(z) \cdot \frac{u_i'(z_j)}{u_i(z_j)}$$

and $\Delta(z) = \det(\psi_{ij}) = \psi_1(z) \ldots \psi_g(z) \det\left(\frac{u_i'(z_j)}{u_i(z_j)}\right)$.

Let $w_i(z) := \frac{u_i'(z)}{u_i(z)}$.

<u>Proposition:</u> $w_1(z), \ldots, w_g(z)$ are linearly independent over K.

Before we give the proof, we state a simple lemma

(5.2.1) Let $R = \{z \in K : |z| = r\}$ and $f(z)$ be an analytic function without zeroes on R. Let

$$f(z) = \sum_{n=-\infty}^{+\infty} a_n z^n$$

and N the unique integer such that

$$|a_N| \cdot r^N > |a_n| \cdot r^n$$

for $n \neq N$. N is called the order of $f(z)$ with respect to the origin.

If $N = 0$, then $\left|\frac{f'(z)}{f(z)}\right| < \frac{1}{r}$ for all $z \in R$.

If $N = \pm 1$, then $\left|\frac{f'(z)}{f(z)}\right| = \frac{1}{r}$ for all $z \in R$.

The proof of this lemma is given by explicit calculation.

If $N = 0$, then $f'(z) = \sum_{n=-\infty}^{\infty} na_n z^{n-1}$ and

$|na_n| \cdot r^{n-1} \leq |a_n| \cdot r^n < \frac{1}{r} < |a_0| \cdot \frac{1}{r}$ and thus $|f'(z)| < |a_0| < \frac{1}{r}$.

But $|f(z)| = |a_0|$ for all $z \in R$ which shows that $|\frac{f'(z)}{f(z)}| < \frac{1}{r}$.

If $N = \pm 1$, then $f(z) = z^N . h(z)$ and the order of $h(z)$ with respect to the origin is 0. Then

$$\frac{f'(z)}{f(z)} = \frac{N \cdot z^{N-1}}{z^N} + \frac{h'(z)}{h(z)} = \frac{N}{z} + \frac{h'(z)}{h(z)} .$$

If $|N| = 1$, then $|\frac{f'(z)}{f(z)}| = \frac{1}{|z|} = \frac{1}{r}$ for all $z \in R$.

If $N = \pm 1$, then $|N| = 1$ and we are done.

<u>Proof of the proposition:</u> We know that

$$\text{ord}_{B_k} u_i(z) = \begin{cases} +1 : & i = k \\ 0 : & i \neq k \end{cases}$$

for $1 \leq i, k \leq g$. The lemma shows that

$$|\frac{u_i'(z)}{u_i(z)}| < \frac{1}{r_k} \quad \text{for } z \in F_k,\ k \neq i$$

$$|\frac{u_i'(z)}{u_i(z)}| = \frac{1}{r_i} \quad \text{for } z \in F_i.$$

Consider a linear combination

$$w(z) = c_1 w_1(z) + \ldots + c_g w_g(z),$$

$c_i \in K$. If i is an index for which

$$|c_i| \cdot \frac{1}{r_i} \geq |c_k| \cdot \frac{1}{r_k}$$

for all k, we pick a point $z \in F_i$.

Then $|c_i w_i(z)| = |c_i| \cdot \frac{1}{r_i}$ and if $k \neq i$ we have

$$|c_k w_k(z)| < |c_k| \cdot \frac{1}{r_k}.$$

Therefore we get

$$|w(z)| = |c_i| \cdot \frac{1}{r_i}.$$

This shows that $w \equiv 0$ only if all $c_i = 0$. q.e.d.

(5.3) We consider now the map

$$\mathfrak{w} : \Omega \to K^g$$

given by $\mathfrak{w}(z) = (w_1(z), \dots, w_g(z))$. Now

$$\mathfrak{w}(\Omega) = \{\mathfrak{w}(z) : z \in \Omega\}$$

is not contained in any $(g - 1)$-dimensional linear subspace of K^g. Because then we would obtain $(c_1, \dots, c_g) \neq (0, \dots, 0)$ such that

$$\sum_{i=1}^{g} c_i w_i(z) = 0$$

for all $z \in \Omega$.

Now the determinant $\Delta(z_1, \dots, z_g) \neq 0$ if and only if the vectors $\mathfrak{w}(z_1), \dots, \mathfrak{w}(z_g)$ are linearly independent. This shows that the analytic function $\Delta(z_1, \dots, z_g)$ is not identically zero.

The analytic mapping $\phi : \Omega^g \to C$ is locally bianalytic outside of the set of zeroes of $\Delta(z_1, \dots, z_g)$, see (4.3).

<u>Proposition:</u> Given points $z_1, \dots, z_r \in \Omega - \{\infty\}$ such that $\mathfrak{w}(z_1, \dots, \mathfrak{w}(z_r)$ are linearly independent over K. Let U be an open subset of $\Omega - \{\infty\}$.

Then there are points $z_{r+1}, \dots, z_g \in U$ such that

$$\mathfrak{w}(z_1), \dots, \mathfrak{w}(z_g)$$

are linearly independent.

<u>Proof:</u> As $\mathfrak{w}(U)$ is not contained in any $(g - 1)$ - dimensional subspace of K^g we find points $x_1, \dots, x_g \in U$ such that

$$\mathfrak{w}(x_1), \dots, \mathfrak{w}(x_g)$$

are a basis of K^g.

Among the $x_1, \dots, x_g$ we can pick a subset $z_{r+1}, \dots, z_g$ such that

$$\mathfrak{w}(z_1), \dots, \mathfrak{w}(z_g)$$

is a basis of K^g.

Proposition: Let U be an open subset of Ω. There are points $z_2, \ldots, z_g \in U$ such that ϕ is locally bianalytic at the point

$$(\infty, \; z_2, \ldots, z_g).$$

Proof: Let i be an index such that $|m_i - m_i'| > r_k, r_k'$ for all k. Let z_1 be a point in F such that

$$|z_1| > |m_k|, \; |m_k'|$$

for all k.

Then $w_i(z_1) \neq 0$.

We have seen in (2.3.2) that

$$u_i(z) = \frac{z - \gamma_i^{-1} a}{z - \gamma_i a} \cdot \prod_{\substack{\gamma \in \Gamma \\ \gamma \neq \gamma \\ \gamma \neq id}} \frac{z - \gamma a}{z - \gamma \gamma_i a}.$$

Therefore

$$\frac{u_i'(z)}{u_i(z)} = \frac{-\gamma_i a + \gamma_i^{-1} a}{(z - \gamma_i a)(z - \gamma_i^{-1} a)} + \sum_{\substack{\gamma \neq id \\ \gamma \neq \gamma_i}} \frac{-\gamma \gamma_i a + \gamma a}{(z - \gamma a)(z - \gamma \gamma_i a)}$$

$$\text{as} \quad \frac{\left(\frac{z - a}{z - b}\right)'}{\left(\frac{z - a}{z - b}\right)} = \frac{\frac{(z - b) - (z - b)}{(z - b)^2}}{\left(\frac{z - a}{z - b}\right)} = \frac{a - b}{(z - a)(z - b)}$$

If $a \in F$, then

$$\gamma a, \; \gamma \gamma_i a \in B_k$$

for the same k, $1 \leq k \leq 2g$, as long as $\gamma \neq id$, $\gamma \neq \gamma_i^{-1}$.

$$\text{Thus} \quad \left| \frac{\gamma a - \gamma \gamma_i a}{(z_1 - \gamma a)(z_1 - \gamma \gamma_i a)} \right| < \frac{r_k}{|z_1|^2}.$$

$$\text{But} \quad \left| \frac{\gamma_i a - \gamma_i^{-1} a}{(z_1 - \gamma_i a)(z_1 - \gamma_i^{-1} a)} \right| = \frac{|m_i - m_i'|}{|z_1|^2}$$

which shows that

$$|w(z_1)| = \frac{|m_i - m_i'|}{|z_1|^2} .$$

Thus $w(z_1) \neq 0$.

This shows that $w_i(z)$ has a zero of order 2 at ∞.

If now $\alpha \in \Gamma$, $\alpha = \begin{pmatrix} a & b \\ c & d \end{pmatrix} \in SL_2(K)$.

Then $u(\alpha z) = \eta_\alpha \cdot u(z)$

$$\frac{du(\alpha z)}{dz} = \frac{du}{dz}(\alpha z) \quad \frac{d\alpha(z)}{dz} = \eta_\alpha \cdot \frac{du(z)}{dz}$$

$$= u'(\alpha z) \quad \frac{1}{(cz + d)^2} = \eta_\alpha \cdot u'(z).$$

Thus $\frac{u'(\alpha z)}{u(\alpha z)} = \eta_\alpha \cdot (cz + d)^2 \cdot u'(z) \frac{1}{\eta_\alpha \cdot u(z)}$

$$= (cz + d)^2 \frac{u'(z)}{u(z)}.$$

As $(cz + d)^2$ has a pole of order 2 at ∞ we find that $\frac{u'(\alpha\infty)}{u(\alpha\infty)}$ is some value $\neq 0$ whenever $\alpha \neq \mathrm{id}$.

Take $\alpha \neq \mathrm{id}$ and $\alpha\infty = a$. Then $\mathfrak{w}(a) \neq 0$ and we find $z_2, \ldots, z_g \in U$ such that

$$\mathfrak{w}(a), \mathfrak{w}(z_2), \ldots, \mathfrak{w}(z_g)$$

are linearly independent.

Thus $\Delta(a, z_2, \ldots, z_g) \neq 0$ and ϕ is locally bianalytic at the point $(a, z_2, \ldots, z_g)$.

If $\alpha \in [\Gamma, \Gamma]$, then

$$\phi(\alpha\infty, x_2, \ldots, x_g) = \phi(\infty, x_2, \ldots, x_g)$$

for all x_i.

This shows that ϕ is locally bianalytic at $(\infty, z_2, \ldots, z_g)$.

Chapter III The geometry of Mumford curves

Introduction.

For a Schottky group Γ the set of ordinary points Ω of Γ is open in $\mathbb{P}(K)$. The aim of this chapter is to give Ω and Ω/Γ the structure of an analytic space over k.

For this purpose we develop in §1 the theory of "rigid analytic spaces" in the sense of John Tate. There seems to be no systematic account of that theory. Section 1 can serve as an introduction to "rigid analytic spaces". We have given complete proofs of all that we need in these lecture notes with one exception namely the triviality of the cohomology of coherent sheaves on an affinoid space. For this isolated result the reader is referred to John Tate [42].

In §1 we treat the following subjects: affinoid algebras; Weierstraß theorems; various finiteness theorems; the spectral norm; rational domains; Grothendieck topology on affinoid spaces; the structure sheaf and its cohomology; definition of analytic spaces over k; contour integration; stable fields.
Much of this is somewhat technical. At some places we have assumed that k is algebraically closed. One can see that this assumption is inessential. The reader is advised to examine the many examples in the text (they give the flavour of the subset) and to skip the more technical proofs.

In section 2 the analytic structure on Ω/Γ is introduced. It is shown that Ω/Γ is an algebraic curve. Such a curve ("parametrised by a Schottky group Γ") is called a Mumford curve. The same methods are used to give an analytic structure on a torus $(k^*)^n/\Lambda$. In Chap VI it will turn out that the Jacobian variety of a Mumford curve is an analytic torus.

Reductions of affinoid and analytic spaces are introduced in §2. It is shown in (2.10.6) that this notion coincides with the reduction maps introduced in Chap I.

It is proved that Ω/Γ has a unique stable reduction. The graph of this stable reduction is equal to the graph $T(\mathcal{L})/\Gamma$ (introduced in Chap I) as will be shown in Chap IV.

For any genus g an explicit construction of the stable reduction of Ω/Γ is given. For $g \leq 3$ all the possibilities for the stable reduction are calculated.

§1 k-analytic spaces

In this rather long section we give an introduction to k-analytic and spaces. Our survey is necessarily incomplete. A more or less complete account would fill a book.

The building blocks for k-analytic spaces are k-affinoid spaces and k-affinoid algebras. Let $z_1,\ldots, z_n$ denote indeterminates. Then $k\langle z_1,\ldots, z_n\rangle = T_n$ (or $T_n(k)$) denotes the ring of all power series $\Sigma a_\alpha z_1^{\alpha_1}\ldots z_n^{\alpha_n}$ with $a_\alpha \in k$ and $\lim a_\alpha = 0$.

The norm on T_n is given by $\|\Sigma a_\alpha z_1^{\alpha_1}\ldots z_n^{\alpha_n}\| = \max |a_\alpha|$. We denote by T_n^o the subring of T_n consisting of the elements $f \in T_n$ with $\|f\| \leq 1$. The set $T_n^{oo} = \{f \in T_n \; \|f\| < 1\}$ is an ideal in T_n^o. The residue ring T_n^o/T_n^{oo} is denoted by $\bar{T}_n$. One easily sees that $\bar{T}_n \cong \bar{k}[z_1,\ldots, z_n]$. Since $\bar{T}_n$ has no zero-divisors, it follows that for $f,g \in T_n$ with $\|f\| = \|g\| = 1$ also $\|fg\| = 1$. And so, in general, $\|fg\| = \|f\|\|g\|$ for all $f,g \in T_n$.

An <u>affinoid algebra</u> A over k (or <u>Tate-algebra</u>) is a k-algebra which is a finite extension of some T_n. That means: there exists a k-algebra homomorphism $T_n \to A$ which makes A into a finitely generated T_n-module.

The Tate-algebra T_n shares many properties with the polynomial ring $\bar{k}[z_1,\ldots, z_n]$. The main tool in working with T_n and other affinoid algebras is the Weierstraß theorem. For its formulation we use the following notations: The map $T_n^o \to \bar{T}_n$ is denoted by $f \mapsto \bar{f}$. An element $f \in T_n$ with $\|f\| = 1$ is called <u>regular in</u> z_n <u>of degree</u> d if $\bar{f} \in \bar{T}_n$ has the form $\lambda z_n^d + \sum_{i<d} c_i z_n^i$ with $\lambda \in \bar{k}$, $\lambda \neq 0$, and $c_o, \ldots, c_{d-1} \in \bar{k}[z_1, \ldots, z_{n-1}]$. Further T_{n-1} denotes the subring $k\langle z_1, \ldots, z_{n-1}\rangle$ of T_n and $T_{n-1}[z_n]$ is also considered as subring of T_n.

(1.1) Theorem: (Weierstraß preparation and division)

1) Division: Let $f \in T_n$ be regular of degree d in z_n. For any $g \in T_n$ there exists unique $q \in T_n$ and $r \in T_{n-1}[z_n]$ of degree $<d$ in z_n such that $g = qf + r$. Moreover $\|g\| = \max(\|q\|, \|r\|)$.

(2) Preparation: Let $f \in T_n$ have norm 1. There exists an automorphism σ of T_n such that $\sigma(f)$ is regular in z_n of some degree.

Proof: (1) The condition on f implies that we can write $f = f_o + D$ where $f_o = \lambda z_n^d + \sum_{i=o}^{d-1} c_i z_n^i$ with $\lambda \in k$, $|\lambda| = 1$ and $c_i \in T_{n-1}$ have norm ≤ 1 and D in T_n has norm <1.

It is rather clear that any $g \in T_n$ can be written as $g = qf_o + r$ with $q \in T_n$; $r \in T_{n-1}[z_n]$ of degree $<d$ in z_n and $\|g\| = \max(\|q\|, \|r\|)$. We want to show the same statement for f instead of f_o. Let $g \in T_n$. Then $g = q_o f_o + r_o = q_o f + r_o + g_1$ where $g_1 = -q_o D$ has norm $\leq \|D\|\|g\|$. We continue with g_1 and write $g_1 = q_1 f + r_1 + g_2$ where $g_2 = -q_1 D$ has norm $\leq \|D\|\|g_1\| \leq \|D\|^2\|g\|$. By induction we find a sequence of elements g_m, q_m, r_m with $\lim \|g_m\| = \lim \|q_m\| = \lim \|r_m\| = 0$. Further $r_m \in T_{n-1}[z_n]$ has degree $<d$ in z_n and $g_m = q_m f + r_m + g_{m+1}$ $(m \geq 0)$. Since T_n is complete with respect to the norm, we find

$$g = \left(\sum_{i=o}^{\infty} q_i\right)f + \left(\sum_{i=o}^{\infty} r_i\right) \text{ and } \|g\| = \max\left(\left\|\sum_{i=o}^{\infty} q_i\right\|, \left\|\sum_{i=o}^{\infty} r_i\right\|\right).$$

So the division by f exists and has the required property about the norms. Suppose that the division by f is not unique. Then $g = q_1 f_1 + r_1 = q_2 f_2 + r_2$ exists with $q_1 \neq q_2$. We may suppose $\|q_1 - q_2\| = 1$ and hence also $\|r_1 - r_2\| = 1$.
The equation $(q_1 - q_2)f = r_2 - r_1$ implies that $\overline{(q_1 - q_2)}\bar{f} = \overline{r_2 - r_1}$

in $\bar{T}_n$. This however contradicts $\text{degree}_{z_n}\ (\overline{r_2 - r_1}) < d$ and $\text{degree}_{z_n}\ \bar{f} = d$.

(2) A linear substitution of the variables $z_i \mapsto \Sigma\, \lambda_{ij} z_j$ with $(\lambda_{ij}) \in GL_n(k^o)$ (i.e.: all $\lambda_{ij} \in k^o$ and the determinant has absolute value 1) induces an automorphism σ of $T_n = k\langle z_1, \ldots, z_n\rangle$ which is given by $\sigma(f) = f(\Sigma\lambda_{1j}z_j, \Sigma\lambda_{2j}z_j, \ldots, \Sigma\lambda_{nj}z_j)$. The automorphism σ induces an automorphism $\bar{\sigma}$ of $\bar{T}_n = \bar{k}[z_1, \ldots, z_n]$ which is given by $z_i \mapsto \Sigma\bar{\lambda}_{ij}z_j$.

Write $\bar{f} = h_o + h_1 + \ldots + h_d$ with h_i homogeneous in $z_1, \ldots, z_n$ of degree i and $h_d \neq 0$. If the residue field $\bar{k}$ is infinite then for a suitable σ, the homogeneous polynomial $\bar{\sigma}(h_d)$ contains the term z_n^d. Hence $\sigma(f)$ is regular in z_n of degree d.

If the field $\bar{k}$ is finite then we have to use other substitutions to make f regular in z_n. The substitution $z_i \mapsto z_i + z_n^{e_i} (1 \leq i < n)$ and $z_n \mapsto z_n$, with $e_1, \ldots, e_{n-1}$ integers ≥ 0, clearly gives an automorphism σ of T_n. Let $\bar{f} = \Sigma a_\alpha z_1^{\alpha_1} \ldots z_n^{\alpha_n} \neq 0$ then

$\overline{\sigma(f)} = \bar{\sigma}(\bar{f}) = \Sigma a_\alpha (z_1 + z_n^{e_1})^{\alpha_1} \ldots (z_{n-1} + z_n^{e_{n-1}})^{\alpha_{n-1}} z_n^{\alpha_n}$.

There is a choice of $e_1, \ldots, e_{n-1}$ such that $a_\alpha \neq 0$ and $a_\beta \neq 0$ and $\alpha \neq \beta$ implies $e_1\alpha_1 + \ldots + e_{n-1}\alpha_{n-1} + \alpha_n \neq e_1\beta_1 + \ldots + e_{n-1}\beta_{n-1} + \beta_n$. For this choice the total degree of $\overline{\sigma(f)}$ is

$N = \max\{e_1\alpha_1 + \ldots + e_{n-1}\alpha_{n-1} + \alpha_n \mid a_\alpha \neq 0\}$. The degree of

$\overline{\sigma(f)}(0, \ldots, 0, z_n) = \Sigma a_\alpha z_n^{(e_1\alpha_1 + \ldots + e_{n-1}\alpha_{n-1} + \alpha_n)}$ is also N.

So $\overline{\sigma(f)}$ has the required form.

Some consequences of the Weierstraß theorem

(1.2) T_n is Noetherian.

(1.3) T_n has unique factorization.

(1.4) Every ideal in T_n is closed.

(1.5) For every ideal I in T_n there is a finite injective map $T_d \to T_n/I$.

The number d is equal to the Krull-dimension of T_n/I.

(1.6) *For every maximal ideal $\underline{m}$ of T_n the field $T_n/\underline{m}$ is a finite extension of k.*

(1.7) *Every affinoid algebra A has the form T_n/I and is a Banach space with respect to the quotient norm. The topology of A does not depend on the presentation $A = T_n/I$.*

Proof: (2) Let I be a non-trivial ideal of T_n. Using Weierstraß' preparation we may suppose that I contains an element f regular in z_n of degree d. Using the division by f, we see that I is generated by f and $J = I \cap T_{n-1}[z_n]$. By induction T_{n-1} and $T_{n-1}[z_n]$ are Noetherian. So the ideal J is finitely generated. Hence also I is finitely generated.

(3) Let f be a non-zero element of T_n. After multiplying g by a constant and a change of coordinates we may suppose that f is regular in z_n of degree d. The division $z_n^d = qf + r$ yields that $z_n^d - r$ is also regular in z_n of degree d. The division $f = q'(z_n^d - r) + r'$ yields $f = qq'f + r'$. The unicity of the division implies $qq' = 1$ and $r' = 0$. Hence $f = q'(z_n^d - r)$ where $q' \in T_n$ is a unit.

By induction the rings T_{n-1} and $T_{n-1}[z_n]$ have already unique factorization. So $f = q'f_1 \ldots f_s$ where the $f_i \in T_{n-1}[z_n]$ are monic in z_n, have norm 1 and are prime elements of $T_{n-1}[z_n]$.

Let g in $T_{n-1}[z_n]$ be monic in z_n, have norm 1 and be prime in $T_{n-1}[z_n]$. Then we have to show that g is also prime in T_n. Let $g = g_1g_2$ be a decomposition in which g_1 is not a unit. Then we may suppose that $\|g_1\| = 1$ and it follows that g_1 is regular in z_n. By the reasoning above we may replace g_1 by a monic polynomial in z_n. The division of g by g_1 is unique and so g_2 also belongs to $T_{n-1}[z_n]$. But since g is prime in $T_{n-1}[z_n]$ we must have g_2 is a unit.

So we have shown that any $f \neq 0$ in T_n is a product of a unit and prime elements. A similar reasoning shows that this decomposition is unique up to the order of the prime factors and units.

(4) Suppose that T_n contains non-closed ideals. Let I be maximal among them. Since T_n is Noetherian I exists. Between I and its closure $\hat{I}$ there are no ideals. Hence $\hat{I}/I \cong T_n/\underline{m}$ where $\underline{m}$ is a maximal ideal of T_n. Any maximal ideal in a Banach algebra with 1 is closed. So $\underline{m}$ is closed. Let $e_2, \ldots, e_a$ generate I and $e_1, e_2, \ldots, e_a$ generate $\hat{I}$. Then $\phi : T_n^a \to \hat{I}$ defined by $(f_1, \ldots, f_a) \mapsto \Sigma f_i e_i$ is continuous and has a closed kernel ker ϕ. Then $T_n^a/\ker \phi \to \hat{I}$ is an isomorphy of Banach spaces. Hence $\underline{m} \oplus T_n^{a-1}/\ker \phi \approx I$ is closed. This contradicts our assumption.

(5) If $I \neq 0$ then we may suppose that I contains an element f regular in z_n of degree d. As in the proof of (3) f is a unit times $g = z_n^d + \sum_{i=0}^{d-1} a_i z_n^i$ with $a_0, \ldots, a_{d-1} \in T_{n-1}^o$. Clearly $g \in I$ and $T_n/(g)$ is a finitely generated T_{n-1}-module with generators $1, z_n, \ldots, z_n^{d-1}$. Let $J = I \cap T_{n-1}$ then $T_{n-1}/J \hookrightarrow T_n/I$ is a finite, injective map. By induction there exists a finite, injective $T_d \hookrightarrow T_{n-1}/J$. This proves the first statement.

Since T_n/I is a finite extension of T_d, the two rings have the same Krull-dimension. The sequence of prime ideals $(z_1, \ldots, z_d) \supset (z_1, \ldots, z_{d-1}) \supset \ldots \supset (z_1) \supset (0)$ in T_d shows Krull-dim $T_d \geq d$. By induction on d, we show that infact Krulldim $T_d = d$. Let $0 \subsetneqq \underline{p}_1 \subsetneqq \cdots \subsetneqq \underline{p}_s$ be a sequence of prime ideals in T_d. Choose a non-zero element $f \in \underline{p}_1$ and a finite, injective map $T_{d-1} \hookrightarrow T_d/(f)$. Then $s \leq 1 +$ Krulldim $T_d/(f)$ and by induction Krulldim $T_d/(f) =$ Krull-dim $T_{d-1} = d - 1$. Hence $s \leq d$.

(6) Let $T_d \hookrightarrow T_n/\underline{m}$ be finite and injective. Since $T_n/\underline{m}$ is a field we must have $d = 0$ and $T_o = k$. This proves the statement.

(7) An affinoid algebra A is equipped with a finite map $\phi : T_n \to A$. Let $e_1, \ldots, e_s$ generate A as a T_n-module. After multiplying the e_i's by a suitable constant we may suppose that e_i satisfies an equation of the type.

$$X_i^{d_i} + \phi(a_{d_i-1}^{(i)})X_i^{d_i-1} + \ldots + \phi(a_o^{(i)}) = 0 \quad (i = 1,\ldots, s)$$

where all $a_j^{(i)}$ are in T_n^o. Put $P_i = X_i^{d_i} + a_{d_i-1}^{(i)} X_i^{d_i-1} + \ldots + a_o^{(i)}$. Then $T_n\langle X_1,\ldots, X_s\rangle/(P_1,\ldots, P_s) \simeq T_n[X_1,\ldots, X_s]/(P_1,\ldots, P_s)$ is mapped surjectively to A by sending X_i to e_i. So we have found a surjective map $T_{n_1} \xrightarrow{\phi_1} A$ with kernel I_1. Since I_1 is closed A becomes a Banach space $(A, \|\ \|_1)$ for the induced norm. Let $T_{n_2} \xrightarrow{\phi_2} A$ be another surjective map. Then we define $\phi_3 : T_{n_1+n_2} \to A$ $(n_3 = n_1 + n_2)$ by $\phi_3(z_i) = \phi_1(z_i)$ for $1 \leq i \leq n_1$ and $\phi_3(z_{n_1+i}) = \phi_2(z_i)$ for $1 \leq i \leq n_2$. The induced map of Banach spaces $(A, \|\ \|_1) \to (A, \|\ \|_3)$ is bijective and continuous. Hence it is a homomorphism. The same holds for $(A, \|\ \|_2) \to (A, \|\ \|_3)$. So the topology on A is independent of the presentation of A as a T_n/I.

(1.8) Let A be an affinoid algebra over k. We associate with A the set X of all maximal ideals of A. The set X is called an <u>affinoid space</u> and is denoted by Sp (A).

Every $f \in A$ can be considered as a function on X, namely $x \mapsto f(x)$ = the image of f in the field A/x. The field A/x is a finite extension of k by virtue of (1.6) and it carries a unique valuation extending the valuation of k. So $|f(x)|$ is well defined for any $x \in X$ and $f \in A$. On X there is a lot of additional structure. First of all X has a topology generated by the sets $\{x \in X \mid |f(x)| \leq 1\}$ with $f \in A$. It is the smallest topology for which all $f \in A$ are continuous functions on X. We will in the sequel often write $\mathcal{O}(X)$ for the algebra A. On $\mathcal{O}(X)$ we have the <u>spectral (semi-) norm</u>

$$\|f\|_{sp} = \sup \{|f(x)| \mid x \in X\}.$$

The norm on A/x induced by A is certainly $\geq |\ |$, the valuation of A/x. Hence we have $\|f\|_{sp} \leq \|f\|$.

The meaning of affinoid spaces, spectral semi-norm etc. will be clarified by the following examples.

(1.8.1) Example

Let $k = K$ (or otherwise stated, suppose that k is algebraically closed), then $A/x = K$ for every maximal ideal x of A and every $f \in A$ is an ordinary function on X with values in K.

Every maximal ideal x of $T_n = K\langle z_1, \dots, z_n\rangle$ is determined by the values $z_1(x), \dots, z_n(x)$. In this way we can identify X with the polydisk $\{(x_1,\dots,x_n) \in K^n \mid |x_i| \leq 1 \text{ for all } i\}$. The topology on X coincides with its topology as subspace of K^n.

The norm on T_n coincides with the spectral norm. We know already that $\|f\|_{sp} \leq \|f\|$. Suppose that $\|f\| = 1$ then $\bar{f}$ is a non-zero element of $\bar{K}[z_1, \dots, z_n]$. Take $(x_1, \dots, x_n) \in K^n$ with all $|x_i| \leq 1$ such that $\bar{f}(\bar{x}_1, \dots, \bar{x}_n) \neq 0$. Then $\overline{f(x_1, \dots, x_n)} = \bar{f}(\bar{x}_1, \dots, \bar{x}_n) \neq 0$ and $|f(x_1, \dots, x_n)| = 1$. Hence $\|f\|_{sp} = 1$.

Let $f \in T_n$ have norm 1 and suppose that $\bar{f} \in \bar{K}[z_1, \dots, z_n]$ is not a constant. Then for suitable $x_2, \dots, x_n \in K$ with $|x_i| \leq 1$ also $\bar{f}(z_1, \bar{x}_2, \dots, \bar{x}_n) \in \bar{K}[z_1]$ is not a constant. Consider $g = f(z_1, x_2, \dots, x_n) \in K\langle z_1\rangle$. Then g is regular in z_1 of some degree d and $g = q(z_1^d + a_{d-1}z_1^{d-1} + \dots + a_0)$ where q is a unit and $a_0, \dots, a_{d-1} \in K^o$. The polynomial has a root $x_1 \in K$ with $|x_1| \leq 1$. So $f(x_1, x_2, \dots, x_n) = 0$. This reasoning implies the following statement:

$$f = a_0 + \sum_{\alpha \neq o} a_\alpha z_1^{\alpha_1} \dots z_n^{\alpha_n} \in K\langle z_1, \dots, z_n\rangle \text{ is invertible}$$

if and only if f has no zeros in $\mathrm{Sp}(T_n)$ if and only if $|a_0| > \max\{|a_\alpha| \mid \alpha \neq 0\}$.

This implies that T_n is in fact the completion with respect to the spectral norm of all rational functions on K^n with poles outside $\{(x_1, \dots, x_n) \mid x_i \in K \text{ and } |x_i| \leq 1\}$.

(1.8.2) Example: Again we assume $k = K$.
Let $A = T_n/I$ where I is a radical ideal (i.e. $f^2 \in I$ implies $f \in I$). Write $T_n = K\langle z_1, \ldots, z_n\rangle$ and $I = (f_1, \ldots, f_s)$ and $X = \operatorname{Sp} A$. Then $X \tilde{\to} \{x = (x_1, \ldots, x_n) \in K^n \mid |x_i| \leq 1$ and all $f_j(x) = 0\}$. This isomorphism is a topological isomorphism given by:

$$x \mapsto (z_1(x), \ldots, z_n(x)).$$

We are faced here with two questions:

a) If $f \in T_n$ is zero on X(identified as subset of K^n) is then $f \in I$?

b) Is A equal to the completion with respect to the spectral norm of X of the ring of rational functions in $z_1, \ldots, z_n$ with poles outside X?

Question a) is equivalent with: Is the spectral semi-norm on A in fact a norm?

Question b) is the same as: Is the spectral semi-norm on A equivalent with a quotient norm on A?

The answer to the questions is "yes". We will give a proof in (1.9).

If the ideal I is not radical then using the results of (1.9) it follows that $\{f \in A \mid \|f\|_{sp} = 0\} = \sqrt{I}/I$ = the ideal of all nilpotent elements of A.

(1.8.3) Example: We assume again $K = k$. In Chapter I we have defined a connected affinoid subspace F of $\mathbb{P}(K)$ as the complement of finitely many open disks in $\mathbb{P}(K)$ with radii in $|K^*|$. We will show that F can be considered as an affinoid space in the terminology of this chapter.

Assume for convenience that $\infty \in F$. Then F is given in $\mathbb{P}(K)$ by the inequalities $|z - a_i| \geq |\pi_i|$ $(i = 1, \ldots, n)$ where $a_i, \pi_i \in K$ and the open disks $|z - a_i| < |\pi_i|$ are disjoint.

Consider the map $\phi : F \to \{(x_1, \ldots, x_n) \in K^n \mid$ all $|x_i| \leq 1\}$ given by $z \mapsto (\frac{\pi_1}{z-a_1}, \ldots, \frac{\pi_n}{z-a_n})$. The map ϕ is injective and its image G is the subset of $\{(x_1, \ldots, x_n) \in K^n \mid$ all $|x_i| \leq 1\}$ given by the equations:

$$E_{ij} = \frac{\pi_i}{a_i - a_j} z_j - \frac{\pi_j}{a_i - a_j} z_i + z_i z_j = 0 \qquad (i \neq j).$$

Let I be the ideal of T_n generated by the E_{ij} and $A = T_n/I$. Then $G = Sp(A)$ and we have identified F with an affinoid space.

Every $f \in T_n$ can uniquely be written as

$$f = f_o + f_1 \quad \text{where } f_1 \in I \text{ and } f_o = a + \sum_{m \geq 1} \sum_{i=1}^{n} a_{im} z_i^m$$

with $a_1, a_{im} \in K$ and limit $a_{im} = 0$. Moreover $\|f\| = \max(\|f_o\|, \|f_1\|)$. One easily verifies this statement for $f \in K[z_1, \ldots, z_n]$. The general case follows by taking limits. The algebra A is given the quotient norm with respect to the surjective map $T_n \to A$ and the map is denoted by $f \mapsto f^*$. From the formula above it follows that $\|f^*\| = \|f_o\|$. We will show that $\|f^*\| = \|f^*\|_{sp}$ where $\| \ \|_{sp}$ denotes the spectral norm on A.

Let $\|f^*\| = 1$ then $\|f_o\| = 1$ and $\max(|a|, |a_{im}|) = 1$. We may assume that $\max(|a|, |a_{1m}|) = 1$. For $z \in F$ with $|z - a_1| = |\pi_1|$ we have:

$$\overline{f^*(\phi(z))} = \bar{a} + \sum_{m \geq 1} \sum_{i=1}^{n} \bar{a}_{i\,m} \left(\overline{\left(\frac{\pi_i}{z - a_i} \right)} \right)^m .$$

Let $t \in \bar{K}$ denote the residue of $\frac{\pi_1}{z - a_1}$ in $\bar{K}$. The residue of $\frac{\pi_i}{z - a_i}$ in $\bar{K}$ is constant if $|\pi_1| < |\pi_i|$ or if $|\pi_i| < |a_1 - a_i|$. If $|\pi_1| = |\pi_i| = = |a_1 - a_i|$ then the residue of $\frac{\pi_i}{z - a_i}$ is equal to $\frac{t}{c_i + d_i t}$ with c_i, d_i non-zero elements of $\bar{K}$. It follows that $\overline{f^*(\phi(z))}$ is a non-trivial function of t. Hence for a good choice of z (or t) we find $\|f^*(\phi(z))\| = 1$. So $\|f^*\| = \|f^*\|_{sp}$.

If one considers A as functions on $F \subset \mathbb{P}(K)$ then A is the completion (w.r.t. supremum-norm on F) of the rational functions on $\mathbb{P}(K)$ with poles outside F.

Moreover we have found a well known decomposition for the functions $f \in A = \mathcal{O}(F)$ namely:

Any $f \in \mathcal{O}(F)$ with $f(\infty) = 0$ can uniquely be written as $f = f_1 + \ldots + f_n$ with f_i convergent on $|z - a_i| \geq |\pi_i|$ and $f_i(\infty) = 0$. In this decomposition $\|f\| = \max(\|f_i\|)$ and $\|f_i\|$ is also equal to $\sup\{|f(z)| \,|\, z \in \mathbb{P}(K);\ |z - a_i| \geq |\pi_i|\}$.

(1.8.4) Example

Let $\rho > 0$ and let A denote the algebra of all series $\Sigma a_n z^n$ with $a_n \in k$ and $\lim |a_n|\rho^n = 0$. One can consider A as an algebra of functions on $\{z \in K \,|\, |z| \leq \rho\}$. The spectral norm on A is given by $\|\Sigma a_n z^n\| = \max |a_n|\rho^n$ as one easily verifies. We will prove the following statement:

A is affinoid over k if and only if some positive power of ρ belongs to $|k^*|$.

Proof: If $\rho^n \in |k^*|$ and $n > 0$, then $T_1 = k\langle X\rangle \to A$ defined by $X \mapsto \frac{1}{\lambda}z^n$, where $|\lambda| = \rho^n$, is a finite map.

On the other hand: if A is affinoid then there exists a finite injective $\phi : T_1 \hookrightarrow A$ since A has Krull-dimension 1.
Let the minimal equation of z over T_1 be

$$z^d + a_1 z^{d-1} + \ldots + a_d = 0 \text{ with } a_1, \ldots, a_d \in T_1.$$

Let μ denote $\max\limits_{1\leq i\leq d} \|a_i\|^{\frac{1}{i}}$. Then $\mu^d \in |k^*|$. For any $x \in Sp(T_1)$ and any root λ of the equation $z^d + a_1(x)z^{d-1} + \ldots + a_d(x) = 0$ there is a maximal ideal y of A with $z(y) = \lambda$ and $y \cap T_1 = x$. This follows from the finiteness of $T_1 \hookrightarrow A$. The $\max\{|\lambda| \,|\, \lambda$ is root of $z^d + a_1(x)z^{d-1} + \ldots + a_d(x) = 0\}$ is equal to $\max\limits_{1\leq i\leq d} |a_i(x)|^{1/i}$.
And so $\|z\|_{sp} = \max\limits_{x\in Sp(T_1)} (\max\limits_{1\leq i\leq d} |a_i(x)|^{1/i}) = \mu$ and $\mu = \rho$.
Hence $\rho^d \in |k^*|$.

Remark: The result above is the reason why we have insisted in Chapter I on closed or open disks with radii in $\sqrt{|k^*|} = \{\rho \in \mathbb{R} \,|\, \rho > 0$ and some positive power of ρ lies in $|k^*|\}$.

(1.9) Theorem: *Let* A *be a reduced affinoid algebra over* k. *In case* k *has characteristic* p *we assume that* $[k : k^p] < \infty$. *Then:*

(1) *The integral closure of* A *in its total quotientring is a finitely generated* A*-module.*

(2) *The spectral semi-norm on* A *is a norm equivalent with any quotient-norm on* A *(i.e. induced by a surjective* $T_n \to A$).

Proof: We note that for any finite extension $A \subset B$ of affionid algebras the spectral (semi-)norm on A coincides with the restriction of the spectral semi-norm on B. This follows from: for every maximal ideal $\underline{m}$ of A there is a maximal ideal $\underline{m}'$ of B with $\underline{m}' \cap A = \underline{m}$.

Let now $\underline{p}_1, \dots, \underline{p}_s$ be the minimal prime ideals of A. Then $A \hookrightarrow A/\underline{p}_1 \oplus \dots \oplus A/\underline{p}_s$. The embedding is an isometry for the spectral norms and a homeomorphism for the quotient norms. The integral closure A^η of A in its total quotientring is equal to $(A/\underline{p}_1)^\eta \oplus \dots \oplus (A/\underline{p}_s)^\eta$. From all this it follows that we may assume in the proof of (1.9) that A has no zero-divisors.

Furthermore we may replace A by any finite extension B and give a proof of statements (1) and (2) for B.

Let $T_d \hookrightarrow A$ be a finite, injective map. Let N_o denote the quotient field of T_d; N_1 the smallest normal field extension of N_o containing A; N_2 the intermediate field satisfying $N_2 \supset N_o$ is purely inseparable and $N_1 \supset N_2$ is a Galoisextension.

$$\begin{array}{ccccc} N_o & \subset & N_2 & \subset & N_1 \\ & & \cap & & \cap \\ & & N_3 & \subset & N_4. \end{array}$$

In the case k has characteristic $p \neq 0$ the field N_2 is contained in $N_3 = N^{\frac{1}{p^a}}$ = the quotient field of the algebra $k^{\frac{1}{p^a}}\langle z_1^{\frac{1}{p^a}}, \dots, z_d^{\frac{1}{p^a}}\rangle$ for some $a \geq 0$.

Let N_4 denote the compositum of N_1 and N_3. Since $k^{p^{\frac{1}{a}}}\langle z_1^{p^{\frac{1}{a}}},\ldots, z_d^{p^{\frac{1}{a}}}\rangle$ is clearly a finite extension of $k\langle z_1,\ldots, z_d\rangle$, it suffices to consider the Galoisextension $N_4 \supset N_3$ generated by one element f of which we may assume $\|f\|_{sp} \leq 1$.

So we have reduced the general case to a special one namely:

Lemma: Let $f = z_n^d + a_1 z_n^{d-1} + \ldots + a_d \in T_n = k\langle z_1,\ldots, z_n\rangle$ be irreducible with coefficients $a_1,\ldots, a_d \in T_{n-1}^o$. Suppose that the quotient field of $A = T_n/(f)$ is a Galois-extension of the quotient field of T_{n-1}. Let B denote the integral closure of A. Then:

(1) B is a finitely generated T_{n-1}-module.

(2) The spectral semi-norm on B is a norm equivalent with any quotient-norm on B.

Proof of the lemma:

Let N_o denote the quotient field of T_{n-1} and N_1 the quotient field of A. By $Tr : N_1 \to N_o$ we denote the ordinary trace, $Tr(a) = \Sigma\sigma(a)$, the sum taken over all σ in the Galoisgroup of N_1 over N_o. The bilinear form $\langle x,y\rangle = Tr(xy)$ on the vectorspace N_1 is non-degenerate.

In fact $\det((\langle z_n^i, z_n^j\rangle)_{i,j=o}^{d-1}) = \Delta$ is equal to the ordinary discriminant of f with respect to the variable z_n. By assumption $\Delta \neq 0$ and $\Delta \in T_{n-1}$.

B^o denote the set of elements of B with spectral norm ≤ 1. It is clear that B and B^o are invariant under the action of the Galoisgroup. In particular $Tr(B^o) \subset B^o \cap N_o$. We have $B \cap N_o = T_{n-1}$ since T_{n-1} has unique factorization and is as a consequence integrally closed. Further $B^o \cap T_{n-1} = T_{n-1}^o$. Hence $Tr(B^o) \subseteq T_{n-1}^o$ and in particular $\Delta \in T_{n-1}^o$. The dual base $e_o,\ldots, e_{d-1}$ of N_1 over N_o defined by $\langle e_i, z_n^j\rangle = \delta_{ij}$ satisfies: $\Delta e_i \in T_{n-1}^o[z_n]$.

Let $b \in B$ have the form $b = \Sigma b_i e_i$ with $b_i \in N_o$. Then $b_j = Tr(bz_n^j) \in Tr(B) \subseteq T_{n-1}$. So $\Delta b \in T_{n-1}[z_n]$. It follows that $B \cong \Delta B$ lies in the finitely generated T_{n-1}-module $T_{n-1}[z_n]$. Hence B is finitely generated as T_{n-1}-module.

Similary $\Delta B^o \subseteq T^o_{n-1}[z_n]$. We choose a presentation $B = T_m/I$ such that the induced norm on B gives on $T_{n-1} \subset B$ the usual spectral norm. Let $\| \|$ denote this norm.

Multiplication by Δ is a continuous injective map $B \to B$ with a closed ideal as image. Hence $B \tilde{\to} \Delta B$ is a homeomorphism and there is a constant $c > 0$ such that $c\|b\| \leq \|\Delta b\| \leq \|b\|$ for all $b \in B$.

Then $\Delta B^o \subseteq T^o_{n-1}[z_n] \subseteq \{b \in B | \|b\| \leq D\}$ where $D = \max (\|1\|, \|z_n\|, \ldots, \|z_n^{d-1}\|)$. Hence $\Delta B^o \subseteq \{b \in B | \|b\| \leq D\} \cap \Delta B \subseteq \{\Delta b | \|b\| \leq c^{-1}D\}$, and $\{b \in B | \|b\| \leq 1\} \subset B^o \subset \{b \in B | \|b\| \leq c^{-1}D\}$. This proves that $\| \|$ and $\| \|_{sp}$ on B are equivalent norms.

(1.10) Remarks:

(1) Theorem (1.9) is also true if k has characteristic p and $[k : k^p] = \infty$. The "purely inseparable" case is however somewhat more difficult to prove. See [16].

(2) If the valuation of k is discrete, then k^o is a Noetherian valuation ring and also $k^o\langle z_1, \ldots, z_{n-1}\rangle$ is Noetherian. From "$\Delta B^o \subset T_{n-1}[z_n]$" in the proof of the lemma above, it follows that B^o is in fact a finitely generated T^o_{n-1}-module.

If the valuation of k is non-discrete then in general B^o is not finitely generated as T^o_{n-1}-module. However, if k is algebraically closed then again B^o is finitely generated over T^o_{n-1}. See [19]. We will discuss this statement in (1.13).

(3) Affinoid algebras have many more nice properties. In fact they are excellent rings. In particular any affinoid algebra has a closed singular locus and satisfies the universal chain condition on prime

ideals. See [2].

(1.11) Remarks on the spectral norm:

(1) Let N be a field with a (non-archimedean) valuation; $\hat{N}$ the completion of N and N_1 the algebraic closure of $\hat{N}$ which carries a unique valuation $|\,|$ extending the valuation on N. Let $P = X^n + a_1X^{n-1} + \ldots + a_n \in N[X]$. Then $\max |a_i|^{\frac{1}{i}} = \max \{|\alpha| \mid \alpha \in N_1$ is a root of $P\}$.

Proof: Write $P = (X - \alpha_1) \ldots (X - \alpha_n)$ with $\alpha_1, \ldots, \alpha_n \in N_1$ and let $|\alpha_1| = \max (|\alpha_i|)$. Put $X = \alpha_1 Y$ then:

$$(Y - 1)(Y - \frac{\alpha_2}{\alpha_1}) \ldots (Y - \frac{\alpha_n}{\alpha_1}) = Y^n + \frac{a_1}{\alpha_1}Y^{n-1} + \ldots + \frac{a_n}{\alpha_1^n}.$$

It follows that $\max |\frac{a_i}{\alpha_1^i}| = 1$ and so $\max |a_i|^{\frac{1}{i}} = |\alpha_1|$.

(2) Let M be a finite extension of N containing some element f which satisfies the irreducible polynomial $P = X^n + a_1X^{n-1} + \ldots + a_n$ over N. Let $v_1, \ldots, v_s$ be the valuations of M extending the valuation $|\,|$ on N. Then $\max\limits_{1 \leq i \leq n} |a_i|^{\frac{1}{i}} = \max\limits_{1 \leq j \leq s} v_j(f)$.

Proof: We may suppose that $M = N(f)$. Every valuation v on M extending $|\,|$ is obtained by embedding N into N_1, the algebraic closure of $\hat{N}$. Such an embedding corresponds to a root $\alpha \in N_1$ of P. Consequently $\max v_j(f) = \max \{|\alpha| \mid \alpha \in N_1$ is a root of $P\}$. Statement (2) now follows from (1).

(3) Let A be an affinoid algebra without zero-divisors. Let $T_d \hookrightarrow A$ be a finite, injective map. A non-zero element $f \in A$ satisfies an irreducible polynomial $P = X^n + a_1X^{n-1} + \ldots + a_n$ with coefficients in T_d. Further $\|f\|_{sp} = \max \|a_i\|^{\frac{1}{i}}$; there is an $x \in X$ with $|f(x)| = \|f\|_{sp}$ and $\|f\|_{sp} \in \sqrt{|k^*|}$.

Proof: P is the irreducible polynomial of f over the quotient field of T_d. Since T_d is integrally closed one finds that all $a_i \in T_d$. For any $y \in Sp(T_d)$ and root λ of the polynomial $X^n + a_1(y)X^{n-1} + \ldots + a_n(y)$ there is a $x \in Sp(A)$ with $x \cap T_d = y$ and $f(x) = \lambda$. Hence $\|f\|_{sp} = \max_{y \in Sp(T_d)} (\max_i |a_i(y)|^{1/i}) = \max \|a_i\|^{1/i}$ and the statements follow.

(4) Let $T_d \overset{\phi}{\hookrightarrow} A$ be a finite, injective map between affinoid algebras, then $\phi^o : T_d^o \to A^o$ is integral. In particular $A^o = \{f \in A \mid \sup \|f^n\| < \infty\}$ where $\| \|$ is some norm on A induced by a surjective $T_m \to A$.

Proof: In proving that $f \in A^o$ is integral over T_d^o we may suppose that A has no nilpotents. Further we may replace A be its integral closure $A^n = A_1 \oplus \ldots \oplus A_s$ in which every A_i is an affinoid algebra without zero-divisors. Hence we may assume that A has no zero-divisors. Then the statement follows from (3).

Choose $T_m \to A$ such that the norm $\| \|$ on A is restricted to $T_d \hookrightarrow A$ the ordinary norm. Since $f \in A^o$ satisfies an equation $f^n + a_1 f^{n-1} + \ldots + a_n = 0$ with $\|a_i\| \leq 1$, $a_i \in T_n$, it follows that $\sup \|f^m\| = \max (1, \|f\|, \ldots, \|f^{n-1}\|)$.

On the other hand, $\sup \|f^m\| < \infty$ implies $\sup \|f^m\|_{sp} < \infty$. Since $\|f^m\|_{sp} = \|f\|_{sp}^m$, it follows that $f \in A^o$.

(5) The spectral semi-norm $\| \|_{sp}$ on any affinoid A satisfies $\|a\|_{sp} = \lim_{n\to\infty} \|a^n\|^{1/n}$.

Proof: Easy consequence of (4).

(6) Let $\phi : A \to B$ be a morphism of affinoid algebras. Then ϕ is continuous. There exists an extension $\psi : A\langle X_1, \ldots, X_s\rangle \to B$ with $\psi(X_i) = b_i$ if and only if $b_1, \ldots, b_s \in B^o$. Moreover if ψ exists then ψ is unique.

Proof: If ψ exists then $\psi(A\langle X_1, \ldots, X_s\rangle^o) \subset B^o$ and so $b_1, \ldots, b_s \in B^o$.

If $b_1, \ldots, b_s \in B^o$ then $\sup \|b_1^{\alpha_1} \ldots b_s^{\alpha_s}\| < \infty$ and $\psi(\Sigma a_\alpha X_1^{\alpha_1} \ldots X_s^{\alpha_s}) = \Sigma \phi(a_\alpha) b_1^{\alpha_1} \ldots b_s^{\alpha_s}$ is well defined and is a k-algebra homomorphism extending ϕ.

We can take $b_1, \ldots, b_s \in B^o$ such that ψ is surjective and we can take a surjective $\chi : T_n \to A$. The resulting surjective map $T_n\langle X_1, \ldots, X_s\rangle \to B$ induces a norm on B equivalent with the given norm. Hence ψ is continuous. Any choice for ψ extending ϕ is determined on $A[X_1, \ldots, X_s]$ and continuous. So there is only one possibility for ψ.

(1.12) We come now to a key result on affinoid algebras. Let A^o (A^{oo} resp.) denote the elements of A with $\| \|_{sp} \leq 1$ ($\| \|_{sp} < 1$ resp.). Let $\bar{A} = A^o/A^{oo}$. This is a $\bar{k} = k^o/k^{oo}$ - algebra.

<u>Theorem:</u> <u>Let</u> $\phi : A \to B$ <u>be a morphism of affinoid algebras.</u> <u>The following statements are equivalent:</u>

1) $\phi : A \to B$ <u>is finite.</u>
2) $\phi^o : A^o \to B^o$ <u>is integral.</u>
3) $\bar{\phi} : \bar{A} \to \bar{B}$ <u>is finite.</u>

<u>Proof:</u> It suffices to show the equivalence of the three statements for the case $A = T_d \hookrightarrow B$. In (1.11) we have already shown 1) $\Rightarrow$ 2). Further 2) implies that $\bar{\phi}$ is an integral extension. If we can show that the total quotient ring of $\bar{B}$ is a finitely generated module over the quotient field of $\overline{T_d}$, then $\bar{B}$ is finitely generated over $\overline{T_d}$.

The reason for this is that the integral closure of $\bar{k}[z_1, \ldots, z_d]$ in any finite field extension of $\bar{k}(z_1, \ldots, z_d)$ is a finitely generated module over $\bar{k}[z_1, \ldots, z_d]$.

In proving the statement we may assume that B has no zero-divisors. Let N be the quotient field of T_d, provided with the valuation $\|\frac{f}{g}\| = \frac{\|f\|}{\|g\|}$ (for any $f, g \in T_d$; $g \neq 0$). Let M be the quotient field of B and $v_1, \ldots, v_s$ the valuations on M extending $\| \|$ on M. As we have seen in (1.11) part (2) and (3), $\| \|$ = max v_i ( ) restricted to B is the spectral norm on B.

Hence $\bar{B}$ lies in the direct sum (i = 1,..., s) of the residue fields of (M, v_i). The residue field (M, v_i) is a finite extension of the residue field of (N, $\| \|$) = $\bar{k}(z_1, \ldots, z_d)$. So we have shown 2) ⇒ 3).

"3) ⇒ 1)". Let $\phi : T_d \to B$ be such that $\bar{\phi} : \bar{T}_d \to \bar{B}$ is finite. Choose $b_1, \ldots, b_n \in B^o$ such that $\phi_n : T_d\langle X_1, \ldots, X_n\rangle \to B$, given by $\phi_n(X_i) = b_i (i = 1, \ldots, n)$, is surjective. The element $\bar{b}_n$ satisfies some equation $X^e + \bar{a}_1 X^{e-1} + \ldots + \bar{a}_e = 0$ with $a_1, \ldots, a_e \in T_d^o$. So $p = b_n^e + a_1 b_n^{e-1} + \ldots + a_e \in B^{oo}$ and some power p^f of p is the image of an element $q \in T_d\langle X_1, \ldots, X_n\rangle$ with $\|q\| < 1$. The kernel of ϕ_n contains the element $(X_n^c + a_1 X_n^{e-1} + \ldots + a_e)^f - q$. It follows that $\phi_{n-1} : T_d\langle X_1, \ldots, X_{n-1}\rangle \to B$ is finite. Let B' be the image of ϕ_{n-1}. By induction on n it follows that $T_d \to B'$ is finite. This ends the proof.

(1.13) <u>Example:</u> A finite algebraic extension ℓ of k can be considered as a morphism $k \to \ell$ of affinoid algebras. If the valuegroup of k is dense in $\mathbb{R}_{>0}$ then $k^o \to \ell^o$ finite implies ℓ^o is a free k^o - module generated by elements of absolute value 1. Hence $[\ell : k] = [\bar{\ell} : \bar{k}]$.

One can make less trivial examples to show that the condition $[\ell : k] = [\bar{\ell} : \bar{k}]$ is necessary for the implication "A $\overset{\phi}{\to}$ B finite then $A^o \xrightarrow{\phi^o} B^o$ finite". This brings us to the following result.

<u>Theorem:</u> <u>If the valuation of</u> k <u>is discrete or if</u> k <u>has the property</u> $[\ell : k] = [\bar{\ell} : \bar{k}]$ <u>for all finite field extensions</u> ℓ <u>of</u> k, <u>then:</u>

$$\phi : A \to B \text{ finite} \qquad \text{implies } \phi^o : A^o \to B^o \text{ finite.}$$

<u>Remark:</u> The proof of this result is rather complicated. We refer to [19]. In (1.20) we will give a proof in case char k = 0 or char k = p ≠ 0 and $[k : k^p] < \infty$.

(1.14) <u>More structure on affinoid spaces:</u>

For our purpose we take the following definition of a Grothendieck topology on a topological space X. It consists of:

1^0. A family $\mathcal{F}$ of open subsets of X satisfying: $\emptyset$, $X \in \mathcal{F}$ and if $U, V \in \mathcal{F}$ then $U \cap V \in \mathcal{F}$.

2^0. For every $U \in \mathcal{F}$ a set Cov (U) of coverings by elements of $\mathcal{F}$. (i.e. $\mathcal{U} = (U_i) \in$ Cov (U) must satisfy: all $U_i \in \mathcal{F}$ and $\cup U_i = U$).

Moreover Cov must satisfy

3^0. $\{U\} \in$ Cov (U) for all $U \in \mathcal{F}$.

4^0. If $\mathcal{U} \in$ Cov (U) and $V \subset U$ with $U, V \in \mathcal{F}$ then $\mathcal{U} \cap V \in$ Cov (V).

5^0. If $\mathcal{U}_i \in$ Cov (U_i) and $(U_i) \in$ Cov (U) then $\cup \mathcal{U}_i \in$ Cov (U).

The elements of $\mathcal{F}$ are often called allowed subsets and the elements of Cov (U) allowed coverings of U.

Let $X = Sp(A)$ be an affinoid space. A subset $R \subset X$ is called a <u>rational domain</u> if there are $f_0, f_1, \ldots, f_n \in A$ which generate the unit ideal and such that

$$R = \{x \in X \mid |f_i(x)| \leq |f_0(x)| \text{ for all } i\}.$$

With R we associate the affinoid algebra

$B = A\langle X_1, \ldots, X_n\rangle/(f_1 - X_1 f_0, \ldots, f_n - X_n f_0)$.

One easily verifies that the obvious map $\phi : A \to B$ induces a homomorphism $Sp(\phi) : Sp(B) \xrightarrow{\sim} R \subset Sp(A) = X$.

So R is itself an affinoid space with respect to some affinoid algebra B. The algebra B satisfies a universal property (which makes it independent of the choice of $f_0, \ldots, f_n$), namely:

"If $A \xrightarrow{\psi} C$ is a morphism of affinoids such that the image of $Sp\,\psi$ lies in $R \subset X = Sp(A)$, then there is a unique morphism $\chi : B \to C$ with $\psi = \chi \circ \phi$".

Indeed, the elements $\psi(f_0), \ldots, \psi(f_n) \in C$ generate C and satisfy $|\psi(f_0)(y)| \geq \max |\psi(f_i)(y)|$ for all $y \in Sp(C)$.

Hence $\psi(f_o)$ is invertible in C and $\psi(f_i)/\psi(f_o)$ are in C^o. The unique map $\chi : B \to C$ is clearly given by $X_i \mapsto \psi(f_i)/\psi(f_o)$ $(i = 1,\ldots, n)$.

The Grothendieck topology on $X = Sp(A)$ is now given by:

$\mathcal{F}$ = the family of all rational domains in X.

Cov (U) = those coverings $\mathcal{U}$ of U containing finitely many $U_1,\ldots, U_n$ with $U_1 \cup \ldots \cup U_n = U$.

Moreover we have a pre-sheaf $\mathcal{O} = \mathcal{O}_X$ on X with respect to the Grothendieck topology, namely $\mathcal{O}(R)$ = the affinoid algebra B belonging to the rational domain R. For any finitely generated A-module M one can form the presheaves $\tilde{M}$ defined by: $R \mapsto M \otimes_A \mathcal{O}(R)$.

Now we come to a basic result on the presheafs $\tilde{M}$. Let $\mathfrak{X} = \{X_1,\ldots, X_n\}$ be an (allowed) covering of X. The Čech-complex of $\tilde{M}$ with respect to this covering is a complex:

$$0 \to C^o \xrightarrow{d^o} C^1 \xrightarrow{d^1} C^2 \xrightarrow{d^2} \ldots \to C^{n-1} \to 0$$

given by $C^p = \bigoplus_{i_o<i_1<\ldots<i_p} \tilde{M}(X_{i_o} \cap X_{i_1} \cap \ldots \cap X_{i_p})$. The coordinates of an element $f \in C^p$ will be denoted by $f(i_o,\ldots, i_p)$. The coboundary map d^p is defined by $d^p(f)(i_o,\ldots, i_{p+1}) = \sum_{j=o}^{p+1} (-1)^j f(i_o,\ldots,\hat{i}_j,\ldots,i_{p+1})$.

The cohomology groups $\check{H}^p(\mathfrak{X}, \tilde{M})$ of the complex are defined as $\ker d^p/\mathrm{im}\, d^{p-1}$. The basic result is:

(1.15) <u>Theorem:</u> <u>For any finite covering</u> $\mathfrak{X}$ <u>of</u> X <u>and any finitely generated</u> $A = \mathcal{O}(X)$ - <u>module</u> M:

$$H^p(\mathfrak{X}, \tilde{M}) = \begin{cases} M & \text{if } p = 0 \\ 0 & \text{if } p \neq 0. \end{cases}$$

<u>Remark:</u> A proof of this statement can be found in [42]. In fact J. Tate works there with a larger collection of subsets of X, namely affinoid subsets. However it turns out (this is the main result of [16]) that any affinoid subset of X is a finite union of rational domains. So the other choice of a Grothendieck topology gives the same theory of sheaves

and cohomology.

The statement (1.15) implies also that $\mathcal{O}$ and $\tilde{M}$ are in fact sheaves for the Grothendieck topology. The sheaf $\mathcal{O}$ is called the structure sheaf of X and $\tilde{M}$ is by definition a coherent sheaf on X. An <u>affinoid space</u> will now mean: $X = Sp(A)$ with its Grothendieck topology and structure sheaf. Rational domains have rather nice properties, e.g.:

(1.16) <u>Lemma:</u> (1) <u>If Y_1 and Y_2 are rational domains then so is $Y_1 \cap Y_2$. Moreover</u> $\mathcal{O}(Y_1 \cap Y_2) \cong \mathcal{O}(Y_1) \hat{\otimes}_A \mathcal{O}(Y_2)$.

(2) <u>If $Y_1 \subset Y_2 \subset X$ are open subsets such that Y_2 is rational in X and Y_1 is rational in Y_2, then Y_1 is rational in X.</u>

<u>Proof:</u> (1) Let Y_1 be given by $|f_o(x)| \geq |f_i(x)|$ $(i = 1, \ldots, n)$ and Y_2 by $|g_o(x)| \geq |g_j(x)|$ $(j = 1, \ldots, m)$ then $Y_1 \cap Y_2$ is given by $|(f_o g_o)(x)| \geq |f_i g_j(x)|$ $(1 \leq i, j \leq n, m)$.

The tensorproduct $\mathcal{O}(Y_1) \hat{\otimes}_A \mathcal{O}(Y_2)$ is the completion of $\mathcal{O}(Y_1) \otimes_A \mathcal{O}(Y_2)$ with respect to the tensorproduct norm. One easily verifies that $A\langle X_1, \ldots, X_n\rangle \hat{\otimes}_A A\langle Y_1, \ldots, Y_m\rangle = A\langle X_1, \ldots, X_n, Y_1, \ldots, Y_m\rangle$ and that $\hat{\otimes}_A$ commutes with the formation of residue rings. Hence

$$\mathcal{O}(Y_1) \hat{\otimes}_A \mathcal{O}(Y_2) = A\langle X_1, \ldots, X_n\rangle / (f_i - X_i f_o) \hat{\otimes}_A A\langle Y_1, \ldots, Y_m\rangle / (g_j - Y_j g_o)$$

$$\cong A\langle X_1, \ldots, X_n, Y_1, \ldots, Y_m\rangle / (f_i - X_i f_o, g_j - Y_j g_o) = \mathcal{O}(Y_1 \cap Y_2).$$

(2) Let Y_2 be given by $|g_o(x)| \geq |g_j(x)|$ for $1 \leq j \leq m$ and Y_1 by $|f_o(y)| \geq |f_i(y)|$ for $1 \leq i \leq n$. The elements $f_o, \ldots, f_n \in A\langle S_1, \ldots, S_m\rangle / (g_j - S_j g_o)$ can be replaced by $f'_o, \ldots, f'_n \in \mathcal{O}(Y_2)$ if $\| f_i - f'_i \|$ are sufficiently small. So we may take $f_o, \ldots, f_n \in A[S_1, \ldots, S_m]$ of total degree $\leq N$. We may replace $f_o, \ldots, f_n$ by $g_o^N f_o, \ldots, g_o^N f_n \in A$.

So we may suppose that $f_o, \ldots, f_n \in A$. For suitable constants $\lambda_o, \ldots, \lambda_m \in k^*$ and all $x \in Y_1$ one has

$$|f_o(x)| \geq |\lambda_i g_i(x)| \quad (i = 0, \ldots, m).$$

Hence $Y_1 = Y_2 \cap Y_3$ where Y_3 is defined by the inequalities:
$|f_o(x)| \geq |f_i(x)|$ $(i = 1, \ldots, n)$ and $|f_o(x)| \geq |\lambda_i g_i(x)|$ $(i = 0, \ldots, m)$.
So using (1) one finds that Y_1 is rational in X.

(1.17) An <u>analytic space X over k</u> is defined as follows.
X has a topology, a Grothendieck topology G and a sheaf $\mathcal{O}_X$ such that there exists a $(X_i) \in \mathrm{Cov}(X)$ with $(X_i, G|X_i, \mathcal{O}_X|X_i)$ is an affinoid space.

(1.18) <u>Examples:</u> In order to simplify we will assume k = K.

(1) $X = \mathrm{Sp}(K\langle z\rangle) \cong \{\lambda \in K \mid |\lambda| \leq 1\}$.
A <u>standard rational</u> subset of X is given by the inequalities
$|z - a| \leq \rho$ and $|z - a_i| \geq \rho_i$ $(i = 1, \ldots, s)$ (of course $\rho, \rho_i \in |K^*|$).
It is a closed disk minus finitely many open disks. One easily verifies the following statements:

a) if S_1, S_2 are standard rational, then $S_1 \cap S_2$ is either empty or again a standard rational.

b) if S_1, S_2 are standard rational with $S_1 \cap S_2 \neq \emptyset$ then $S_1 \cup S_2$ is a standard rational.

c) every finite union S of standard rationals can uniquely be written as a disjoint union of finitely many standard rationals.

We claim the following:
"Every rational domain in X is a finite union of standard rationals".
Indeed: let the rational domain be given by
$|f_o(x)| \geq |f_i(x)|$ $(i = 1, \ldots, n)$. A slight change of $f_o, \ldots, f_n$ does not change the set and we may suppose that f_i and f_j have no common zeros for $i \neq j$. So we have only to study a single inequality $|f_o(x)| \geq |f_1(x)|$. Using the Weierstraß theorem we may suppose that f_o and f_1 are polynomials in z with all their roots in $\{\lambda \in K \mid |\lambda| \leq 1\}$.
The inequality is now $|(z - \alpha_1)\ldots(z - \alpha_n)| \geq |\lambda||(z - \beta_1)\ldots(z - \beta_m)|$ with $\alpha_1, \ldots, \alpha_n, \beta_1, \ldots, \beta_m \in K^o$ and $\lambda \in K^*$.

A somewhat cumbersome calculation and induction on n, m gives the desired result. The converse: "any finite union of standard rationals is itself rational" will be shown in (2.6).

(2) $X = \mathbb{P}(K)$.
The two subsets $X_1 = \{z \mid |z| \leq 1\}$ and $X_2 = \{z \mid |z| \geq 1\}$ are both identified with $Sp(K\langle z\rangle)$. An allowed subset of X is a subset Y such that $Y \cap X_1$ and $Y \cap X_2$ are rational in X_1 and X_2. A standard rational subset in X is the complement of finitely many open disks. It is certainly an allowed subset. Using (1), every allowed subset of X is a finite union of standard rational subsets.

A covering is allowed if it has a finite subcovering. For any standard rational $F \subset \mathbb{P}(K)$ the algebra $\mathcal{O}(F)$ is described in (1.8.3).

The Čech-cohomology of $\mathbb{P}(K)$ w.r.t. the sheaf $\mathcal{O}$ can be calculated with the covering $\{X_1, X_2\}$ since $\mathcal{O}$ has trivial cohomology on $X_1, X_2, X_1 \cap X_2$. The complex is:

$$0 \to \mathcal{O}(X_1) \oplus \mathcal{O}(X_2) \to \mathcal{O}(X_1 \cap X_2) \to 0$$

and more explicitely

$$0 \to K\langle z\rangle \oplus K\langle z^{-1}\rangle \to K\langle z, z^{-1}\rangle \to 0$$

where the map is given by $(f_1, f_2) \mapsto f_1 - f_2$.
It follows that $H^0(\mathbb{P}(K), \mathcal{O}) = K$ and $H^p(\mathbb{P}(K), \mathcal{O}) = 0$ for $p \neq 0$.

(3) $\Omega = \mathbb{P}(K) - \mathcal{L}$, where $\mathcal{L}$ is a compact set.
Ω is given the induced topology. A subset $F \subset \Omega$ is allowed if either $F = \Omega$ or F is a rational subset.

A covering $\{F_i\}$ of Ω is allowed if any rational subset $F \subset \Omega$ is contained in the union of finitely many F_i.

In particular, take $\pi \in K$, $0 < |\pi| < 1$, and $n \geq 1$. Then $\mathcal{L}$ is contained in finitely many open disks of radii $|\pi^n|$. Call its complement F_n. Then $F_n \subset F_{n+1}$ and $\{F_n \mid n \in \mathbb{N}\}$ is an allowed covering of Ω.

We define $\mathcal{O}(\Omega) = \varprojlim \mathcal{O}(F_n)$ = the functions on Ω which are uniform limits of rational functions on each affinoid F in Ω. We claim the following:

For any allowed covering $\mathfrak{X}$ of Ω

$$\check{H}^p(\mathfrak{X},\mathcal{O}) = \begin{cases} \mathcal{O}(\Omega) & \text{for } p = 0 \\ 0 & \text{for } p \neq 0. \end{cases}$$

Since affinoids have no cohomology it suffices to use the covering $\{F_n\}_{n\in\mathbb{N}}$.

The Čech-complex of that covering is:

$$0 \to \Pi\mathcal{O}(F_n) \overset{d^o}{\to} \prod_{n_o<n_1} \mathcal{O}(F_{n_o} \cap F_{n_1}) \overset{d^1}{\to} \Pi\, \mathcal{O}(F_{n_o} \cap F_{n_1} \cap F_{n_2}) \to \cdots$$

Now $\ker d^1 \cong \Pi\mathcal{O}(F_n)$ via the map $f \mapsto (f(n,n+1))_{n\in\mathbb{N}}$ and the resulting $\delta : \Pi\mathcal{O}(F_n) \to \Pi\mathcal{O}(F_n)$ has the form:

$$(f_1, f_2, f_3,\ldots) \mapsto (f_1 - f_2, f_2 - f_3, f_3 - f_4,\ldots).$$

Clearly $\ker \delta = \varprojlim \mathcal{O}(F_n) = \mathcal{O}(\Omega)$. Let $(a_1, a_2, a_3,\ldots,)$ be given. There are rational function $b_1, b_2,\ldots,$ with poles in $\mathcal{L}$ such that $\|a_i - b_i\| \leq |\pi|^i$ on F_i.

Put $(a_1, a_2,\ldots) = (b_1, b_2,\ldots) + (c_1, c_2,\ldots)$. Then $(b_1, b_2, b_3,\ldots) = \delta(0, -b_1, -b_1-b_2, -b_1-b_2-b_3,\ldots)$ and $(c_1, c_2, c_3,\ldots) = \delta(\sum_{n\geq 1} c_n, \sum_{n\geq 2} c_n, \sum_{n\geq 3} c_n,\ldots)$. Hence δ is surjective and $\check{H}^1(\mathfrak{X},\mathcal{O}) = 0$. In a similar way one shows that $\check{H}^p(\mathfrak{X},\mathcal{O}) = 0$ for all $p \geq 2$.

(4) Contour-integration in $\mathbb{P}(K)$:

Let B be an open disk in K with corresponding closed disk B^+. We call $\partial B = B^+ - B$ the "boundary" of B. The algebra $\mathcal{O}(\partial B) \cong K\langle T, T^{-1}\rangle = K\langle T, S\rangle / (TS-1)$. An invertible element of $\mathcal{O}(\partial B)$ has the form $f = T^m(\sum_{n\in\mathbb{Z}} a_n T^n)$ where $|a_0| > \max_{n\neq o} |a_n|$.

We call $m = \mathrm{ord}_T(f)$ the order of f with respect to T. If we normalize f such that $\|f\| = 1$ then $\bar{f} \in \overline{\mathcal{O}(\partial B)} = \bar{k}[T, T^{-1}]$ has the form λT^m with $\lambda \in \bar{k}$, $\lambda \neq 0$. Hence up to its sign $\mathrm{ord}_T(f)$ is independent of the choice of T.

The sign is determined if one gives ∂B an orientation. So after a choice of the orientation we can write $\mathrm{ord}_{\partial B}(f) = m$. For a differential form ω on ∂B, i.e. ω has the form $\sum_{n=-\infty}^{\infty} a_n T^n dT$, we define its residue (w.r.t. the variable T) by $\mathrm{Res}_T(\omega) = a_{-1}$. In order to show that $\mathrm{Res}_T(\omega)$ depends only on ∂B and its orientation we have to show that $\mathrm{Res}_S(T^m \frac{dT}{T}) = 1$ for $m = 0$ and $= 0$ for $m \neq 0$, where $T = \lambda S(1 + \sum_{n \neq o} a_n S^n)$ with $|\lambda| = 1$ and $|a_n| < 1$.

It suffices to verify the cases $m \geq 0$ and only finitely many $a_n \neq 0$.

For $m = 0$ the verification is easy, since we can write
$T = \mu S \prod_{i=1}^{a} (1 + \alpha_i S) \prod_{j=1}^{b} (1 + \beta_j S^{-1})$ with $|\mu| = 1$ and $|\alpha_i| < 1$, $|\beta_j| < 1$.

Then $\frac{dT}{T} = (1 + \Sigma \frac{\alpha_i}{1-\alpha_i S} - \Sigma \frac{\beta_j S^{-2}}{1+\beta_j S^{-1}}) \frac{dS}{S}$ and clearly $\mathrm{Res}_S (\frac{dT}{T}) = 1$.

For $m > 0$ and K of characteristic zero, one has $T^m \frac{dT}{T} = \frac{1}{m} dT^m =$
$= \frac{1}{m} d(\Sigma b_n S^n) = \frac{1}{m} \Sigma n b_n S^n \frac{dS}{S}$. So clearly $\mathrm{Res}_S (T^m \frac{dT}{T}) = 0$. If K has characteristic $\neq 0$ one has

$$T^m \frac{dT}{T} = \lambda^m S^m (1 + \sum_{n \neq o} a_n S^n)^{m-1} (1 + \sum_{n \neq o} (n+1) a_n S^n) \frac{dS}{S}.$$

Replace every $a_n \neq o$ by a variable x_n. Then the constant term of $S^m (1 + \sum_{n \neq o} x_n S^n)^{m-1} (1 + \sum_{n \neq o} (n+1) x_n S^n)$ is a polynomial P in the $\{x_n\}$ with coefficients in $\mathbb{Z}$.

For any values of the x_n in a field of characteristic zero, the polynomial P has value zero. Hence $P = 0$ and as a consequence $\mathrm{Res}_S (T^m \frac{dT}{T}) = 0$.

For a rational function f on $\mathbb{P}(K)$ and a point $p \in \mathbb{P}(K)$ we write $\text{ord}_p(f)$ for the order of f at the point p. For a rational differential form ω on $\mathbb{P}(K)$ (i.e. $\omega = f dz$ where f is a rational function), the residue of ω at p is denoted by $\text{Res}_p(\omega)$.

If f has no zeros or poles on ∂B and ω has no poles on ∂B, then one easily verifies that:

$$\text{ord}_{\partial B}(f) = \sum_{p \in B} \text{ord}_p(f) \text{ and } \text{Res}_{\partial B}(\omega) = \sum_{p \in B} \text{Res}_p(\omega).$$

Let now $F \subset \mathbb{P}(K)$ be the complement of open disks $B_1, \ldots, B_n$. Suppose that $\infty \in F$ and that the closed disks $B_1^+, \ldots, B_n^+$ are disjoint. The interior $\mathring{F}$ of F will be $\mathbb{P}(K) - (B_1^+ \cup \ldots \cup B_n^+)$ and the boundary $\partial F = F - \mathring{F}$ is the disjoint union of $\partial B_1, \ldots, \partial B_n$.

For a rational f without poles or zeros on ∂F and a rational ω without poles on ∂F one has:

$$\sum_{p \in \mathring{F}} \text{ord}_p(f) + \sum \text{ord}_{\partial B_i}(f) = 0 \text{ and}$$

$$\sum_{p \in \mathring{F}} \text{Res}_p(\omega) + \sum \text{Res}_{\partial B_i}(\omega) = 0.$$

By taking limits, the formulas also hold for any meromorphic f on F (no zeros or poles on ∂F) and meromorphic ω (no poles on ∂F). The formulas represent the non-archimedean analogue of the complex "Contour-integration".

(5) Algebraic varieties as k-analytic spaces:

In order to simplify the exposition we suppose that k is algebraically closed. Obvious changes have to be made for fields that are not algebraically closed. An affine algebraic variety X is the set of maximal ideals of a ring R having the form $k[z_1, \ldots, z_n]/(f_1, \ldots, f_r)$. The set X can be identified with $\{x = (x_1, \ldots, x_n) \in k^n \mid f_1(x) = \ldots = f_r(x) = 0\}$. Take $\pi \in k$ with $0 < |\pi| < 1$ and $m \geq 0$. Put $z_i = \pi^{-m} Y_i$ and $g_j = f_j(\pi^{-m} Y_1, \ldots, \pi^{-m} Y_m)$. Then $A_m = k\langle Y_1, \ldots, Y_n\rangle/(g_1, \ldots, g_r)$ is an affinoid algebra and $X_m = Sp(A_m)$ can be identified with the $x = (x_1, \dots, x_n) \in X$ satisfying $|x_i| \leq |\pi|^{-m}$ for all i.

The Grothendieck topology on X is given as follows: an allowed subset of X is either X itself or a rational domain in some X_m. A covering $\mathfrak{X}$ (of an allowed set) is allowed if any allowed affinoid, which is contained in the union, is already contained in the union of finitely many elements of $\mathfrak{X}$.

The sheaf $\mathcal{O}$ on X is given by: $\mathcal{O}(X_m) = A_m$ and $\mathcal{O}(F) = A_m\langle x_1, \dots, x_s\rangle / (f_1 - x_1 f_o, \dots, f_s - x_s f_o)$ if $F \subset X_m$ is given by the inequalities $|f_o(x)| \geq \max |f_i(x)|$. Further $\mathcal{O}(X) = \varprojlim \mathcal{O}(X_m)$. One can easily verify that X is a k-analytic space in this manner and that the analytic structure on X does not depend on the chosen embedding $X \hookrightarrow k^n$.

Let now X denote a (separated) algebraic variety over k.
A proper subset F of X is allowed if F is an affinoid subset of some open affine Y in X (as defined above).

Let F_i be an affinoid subset of $Y_i (1 = 1,2)$, where Y_1, Y_2 are open affine subsets of K. The coordinate rings R_i of Y_i have a presentation $R_1 = k[X_1, \dots, X_n]/(f_1, \dots, f_a)$ and $R_2 = k[Y_1, \dots, Y_m]/(g_1, \dots, g_b)$ such that F_1 is defined by "$|X_i| \leq 1$" and F_2 by "$|Y_j| \leq 1$".

Since X is separated, also $Y_1 \cap Y_2$ is affine and $R_1 \otimes_k R_2$ maps surjectively on its coordinate ring R. Hence R has the form $k[X_1, \dots, X_n, Y_1, \dots, Y_m]/I$ where the ideal I contains the elements $f_1, \dots, f_a, g_1, \dots, g_b$. The set $F_1 \cap F_2$ is given by the inequalities $|X_i| \leq 1$ and $|Y_j| \leq 1$ and as a consequence, it is affinoid in $Y_1 \cap Y_2$. Further $\mathcal{O}(F_1) \hat{\otimes}_k \mathcal{O}(F_2)$ maps surjectively to $\mathcal{O}(F_1 \cap F_2)$ and one can show that $F_1 \cap F_2$ is a rational domain in F_1.

In particular, if $F_1 = F_2$ and $Y_1 \neq Y_2$ it follows that $\mathcal{O}(F_1) = \mathcal{O}(F_2)$. This means that the affinoid algebra $\mathcal{O}(F)$ of F does not depend on the choice of open affine Y in which F is affinoid.

Again we call a covering $\mathfrak{X}$ allowed if any allowed affinoid, which is contained in the union, is already contained in a finite union of elements of $\mathfrak{X}$.

The algebraic variety X is supposed to be of finite type over k. So $X = X_1 \cup \dots \cup X_s$ where $X_1, \dots, X_s$ are open affine subsets. Let $X_{i,m}$ ($i = 1, \dots, s$ and $m \geq 0$) denote the affinoid subset of X_i, introduced earlier. Some calculation shows that $\{X_{i,m} \mid \text{all } i,m\}$ is an allowed covering of X. We define $\mathcal{O}(X)$ to be the kernel of $\Pi\mathcal{O}(X_{i,m}) \to \Pi\mathcal{O}(X_{i,m} \cap X_{j,n})$. Then we have on X the structure of an analytic space over k.

For a projective variety $X \subset \mathbb{P}^n(K)$ given by homogeneous equations $p_1, \dots, p_s$ in $z_0, \dots, z_n$ one can make the procedure above more explicit. The set X_i ($i = 0, \dots, n$) given as $\{\ulcorner z_0, \dots, z_n \urcorner \in X \mid |z_i| \geq \max |z_j|\}$ is clearly an allowed subset. Its affinoid algebra $\mathcal{O}(X_i)$ is equal to $k\langle T_0, \dots, \hat{T}_i, \dots, T_n\rangle / (p_j(T_0, \dots, T_{i-1}, 1, T_{i+1}, \dots, T_n))$. If X is irreducible and reduced the $\mathcal{O}(X_i)$ consists of the limits of the rational functions on X with poles outside X_i (w.r.t the supremum-norm on X_i). Further one can verify that $\mathcal{O}(X) = k$.

In particular <u>if X is a complete, non-singular and irreducible curve over k</u> with function field k(X), then for any $f \in k(X)$ with $f \neq 0$ the set $Z_f = \{x \in X \mid |f(x)| \leq 1\}$ is an allowed affinoid subset of X. Any allowed affinoid subspace of X is a rational domain in some Z_f.

The affinoid algebra $\mathcal{O}(Z)$ on $Z = Z_f$ can be described as follows. Let $x_1, \dots, x_s$ be the poles of f and let R denote the ring of regular functions on $X - \{x_1, \dots, x_s\}$. Then R is the integral closure of $k[f]$ in k(X). Further $\mathcal{O}(Z)$ is the completion of R with respect to the spectral norm on Z. This is a finitely generated $k\langle f\rangle \cong k\langle T\rangle$ - module and in fact isomorphic to $k\langle f\rangle \otimes_{k[f]} R$. (more details in (1.20)). Let $\mathcal{O}^a$ denote the algebraic sheaf on X. The covering $X_+ = \{x \in X$ with $t(x) \neq \infty\}$; $X_- = \{x \in X \mid t(x) \neq 0\}$ can be used to calculate the cohomology group of X. So $H^p(\mathcal{O}^a)$ are the cohomology groups of the complex $0 \to \mathcal{O}^a(X_+) \oplus \mathcal{O}^a(X_-) \to \mathcal{O}^a(X_+ \cap X_-) \to 0$. Results: $H^o(\mathcal{O}^a) = k$, $H^1(\mathcal{O}^a) = k^g$ with g = genus of X and $H^p(\mathcal{O}^a) = 0$ for $p > 1$.

The affinoid covering $\tilde{X}_+ = \{x \in X \,|\, |t(x)| \leq 1\}$; $\tilde{X}_- = \{x \in X \,|\, |t(x)| \geq 1\}$ can be used to calculate the cohomology of the analytic sheaf. The complex is $0 \to (\mathcal{O}^a(\tilde{X}_+) \hat{\otimes} k\langle t\rangle) \oplus (\mathcal{O}^a(\tilde{X}_-) \hat{\otimes} k\langle t^{-1}\rangle) \to \mathcal{O}^a(\tilde{X}_+ \cap \tilde{X}_-) \hat{\otimes} k\langle t,t^{-1}\rangle \to 0$ This has clearly the same cohomology groups. Hence $\check{H}^o(X,\mathcal{O}) = k$; $\check{H}^1(X,\mathcal{O}) = k^g$ with g = genus of X and $\check{H}^p(X,\mathcal{O}) = 0$ for $p > 1$.

(1.19) At this point we interrupt our discussion of examples of analytic spaces and we study in more detail the spectral norm on a complete, non-singular, irreducible curve.

A valued field k is called <u>stable</u> if $|k^*| = \sqrt{|k^*|}$ and for every finite field extension ℓ of k one has $[\ell : k] = [\bar{\ell} : \bar{k}]$. Let k be stable and let the field F be a finite extension of k(t). On k(t) we use the valuation $\| \|$ given by $\|a_0 + a_1 t + \ldots + a_n t^n\| = \max |a_i|$. On F we use the spectral norm $\| \| = \| \|_{sp}$ with respect to this valuation on k(t).

Let R denote the integral closure of k[t] in F. It is a finitely generated free k[t]-module. Let Z denote the set of maximal ideals x of R satisfying $|t(x)| \leq 1$. And let k(x) denote R/x (with its unique valuation).

Then $\phi : R \to \prod_{x \in X} k(x)$ given by $f \mapsto (f(x))_{x \in X}$

is an isometry for the spectral norm on R. Moreover for any $f \in R$ there is an element $x \in Z$ with $|f(x)| = \|f\|$.

(1) <u>Lemma: Every finite dimensional subspace</u> V <u>of</u> R <u>has an orthonormal base.</u>

<u>Proof:</u> The statement means that there are elements $v_1, \ldots, v_n$ in V, forming a base, and such that

$$\|\lambda_1 v_1 + \ldots + \lambda_n v_n\| = \max(|\lambda_i|) \text{ for all } \lambda_1, \ldots, \lambda_n \in k.$$

The condition on k implies that every $k(x)$ has an orthonormal base. Let $\mathcal{L}$ denote the set of linear functionals on V given by $V \subset R \to k(x) \xrightarrow{\ell} k$, where ℓ is the projection of $k(x)$ on some coordinate w.r.t. an orthnormal base of $k(x)$.

V has the following property: $\|v\| = \max\{|\ell(v)| \mid \ell \in \mathcal{L}\}$. Take $v_1 \in V$ with $\|v_1\| = 1$ and $\ell_1 \in \mathcal{L}$ with $|\ell_1(v_1)| = 1$. Let $W = \ker \ell_1$ and $\mathcal{L}' = \{\ell - \alpha_\ell \ell_1 \mid \ell \in \mathcal{L}\}$ where each $\alpha_\ell \in k^o$ is chosen such that $\ell - \alpha_\ell \ell_1(v_1) = 0$. Then $\|\lambda v_1 + w\| = \max(|\lambda|, \|w\|)$ for any $\lambda \in k$ and $w \in W$. Further W has the same property as V w.r.t. the set of functionals $\mathcal{L}'$. By induction on the dimension we find that V has an orthonormal base.

Let now X denote the algebraic curve, corresponding to the field F. Let $D = \Sigma n_i x_i$ denote the pole-divisor of the function t on X. Then $\deg(D) = [F : k(t)]$.

Let $R(n)$ denote the subset of R given by $R(n) = \{f \in F \mid \text{div}(f) \geq -nD\}$. Then $R(n)$ is finite-dimensional and its dimension (for n big) is given by the Riemann-Roch formula:

$$\dim R(n) = \deg(nD) + 1 - g.$$

Hence $\lim_{n\to\infty} \frac{1}{n} \dim R(n) = \deg(D) = [F : k(t)] =$ the rank of R. For the extension $\bar{k}[t] \to \bar{R}$ (= R^o/R^{oo}, usual) we find the same formula:

$$\lim_{n\to\infty} \frac{1}{n} \bar{R}(n) = \text{the rank of } \bar{R} \text{ as } \bar{k}[t]\text{-module}.$$

Further $f \in R(n)$ if and only if $f \in R$ and $t^{-n}f$ is integral over $k[t^{-1}]$ (and similar for $\bar{R}(n)$).

It follows that $\overline{R(n)} \subset \bar{R}(n)$. According to (1) one has $\dim R(n) = \dim \overline{R(n)}$.

As a consequence we find: rank of $R \leq$ rank of $\bar{R}$. The other inequality is obvious; so we have shown

Lemma (2). rank R = rank $\bar{R}$.

Take $e_1, \ldots, e_s \in R^o$ such that $\bar{e}_1, \ldots, \bar{e}_s$ forms a free base of $\bar{R}$ over $\bar{k}[t]$. Then $\Sigma k[t]e_i \subseteq R \subseteq \Sigma k(t)e_i$ and $\|\Sigma a_i e_i\| = \max \|a_i\|$ for all $a_1, \ldots, a_s \in k(t)$.

Let R_p denote $R \cap k(t)e_1 + \ldots + k(t)e_p$. The projection of R_p on the last coordinate has the form $k[t]\frac{1}{a_p} e_p$ where $a_p \in k[t]$ has norm 1. We claim that the image of R_p^o is $k^o[t] \frac{1}{a_p} e_p$ and that $\bar{a}_p \in \bar{k}$.

For $p = 1$ this is clear. We may change e_1 into $\frac{1}{a_1} e_1$, or what amounts to the same, we may assume $a_1 = 1$. Now we consider the case $p = 2$.

Write $a_2 = a_2^+ a_2^-$ where a_2^+ contains the roots of a_2 with absolute value > 1 and a_2^- the roots with absolute value ≤ 1. Let $be_1 + \frac{1}{a_2} e_2$ belong to R_2. Then $a_2 b = c \in k[t]$. Write $c = qa_2^- + r$ where degree $(r) <$ degree (a_2^-). The element $\frac{1}{a_2^-} (re_1 + e_2)$ belongs to R_2. If $\|r\| > 1$ then we choose a $\lambda \in k$ with $\|\lambda r\| = 1$. Then $\frac{\lambda r}{a_2^-} e_1 + \frac{\lambda}{a_2^-} e_2 \in R_2^o$. Hence $\overline{\frac{\lambda r}{a_2^-}} \in \bar{k}[t]$.

This is a contradiction, since $0 \leq$ degree $(\bar{\lambda} r) <$ degree $(\overline{a_2^-})$. Hence $\|r\| \leq 1$ and $\frac{1}{a_2^-} (re_1 + e_2) \in R_2^o$ and $\frac{1}{\overline{a_2^-}} \in \bar{k}[t]$. Hence $\overline{a_2^-} \in \bar{k}$ and $a_2 \in k$. So we can write $a_2 = 1 + q$ with $q \in k^o[t]$ and $\|q\| < 1$.

The element $\frac{1}{a_2} (ce_1 + e_2)$ belongs to R_2. Write $\frac{c}{a_2} = \frac{ca_2}{a_2} - \frac{qc}{a_2}$. Then also $\frac{1}{a_2} (-qce_1 + e_2) \in R_2$. Repeating this one obtains $\frac{1}{a_2} ((-q)^d ce_1 + e_2) \in R_2$. For d big enough, the element belongs to R_2^o and we have shown that the image of R_2^o is $k^o[t]\frac{1}{a_2}e_2$. We may change e_2 into $\frac{1}{a_2} ((-q)^d ce_1 + e_2)$, or what is the same, assume that $a_2 = 1$.

It is clear how to continue this process. We have therefore proved the claim and moreover shown:

Corollary (3): R has a free base $e_1, \ldots, e_s$ such that $e_1, \ldots, e_s \in R^o$ and their images $\bar{e}_1, \ldots, \bar{e}_s$ form a free base of $\bar{R}$ over $\bar{k}[t]$. The elements $e_1, \ldots, e_s$ are also a free base of R^o as $k^o[t]$-module.

Corollary (4): The field $k(t)$ with the valuation given by $\|\Sigma a_i t^i\| = \max \|a_i\|$ is stable.

Corollary (5): The quotient field of T_n is stable with respect to the spectral norm on T_n (provided that $[k : k^p] < \infty$ if char $k = p \neq 0$).

Proof: For a valued field N which is not complete, we define stable as follows: 1) $|N^*| = \sqrt{|N^*|}$

2) for every finite field extension M of N we give M the spectral norm and as usual $\bar{M} = M^o/M^{oo}$. Then we require that $[M : N] = [\bar{M} : \bar{N}]$.

So result (4) follows from lemma (2). The result (5) follows from a stepwise use of (4) and the following lemma.

Lemma (6): Suppose that the valued field N has either characteristic zero or has characteristic $p \neq 0$ and satisfies $[N : N^p] = [\hat{N} : \hat{N}^p] < \infty$. Then:

N is stable if and only if $\hat{N}$ is stable.

Proof: "⇒" Let M be a finite extension of $\hat{N}$. For some $n \geq 0$ the field $M\hat{N}^{1/p^n}$ is separable over $\hat{N}^{1/p^n}$. So we may as well suppose that M is separable over $\hat{N}$, since $\hat{N}^{1/p^n}$ has an orthonormal base over $\hat{N}$.

Let $P = T^d + a_1T^{d-1} + \ldots + a_d$ be the minimal polynomial of a generator of M over N. For $a_i^* \in N$ with $|a_i - a_i^*|$ sufficiently small, the polynomial $Q = T^d + a_1^*T^{d-1} + \ldots + a_d^*$ is irreducible over $\hat{N}$ and has a root f in M. Let M_1 be the finite extension of N generated by f. Since Q is irreducible over $\hat{N}$ it follows that the absolute value of N has a unique extension to M_1. So $M_1 = M$ and $[\bar{M} : \bar{\hat{N}}] = [\bar{M}_1 : \bar{N}] = [M_1 : N] = [M : \hat{N}]$ proves the statement.

"⇐" Again we have only to consider finite separable extensions M of N. Then $M \otimes_N \hat{N}$ is a finite direct sum of field extensions of $\hat{N}$. Hence $M \otimes_N \hat{N}$ has an orthonormal base over $\hat{N}$. The mapping $M \to M \otimes_N \hat{N}$ is an isometry for the spectral norm. Hence M has an orthonormal base over N and $[\bar{M} : \bar{N}] = [M : N]$.

(1.20) Proof of Theorem (1.13), under the hypothesis

$$\text{char } k = 0 \text{ or char } k = p \neq 0 \text{ and } [k : k^p] < \infty.$$

We have to show the following result:

Theorem: If k is stable and $\phi : A \to B$ is a finite morphism of k-affinoid algebras then $\phi^o : A^o \to B^o$ is finite.

Proof: We may suppose that $A = T_d$ and that B has no nilpotents. According to (1.19.5) the dimension of the total quotient ring $\bar{B}$ over $\bar{k}(z_1, \ldots, z_d)$ is equal to the dimension of the total quotient ring of B over the quotient field N of T_d. Let $e_1, \ldots, e_s \in B^o$ be such that $\bar{e}_1, \ldots, \bar{e}_s$ is a base over $\bar{k}(z_1, \ldots, z_d)$. Then $B \subset Ne_1 + \ldots + Ne_s$ and $\|\Sigma n_i e_i\| = \max \|n_i\|$ for all $n_1, \ldots, n_s \in N$. After multiplying B with a non-zero element in T_d, we may suppose that $B \subseteq \Sigma T_d e_i$.

The space $\Sigma T_d e_i$ has an orthonormal base $X^\alpha e_i$ ($i = 1, \ldots, s$ and $\alpha = (\alpha_1, \ldots, \alpha_d) \in \mathbb{N}_0^d$). Choose $f_1, \ldots, f_m \in B^o$ such that $\bar{f}_1, \ldots, \bar{f}_m$ generate $\bar{B}$ as $\bar{T}_d$ -module. Choose $\pi \in k$, $0 < |\pi| < 1$, and $|\pi|$ sufficiently close to 1 such that the coefficients of $f_1, \ldots, f_m$ with respect to the orthonormal base $\{X^\alpha e_i\}$ lie in $R = k_1^o + \pi k^o$ in which k_1 is a suitably chosen discrete subfield of k.

Choose a subset S of $\{X^\alpha f_i \mid i = 1, \ldots, m;\ \alpha \in \mathbb{N}_0^d\}$ and a subset T of $\{X^\alpha e_i \mid i = 1, \ldots, s \text{ and } \alpha \in \mathbb{N}_0^d\}$ with the properties: $\{\bar{s} \mid s \in S\}$ is a base of the $\bar{k}$-vectorspace spanned by $\{\overline{X^\alpha f_i}\}$ and $\{\bar{s} \mid s \in S\} \cup \{\bar{t} \mid t \in T\}$ is a base of $\Sigma \bar{T}_d \bar{e}_i$.

Lemma: $S \cup T$ is an orthonormal base of $\Sigma T_d e_i$.

Proof: The set is certainly orthonormal. To show that it is an orthonormal base it suffices to show: any $f \in \Sigma T_d e_i$ with $\|f\| \leq 1$ can be written $f = \sum_{v \in S \cup T} \lambda_v v + g$ with $\|g\| \leq |\pi|$ and all $|\lambda_v| \leq 1$. It suffices to show this for $X^\alpha e_i$.

We know $\overline{X^\alpha e_i} = \sum_{v \in S \cup T} \mu_v \bar{v}$ with $\mu_v \in \bar{k}_1$. Hence $\| X^\alpha e_i - \Sigma \lambda_v v \| < 1$ for suitable $\lambda_v \in k_1^o \subset R$. But the element $X^\alpha e_i - \Sigma \lambda_v v$ has with respect to the original base of $\Sigma T_d e_i$ all its coefficients in R. Hence

$$\| X^\alpha e_i - \Sigma \lambda_v v \| \leq |\pi| .$$

We continue the proof of the theorem. Any element $f \in B$ with $\| f \| \leq 1$ can be written as $f = \sum_{s \in S} \lambda_s s + \sum_{t \in T} \mu_t t$ with all $|\lambda_s| \leq 1$, $|\mu_t| \leq 1$. The part $\sum_{s \in S} \lambda_s s$ belongs to B. So also $\Sigma \mu_t t = g \in B$. If $\| g \| = |\lambda| \neq 0$ then $\frac{1}{\lambda} g = \Sigma \frac{1}{\lambda} \mu_t t \in B$ and $0 \neq \Sigma (\overline{\frac{\mu_t}{\lambda}}) \bar{t} \in \bar{B}$. This is a contradiction. It follows that S is an orthonormal base for B and B^o is generated as T_d^o-module by $f_1, \ldots, f_m$.

<u>So we have proved a little more than the theorem, namely: If $A \to B$ is finite and $f_1, \ldots, f_m \in B^o$ are such that $\bar{f}_1, \ldots, \bar{f}_m$ generate $\bar{B}$ as $\bar{A}$-module, then $f_1, \ldots, f_m$ generate B^o as A^o-module.</u>

(1.21) <u>Remarks on the cohomology:</u>

On an affinoid space $X = \mathrm{Sp}A$ we have associated with every finitely generated A-module M a sheaf $\tilde{M}$ on X. For a point $x \in X$ the stalk $\tilde{M}_x = \varinjlim \{M(U) | x \in U\} = M \otimes \mathcal{O}_x$, is a finitely generated module over $\mathcal{O}_x = \varinjlim \{\mathcal{O}(U) | x \in U\}$. The ring $\mathcal{O}_x$ is the local analytic ring consisting of the germs of analytic functions at x. Its completion $\hat{\mathcal{O}}_x$ is isomorphic to $\hat{A}_x$. It follows that $M = 0$ if and only if all $\tilde{M}_x$ are zero. Moreover $\check{H}^i(X, \tilde{M}) = 0$ for $i \geq 1$ and $= M$ for $i = 0$. From this it follows that the following three conditions on a sequence $0 \to M_1 \to M_2 \to M_3 \to 0$ of finitely generated A-modules are equivalent:

(1) $0 \to M_1 \to M_2 \to M_3 \to 0$ is exact.

(2) $0 \to \tilde{M}_1 \to \tilde{M}_2 \to \tilde{M}_3 \to 0$ is exact.

(3) $0 \to \tilde{M}_{1,x} \to \tilde{M}_{2,x} \to \tilde{M}_{3,x} \to 0$ is exact for all $x \in X$.

That implies that the sheaf-cohomology (defined as the derived functors of $\tilde{M} \mapsto \check{H}^0(X,\tilde{M}) = H^0(X,\tilde{M})$) for $\tilde{M}$ on X is also trivial.

For any analytic space X over k, we call a sheaf <u>coherent</u> if for any affinoid $Y \subset X$ the sheaf $\mathcal{S}/Y$ is isomorphic to $\tilde{M}$ for some finitely generated $\mathcal{O}(Y)$-module M.

Using Leray's theorem one finds that for any allowed covering $\mathfrak{X}$ of X, such that all finite intersections of elements in $\mathfrak{X}$ are affinoid, and any coherent sheaf $\mathcal{S}$ on X the cohomology groups $H^p(X,\mathcal{S})$ and $\check{H}^p(\mathfrak{X},\mathcal{S})$ agree.

In particular sheaf cohomology and Čech-cohomology for coherent sheaves are the same and we can drop the $\check{}$ on the H.

For a projective variety $X \subset \mathbb{P}^n$ one has the following <u>GAGA-Theorem.</u>

For any coherent algebraic sheaf M on X there is a corresponding analytic coherent sheaf M_{an} on X. The correspondence $M \mapsto M_{an}$ preserves exactness and the formation of cohomology groups. Any analytic coherent sheaf is isomorphic to some M_{an}.

We will not give the proof (it requires the standard techniques of faithfully flat ring-extensions), but merely describe the correspondence $M \mapsto M_{an}$. Let $X_0, \ldots, X_n$ denote the affine covering of X given by $X_i = \{\ulcorner z_0, \ldots, z_n \urcorner \in X \mid z_i \neq 0\}$ and let $\tilde{X}_0, \ldots, \tilde{X}_n$ denote the affinoid covering given by $\tilde{X}_i = \{\ulcorner z_0, \ldots, z_n \urcorner \in X \mid |z_i| \geq \max |z_j|\}$.

An algebraic coherent sheaf M consists of

(1) finitely generated $\mathcal{O}^a(X_i)$-modules M_i.

(2) isomorphisms $\phi_{ji} : M_i \otimes \mathcal{O}^a(X_i \cap X_j) \tilde{\to} M_j \otimes \mathcal{O}^a(X_i \cap X_j)$.

(3) on triple intersections the compatability "$\phi_{kj}\phi_{ji} = \phi_{ki}$".

The analytic sheaf M_{an} is now given by

(1) the $\mathcal{O}(\tilde{X}_i)$-modules $M_i \otimes_{\mathcal{O}^a(X_i)} \mathcal{O}(\tilde{X}_i)$ and has the equivalence of properties (2) and (3) on $\tilde{X}_i \cap \tilde{X}_j$ and $\tilde{X}_i \cap \tilde{X}_j \cap \tilde{X}_k$.

One finds at once that $M \mapsto M_{an}$ preserves exactness by looking at the stalks. The proof that $H^p(X,M) \cong H^p(X,M_{an})$ is quite easy. With some commutative algebra one can show that any coherent analytic sheaf is isomorphic to some M_{an}. Finally as in the $\mathbb{C}$-case, it follows that any analytic subset of $\mathbb{P}^n$ (i.e. locally for the Grothendieck topology given as zeros of analytic functions) is in fact algebraic.

§2 The construction of Ω/Γ for a Schottky group Γ.

(2.1) The ordinary points of a Schottky group Γ of rank g form a subset Ω of $\mathbb{P}(K)$. This set Ω has a structure as analytic space, given in (1.18.3). On the quotient $X = \Omega/\Gamma$ we put a natural structure of analytic space with the help of the canonical map $p: \Omega \to \Omega/\Gamma$.

The Grothendieck topology on X is defined as follows: a subset $Y \subsetneq X$ is allowed (or affinoid) if there exists an affinoid $U \subset \Omega$ such that $p : U \to Y$ is bijective. The structure sheaf $\mathcal{O}_X = \mathcal{O}$ on X is given by $\mathcal{O}(Y) = \mathcal{O}(U)$. (This is clearly independent of the choice of U).

We want to show that X is a separated analytic space. "Separated" means formally: the diagonal $\Delta_X \subset X \times X$ is a closed analytic subset. Equivalently, for every pair of affinoids Y_1, Y_2 in X the set $\Delta_X \cap (Y_1 \times Y_2)$ is given as zero set of some ideal $I \subset \mathcal{O}(Y_1 \times Y_2)$ and $\mathcal{O}(Y_1 \cap Y_2) = \mathcal{O}(\Delta_X \cap (Y_1 \times Y_2)) = \mathcal{O}(Y_1 \times Y_2)/I$. Since $\mathcal{O}(Y_1 \times Y_2) \cong \mathcal{O}(Y_1) \hat{\otimes}_k \mathcal{O}(Y_2)$ the condition translates into the following: $Y_1 \cap Y_2$ is affinoid and $\mathcal{O}(Y_1) \hat{\otimes} \mathcal{O}(Y_2) \to \mathcal{O}(Y_1 \cap Y_2)$ is surjective.

Let $p : U_i \to Y_i$ $(i = 1,2)$ be bijective and let U_1, U_2 be affinoids in Ω. There are only finitely many $\gamma_1, \ldots, \gamma_n \in \Gamma$ with $\gamma U_2 \cap U_1 \neq \emptyset$. Hence $\bigcup_{i=1}^{n} U_1 \cap \gamma_i U_2$ is affinoid and maps bijectively to $Y_1 \cap Y_2$. An easy computation yields $\mathcal{O}(U_1) \hat{\otimes} \mathcal{O}(U_2)$ maps surjectively to $\mathcal{O}(\bigcup_{i=1}^{n} U_1 \cap \gamma_i U_2) = \bigoplus_{i=1}^{n} \mathcal{O}(U_1 \cap \gamma_i U_2)$. (N.B. the map $\mathcal{O}(U_2) \to \mathcal{O}(U_1 \cap \gamma_i U_2)$ is defined by $\mathcal{O}(U_2) \xrightarrow{\sim} \mathcal{O}(\gamma_i U_2)$ followed by the restriction map).

So we have verified: X is a separated analytic space.

In order to compute $H^0(X,\mathcal{O}) = \mathcal{O}(X)$ and $H^1(X,\mathcal{O})$ we construct an affinoid covering of X. Let $F = \mathbb{P} - (B_1 \cup \ldots \cup B_{2g})$ be a fundamental domain for Γ. The B_i's are supposed to be open disks in K such that the corresponding closed disks B_i^+ are still disjoint. The generators $\gamma_1, \ldots, \gamma_g$ of Γ identify the "boundary" of B_i with that of B_{i+g} (identification by γ_i with $i = 1, \ldots, g$).

We choose open disks $\tilde{B}_i$ with $\tilde{B}_i \supset B_i^+$ and such that the corresponding closed disks $\tilde{B}_i^+$ are still disjoint. Let $G = \mathbb{P} - (\tilde{B}_1 \cup \ldots \cup \tilde{B}_{2g})$. Then pG, $p(\tilde{B}_i^+ - B_i)$ $(i = 1, \ldots, 2g)$ forms an affinoid covering of X. The Čech-complex of $\mathcal{O}$ with respect to this covering is:

$$0 \to \mathcal{O}(pG) \oplus \sum_1^{2g} \mathcal{O}(p(\tilde{B}_i^+ - B_i)) \to \sum_1^{2g} \mathcal{O}(pG \cap p(\tilde{B}_i^+ - B_i)) \oplus \sum_{i=1}^{g} \mathcal{O}(p(\tilde{B}_i^+ - B_i) \cap p(\tilde{B}_{i+g}^+ - B_{i+g})) \to 0.$$

We make the following convention: for any affinoid $U \subset \mathbb{P}$ containing ∞, $\mathcal{O}(U)_\infty = \{f \in \mathcal{O}(U) \mid f(\infty) = 0\}$. Using the result in (1.8.3) we can rewrite the complex as

$$0 \to k \oplus \sum_1^{2g} \mathcal{O}(\mathbb{P} - \tilde{B}_i)_\infty \oplus \sum_1^{2g} (\mathcal{O}(\mathbb{P} - B_i)_\infty \oplus \mathcal{O}(\tilde{B}_i^+)) \overset{d}{\to} \sum_{i=1}^{2g} (\mathcal{O}(\mathbb{P} - \tilde{B}_i)_\infty \oplus \mathcal{O}(\tilde{B}_i^+)) \oplus \sum_{i=1}^{2g} \mathcal{O}(\mathbb{P} - \tilde{B}_i)_\infty \oplus k^g \to 0.$$

The map d is easily seen to be bijective if one leaves the terms k and k^g out. Hence $H^0(X,\mathcal{O}) = \mathcal{O}(X) = k$; $H^1(X,\mathcal{O}) = k^g$ and $H^p(X,\mathcal{O}) = 0$ for $p > 1$. We come now to the following result.

(2.2) Theorem: $X = \Omega/\Gamma$ is analytically isomorphic to a complete, non--singular and irreducible curve of genus g.

Proof: A function f on X is called meromorphic if the restriction of f to any affinoid Y in X lies in the total quotient ring of $\mathcal{O}(Y)$. Let D be a positive divisor on X i.e. $\sum_{i=1}^{s} n_i x_i$ with $n_i \geq 0$ and $x_i \in X$. The vector space of $f \in \mathcal{M}$ (= the field of meromorphic functions on X), satisfying div $(f) \geq -D$ is denoted by $L(D)$. The symbol div (f) denotes the divisor of f and is $\sum_{x\in X} \mathrm{ord}_x (f)x$. Let us, for convenience suppose that $x_1,\ldots, x_s \in pG$. Choose a meromorphic function f on pG, with div $(f) \geq -D$. This gives a 1-cocycle for the chosen covering above, namely: components = f on $pG \cap p(\tilde{B}_i^+ - B_i)$ and 0 for the other components. Using that H^1 has dimension g and H^0 has dimension 1, one finds the inequality $\deg (D) + 1 - g \leq \dim L(D) \leq \deg (D) + 1$, where $\deg (D) = \Sigma n_i$.

The formula holds clearly for any divisor $D \geq 0$, since we can change the covering of X a little such that the points of D do not lie on any boundary. If deg (D) is sufficiently large then L(D) contains a non--constant element t. We want to show that $\mathcal{M} \supset K(t)$ is a finite algebraic extension. If some element $s \in \mathcal{M}$ is transcendental over K(t) then after a change of D we may suppose $s, t \in L(D)$. The vectorspace L(nD) (with $n > 0$) contains all monomials $t^a s^b$ with $a + b \leq n$. So $L(nD) \geq \frac{(n+1)(n+2)}{2}$. This contradicts our inequality. Let $k(t) \subset F \subset \mathcal{M}$ such that F is an extension of degree m. Then R, the integral closure of k[t] in F is a free module over k[t] with free base $e_1,\ldots, e_m$. The elements e_i lie in some $L(n_0 D)$. Hence L(nD) contains all $t^\alpha e_i$ with $\alpha \leq n - n_0$. So $\dim L(nD) \geq m(n - n_0)$ and $m \leq \lim_{n\to\infty} \frac{1}{n} \dim L(nD)$, which is equal to deg (D).

So $\mathcal{M}$ is a finite algebraic extension of K(t). Finally we want to establish an isomorphism between X and the non-singular, complete irreducible curve Y associated with $\mathcal{M}$. As an abstract set Y is equal to the set of discrete valuations of $\mathcal{M}$, which are trivial on K. An open, non-empty, set $U \subset Y$ is by definition the complement of a finite subset of X and $\mathcal{O}^a(U)$ = the regular functions on U are the elements of $\mathcal{M}$ with poles outside U (i.e. $v(f) \geq 0$ for all $v \in U$).

Every point $x \in X$ gives the discrete valuation $\mathrm{ord}_x(\ )$ on $\mathcal{M}$. So we have a map $\theta : X \to Y$. It is rather clear that θ is injective. Suppose that $v_0 \in Y$ does not lie in the image of θ. Then there exists a non-

-constant $f \in \mathcal{M}$ with $v(f) \geq 0$ for all $v \neq v_0$. But $f \in \mathcal{O}(X) = k$, yields a contraction. So Θ is bijective. As we have shown in (1.18.5) the analytic structure on Y is given by sets of the form $V = \{y \in Y \mid |t(y)| \leq 1\}$ where t varies and $|t(y)|$ denotes the absolute value of the residue of t in the residue field (= K) of the discrete valuation y. One easily verifies that $\Theta^{-1}(V)$ is affinoid in X and carries the same algebra of analytic functions. So $X \to Y$ is an analytic isomorphism. The rest follows from (1.18.5).

(2.3) Remarks:

(1) Let X and Y be reduced analytic spaces over k. A morphism $\phi : X \to Y$ is a map having the properties

a) there are allowed affinoid coverings $(X_i)_{i \in I}$ of X and $(Y_j)_{j \in J}$ of Y such that $\phi(X_i) \subset Y_{i*}$ for some $i^* \in J$.

b) $\phi : X_i \to Y_{i*}$ is induced by a k-algebra homomorphism $\mathcal{O}(Y_{i*}) \to \mathcal{O}(X_i)$.

(N.B. for non-reduced analytic spaces a more complicated definition of morphism is needed. However we will only deal with reduced analytic spaces).

Clearly the map $p : \Omega \to \Omega/\Gamma$ is a morphism. Moreover Ω/Γ is the "categorical quotient". This means, every morphism $\phi : \Omega \to X$ with the property $\phi \circ \gamma = \phi$ for all $\gamma \in \Gamma$ factors uniquely as $\phi = \tilde{\phi} \circ p$ where $\tilde{\phi} : \Omega/\Gamma \to X$ is a morphism.

(2) A more sophisticated proof of (2.2) would be: For a divisor $D \geq 0$ of sufficiently high degree, the map $\phi_D : X \to \mathbb{P}(L(D))$ which is given by $x \mapsto (f \mapsto f(x)) \in \mathbb{P}(L(D))$, is an analytic embedding. Using "GAGA" one obtains that $\phi_D(X)$ is an algebraic curve.

(3) The Riemann-Roch theorem for the curve $X = \Omega/\Gamma$ can be given a simple analytic proof.

(4) A curve of the type Ω/Γ will be called a Mumford-curve.

(2.4) Analytic tori:

The example of an analytic torus over k is rather useful since it turns out in Chap. VI that the Jacobi-variety of a Mumford curve is an analytic torus. The construction of the analytic tori is very similar to the construction of Ω/Γ in (2.1).

A sugbroup Λ of $G = (k^*)^n$ is called discrete if every affinoid in G meets only finitely many elements of Λ. The map $\ell : G \to \mathbb{R}^n$ given by $\ell(x_1,\dots, x_n) = (-\log|x_1|,\dots, -\log|x_n|)$ is a group homomorphism. It is easily seen that Λ is discrete if, and only if, $\ell(\Lambda)$ is a discrete subgroup of $\mathbb{R}^n$ and ker (ℓ/Λ) is finite. We are interested in the case where Λ has maximal rank and Λ has no torsion elements. In other words $\Lambda \tilde{\to} \ell(\Lambda)$ and $\ell(\Lambda) \cong \mathbb{Z}^n$ is a lattice in $\mathbb{R}^n$.

The analytic torus defined by Λ is $T = G/\Lambda$. Using the canonical map $p : G \to G/\Lambda = T$ we will define a structure of analytic space over k on T. The topology on T is the usual quotient topology. A proper subset Y of T is an allowed affinoid if there exists some allowed affinoid U in G such that $p : U \to Y$ is bijective. We define the sheaf by $\mathcal{O}(X) = k$ and $\mathcal{O}(Y) = \mathcal{O}(U)$. (This does not depend on the choice of U).

In order to show that T is a separated analytic space we have to show that for allowed affinoids $Y_1, Y_2 \subset T$ also $Y_1 \cap Y_2$ is allowed affinoid and that the diagonal Δ_T is a closed analytic subset of $T \times T$.

Let $p:U_i \to T_i$ $(i = 1,2)$ be bijective and U_i allowed affinoid in G. There are only finitely many $\lambda \in \Lambda$ (say $\lambda_1,\dots, \lambda_s$) with $U_1 \cap \lambda U_2 = \emptyset$. So $p : V = \bigcup_{i=1}^{s} (U_1 \cap \lambda_i U_2) \to Y_1 \cap Y_2$ is bijective. The set V is a disjoint union of rational domains in G. Using the rather technical proposition (2.5) it follows that V is also a rational domain, hence V and $Y_1 \cap Y_2$ are allowed.

Let $\varepsilon \in \sqrt{|k^*|}$, $\varepsilon > 1$ and ε close enough to 1.
Put $H = \{(x_1,\dots, x_n) \in (k^*)^n | \varepsilon^{-1} \leq |x_i| \leq \varepsilon \text{ for all } i\}$.

We can translate H over an element $z \in (k^*)^n$ and obtain the set zH. Finitely many $p(z^{(1)}H),\dots, p(z^{(s)}H)$ cover already T. Moreover $p(z^{(i)}H) \cap p(z^{(j)}H) = p(z^{(i)}H \cap z^{(j)}\lambda H)$ for a suitable $\lambda \in \Lambda$. Hence the map $\mathcal{O}(p(z^{(i)}H)) \hat{\otimes} \mathcal{O}(p(z^{(j)}H)) \to \mathcal{O}(p(z^{(i)}H) \cap p(z^{(j)}H))$ is surjective. It follows that the diagonal Δ_T in $T \times T$ a closed analytic subset. (A subset Y of X is called a closed analytic subset if for an allowed affinoid covering (X_i) of X each $Y \cap X_i$ is the zero set of an ideal in $\mathcal{O}(X_i)$.).

One can now easily verify the following result.

Proposition: $T = G/\Lambda$ is an analytic space over k. It is the categorical quotient of G by Λ.

(2.5) We have to prove a rather technical result, namely:

Proposition: Let Y_1, Y_2 be disjoint rational domains in an affinoid space X, then $Y_1 \cup Y_2$ is a rational domain and $\mathcal{O}(Y_1 \cup Y_2) = \mathcal{O}(Y_1) \oplus \mathcal{O}(Y_2)$.

Proof: We remark first that the second statement is rather obvious. Indeed from (1.15) it follows that $0 \to \mathcal{O}(Y_1 \cup Y_2) \to \mathcal{O}(Y_1) \oplus \mathcal{O}(Y_2) \to 0$ is exact.

The diagonal Δ_X in $X \times X$ does not intersect $Y_1 \times Y_2$. The ideal of Δ_X is generated by the set $\{f \otimes 1 - 1 \otimes f \mid f \in \mathcal{O}(X)\} \subset \mathcal{O}(X) \hat{\otimes} \mathcal{O}(X)$. For some $f \in \mathcal{O}(X)^o$, the function $f \otimes 1 - 1 \otimes f$ has no zeros on $Y_1 \times Y_2$. This means that $f(Y_1) \cap f(Y_2) = \emptyset$. Now we need two lemmas.

Lemma (2.6): Any finite union of rational domains in $\{z \in K \mid |z| \leq 1\}$ is again a rational domain.

Lemma (2.7): Given an affinoid space X, a rational domain Y in X and $f \in \mathcal{O}(X)^o$, then $f(Y)$ is a union of a finite set and a rational domain in $\{z \in K \mid |z| \leq 1\}$.

Using the two lemmata we find disjoint rational domains V_1, V_2 in $\{z \in K \mid |z| \leq 1\}$ with $V_i \supset f(Y_i)$. The union $V_1 \cup V_2$ is again rational and $f^{-1}(V_1 \cup V_2)$ is a rational domain in X. $\mathcal{O}(f^{-1}(V_1 \cup V_2)) = \mathcal{O}(f^{-1}V_1) \oplus \mathcal{O}(f^{-1}V_2)$ since $f^{-1}V_1 \cap f^{-1}V_2 = \emptyset$. It suffices to show that $Y_1 \cup Y_2$ is rational in $f^{-1}(V_1 \cup V_2)$. Let Y_1 be given by $|f_o| \geq \max |f_i|$ with all $f_i \in \mathcal{O}(f^{-1}V_1 \cup V_2)$ and let Y_2 be given by $|g_o| \geq \max |g_j|$.

Each element $a \in \mathcal{O}(f^{-1}V_1 \cup V_2)$ will be written as (a^1, a^2) with $a^1 \in \mathcal{O}(f^{-1}V_1)$ and $a^2 \in \mathcal{O}(f^{-1}V_2)$.

The subset $Y_1 \cup Y_2$ of $f^{-1}(V_1 \cup V_2)$ is rational since it is given by the inequalities $|(f_o^1, g_o^2)| \geq \max\limits_{i,j} (|(f_i^1, g_j^2)|)$.

(2.6) <u>Lemma: Any finite union of rational domains in</u> $\{z \in K \mid |z| \leq 1\}$ <u>is itself a rational domain.</u>

<u>Proof:</u> According to (1.18) we have to show that a finite union of standard domains (i.e. given by $|z - a_0| \leq \rho_o$ and $|z - a_i| \geq \rho_i$ $(i = 1, \ldots, s)$) is a rational domain.

A standard domain F is the intersection of the domains $|z - a_0| \leq \rho_o$ and $|z - a_i| \geq \rho_i$ $(i = 1, \ldots, s)$. Since intersections of rational domains are rational, it is enough to consider a finite union $V_1 \cup \ldots \cup V_s \cup U_1 \cup \ldots \cup U_t$ where V_i is given by an inequality $|z - b_i| \geq \mu_i$ and in which U_j is given by $|z - c_j| \leq \nu_j$.

The union V of $V_1, \ldots, V_s$ is the complement in $\{z \in K \mid |z| \leq 1\}$ of finitely many disjoint open disks. Clearly V is rational and can be written as $W_1 \cap \ldots \cap W_n$ in which each W_i is the complement in $\{z \in K \mid |z| \leq 1\}$ of one open disk. Hence it is enough to show that $W \cup U_1 \cup \ldots \cup U_s$ is rational where:

$W = \{z \in K \mid |z| \leq 1\} - B$ with B an open disk.

$U_1, \ldots, U_s$ are closed disks, disjoint and all of them contained in B.

We prefer now to work in $\mathbb{P}(K)$. The set W is rational in $\tilde{W} = \mathbb{P}(K) - B$. We have to show that $\tilde{W} \cup U_1 \cup \ldots \cup U_s$ is rational. After taking a point $a \in B - (U_1 \cup \ldots \cup U_s)$ as ∞ in $\mathbb{P}(K)$ we have reduced (2.6) to showing that a finite union of (disjoint) closed disks in $\{z \in K \mid |z| \leq 1\}$ is rational. This is not completely obvious. In the case of two closed disks $B_1 = \{z \in k \mid |z| \leq |\pi_1|\}$ and $B_2 = \{z \in K \mid |z - 1| \leq |\pi_2|\}$ with $|\pi_1|, |\pi_2| < 1$, one takes some $\pi \in K$ with $1 > |\pi| \geq \max(|\pi_1|, |\pi_2|)$. The set $C = \{z \in K \mid |z(z - 1)| \leq |\pi|\}$ is rational and it is the disjoint union of $C_1 = \{z \in K \mid |z| \leq |\pi|\} \supset B_1$ and $C_2 = \{z \in K \mid |z - 1| \leq |\pi|\} \supset B_2$. So $\mathcal{O}(C) = \mathcal{O}(C_1) \oplus \mathcal{O}(C_2)$ and B_i is rational in C_i $(i = 1,2)$. It follows as before in the proof of (2.5), that $B_1 \cup B_2$ is rational in $C_1 \cup C_2$ and we are done.

Now the general case. Let $B_1, \ldots, B_s$ be closed disks with centers $b_1, \ldots, b_s$. Suppose that they are disjoint and that $\max\{|c - d| \mid c, d \in B_1 \cup \ldots \cup B_s\} = 1$. After renumbering the disks we have $|b_i - b_j| = 1$ for $i \neq j$; $i, j \leq t$ $(t \geq 2)$ and for any $i > t$ there exists a $j \leq t$ with $|b_i - b_j| < 1$. Define C by the inequality $|(z - b_1) \ldots (z - b_t)| \leq |\pi|$ in which $|\pi| < 1$ and $|\pi|$ is sufficiently close to 1. Obviously C is rational and it is the disjoint union of the closed disks C_i given by $|z - b_i| \leq |\pi|$. Hence C contains all B_i. Since $\mathcal{O}(C) = \mathcal{O}(C_1) \oplus \ldots \oplus \mathcal{O}(C_t)$ it is enough to show that the union $(C_i \cap B_1) \cup \ldots \cup (C_i \cap B_s)$ is rational in C_i. This is a union of less than s disks. The statement follows then by induction.

(2.7) Lemma: Given an affinoid space X, a rational domain Y in X and a function $f \in \mathcal{O}(X)^o$. The set $f(Y)$ is a union of a rational domain in $\{z \in K \mid |z| \leq 1\}$ and a finite subset.

Proof: We may of course suppose that $\mathcal{O}(X)$ has no nilpotents and that $X = Y$. Let $T_d = k\langle z_1, \ldots, z_d\rangle \to \mathcal{O}(X)$ be a finite injective map. The minimal polynomial of f over T_d (or its quotient field) has the form $P = T^n + a_1 T^{n-1} + \ldots + a_n$, with $a_1, \ldots, a_n \in T_d^o$.

Then $\lambda \in f(X)$ if and only if $\lambda^n + a_1\lambda^{n-1} + \ldots + a_n \in T_d$ has some zero in Sp (T_d).

We write formally $P = \sum_\alpha P_\alpha(T) z_1^{\alpha_1} \ldots z_d^{\alpha_d}$ where each $P_\alpha(T)$ is a polynomial in T with coefficients in k. The set f(X) in $\{z \in K \mid |z| \leq 1\}$ is given by the inequalities $|P_o(T)| \leq \max_{\alpha \neq o} |P_\alpha(T)|$.

The set of all the polynomials $\{P_\alpha \mid \alpha \geq 0\}$ has g.c.d. Q. Put $P_\alpha = Q\,Q_\alpha$. The polynomials Q_α generate the unit ideal in $k[T]$. Since the coefficients of the polynomials tend to zero, any Q_α with $|\alpha| > N$ can be written as a combination $\sum_{|\beta| \leq N} \pi f_\beta Q_\beta$ with $|\pi| < 1$ and $f_\beta \in k^o[T]$.

As a consequence, f(X) consists of the zeros of Q and the subset of $\{z \in K \mid |z| \leq 1\}$ given by $|Q_o(T)| \leq \max\{|Q_\alpha(T)| \mid |\alpha| \leq D\}$. According to (2.6) the last set is a rational domain in $\{z \in K \mid |z| \leq 1\}$.

(2.8) Remark: The union of two rational domains Y_1, Y_2 in an affinoid X need not be rational, nor even affinoid.

Take for example $X = \{(z_1, z_2) \in K^2 \mid |z_i| \leq 1\}$ and $Y_1 = \{(z_1, z_2) \in X \mid |z_1| \leq |\pi|\}$; $Y_2 = \{(z_1, z_2) \in X \mid |z_2| \leq |\pi|\}$, (with $0 < |\pi| < 1$). If $F = Y_1 \cup Y_2$ were affinoid then according to (1.15) $0 \to \mathcal{O}(F) \to \mathcal{O}(Y_1) \oplus \mathcal{O}(Y_2) \overset{d}{\to} \mathcal{O}(Y_1 \cap Y_2) \to 0$ is exact. One easily computes that the kernel of d consists of all power series $\Sigma a_{nm} z_1^n z_2^m$ with $\lim \pi^n a_{n,m} = \lim \pi^m a_{n,m} = 0$. Those power series converge on the rational subset $|z_1 z_2| \leq |\pi|$. Hence Sp $(\mathcal{O}(F)) \neq Y_1 \cup Y_2$ and we arrive at a contradiction.

(2.9) Reductions of analytic spaces:

With every analytic space X over k we want to associate an algebraic variety $\bar{X}$ over $\bar{k}$ and a map $R : X \to \bar{X}$. For affinoid spaces X = SpA there is an obvious candidate namely $\bar{X} = \text{Max}(\bar{A})$ which is the maximal ideal space of $\bar{A} = A^o/A^{oo}$.

Every maximal ideal x of A induces a k-algebra homomorphism $\phi : A \to \ell$, where ℓ is some finite extension of k. From this we find a $\bar{\phi} : \bar{A} \to \bar{\ell}$ which has as kernel a maximal ideal $\underline{m}$ of $\bar{A}$, since $\bar{\ell}$ is a finite extension of $\bar{k}$. The map $R : X \to \bar{X}$ is given by $R(x) = \underline{m}$ and $X \to \bar{X}$ is called the canonical reduction of the affinoid space X.

(2.9.1) Lemma: The map $R : X \to \bar{X}$ is surjective.

Proof: For a finite and injective map $\phi : T_d \to A$ also $\bar{\phi} : \bar{T}_d \to \bar{A}$ is finite. A maximal ideal $\underline{m}$ of $\bar{A}$ has as inverse image $\bar{\phi}^{-1}(\underline{m})$ which is a maximal ideal of $\bar{T}_d$. After a slight change of the map ϕ one can arrange such that $\bar{\phi}^{-1}(\underline{m}) = (z_1, \ldots, z_d)$ in which $\bar{T}_d = \bar{k}[z_1, \ldots, z_d]$. Let I be the ideal $(z_1, \ldots, z_d)\,A$ in A. The obvious map $\bar{A} \to \overline{A/I}$ has as kernel $\sqrt{\bar{I}}$ = the radical of the ideal $\bar{I} = \{\bar{f} \mid f \in I \cap A^o\}$. The maximal ideal $\underline{m}$ contains $\bar{I}$ and gives a maximal ideal of $\bar{A}/\sqrt{\bar{I}}$. Since A/I and $\overline{A/I}$ have finite dimension over k, this maximal ideal extends to some maximal ideal M of $\overline{A/I}$. We have in this way reduced the problem to the statement: every maximal ideal of $\overline{A/I}$ is the image of a maximal ideal of A/I (or $A/\sqrt{I}$).

But $A/\sqrt{I}$ is finite dimensional and must have the form $A/\sqrt{I} = \ell_1 \oplus \ldots \oplus \ell_s$ where each ℓ_i is a finite extension of k. As a consequence $\overline{A/I} = \bar{\ell}_1 \oplus \ldots \oplus \bar{\ell}_s$ and the map from the maximal ideals of A/I to those of $\overline{A/I}$ is bijective.

(2.9.2) Examples: For convenience we assume again that k is algebraically closed. In the general case one has to make obvious changes. The canonical reduction of T_d, namely:
$R : Sp(T_d) = \{(z_1, \ldots, z_d) \in k^d \mid \text{all } |z_i| \leq 1\} \to \text{Max}\,(\bar{k}^d) = \bar{k}^d$ is the obvious map $(z_1, \ldots, z_d) \mapsto (\bar{z}_1, \bar{z}_2, \ldots, \bar{z}_d)$.

The reduction of an affinoid F in $\mathbb{P}(K)$, given as the complement of disjoint open disks $B_1, \ldots, B_s$, depends on the geometric position of the open disks.

Let F contain ∞ and suppose that the closed disks $B_1^+, \ldots, B_s^+$ are still disjoint. The example (1.8.3) shows that $\overline{\mathcal{O}(F)} = \bar{k}[z_1, \ldots, z_s]/(z_i z_j)_{i \neq j}$. Consequently $\bar{F}$ consists of s irreducible components, every component $\cong \bar{k}$. The s components intersect "quasi-normally" at one point p. We will mean by "quasi-normal" that the complete local ring of $\bar{F}$ at p has the form $\bar{k}[[X_1, \ldots, X_n]]/(X_i X_j)_{i \neq j}$.

A special case is: $F = \{z \mid |\pi| \leq |z| \leq 1\}$ in which $0 < |\pi| < 1$. Then $\bar{F}$ consists of two affine lines meeting normally.

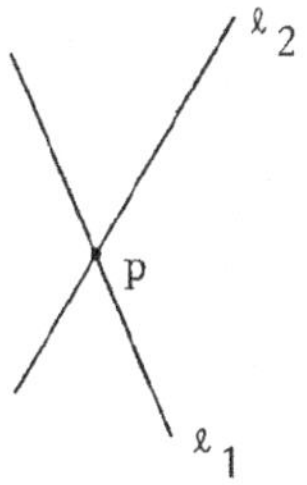

The inverse image of p is $R^{-1}(p) = \{z \mid |\pi| < |z| < 1\}$. The inverse image of $\ell_1 - \{p\}$ is $\{z \in F \mid |z| = 1\}$ and the inverse image of $\ell_2 - \{p\}$ is $\{z \in F \mid |z| = |\pi|\}$.

Another extreme position for the $B_1, \ldots, B_s$ is: the radii of the disks are 1; the centers $b_1, \ldots, b_s$ have absolute value 1. The reduction $\bar{F}$ is a Zariski-open subset of $\bar{k}$ and in fact equal to $\bar{k} - \{\bar{b}_1^{-1}, \ldots, \bar{b}_s^{-1}\}$.

In the general case (1.8.3) gives the following formula for $\overline{\mathcal{O}(F)}$, namely $\overline{\mathcal{O}(F)} = \bar{k}[z_1, \ldots, z_s]/(z_i z_j + b_{ij} z_j + c_{ij} z_i)_{i<j}$ in which $b_{ij}, c_{ij} \in \bar{k}$. The reduction $\bar{F}$ consists of finitely many components $A_1, \ldots, A_t$. Each A_i is a Zariski-open subset of $\bar{k}$ and every intersection of components is "quasi-normal".

(2.9.3) <u>Example:</u> The non-singular elliptic curve $y^2 = x(x-1)(x-\pi)$, with $0 < |\pi| < 1$, has $F = \{(x,y) \in K^2 \mid |x| \leq 1,\ |y| \leq 1$ and $y^2 = x(x-1)(x-\pi)\}$ as allowed affinoid subset. The affinoid algebra $\mathcal{O}(F) = k\langle X, Y\rangle/(Y^2 - X(X-1)(X-\pi))$ has a quotient norm induced by $k\langle X, Y\rangle \to \mathcal{O}(F)$. The quotient norm happens to be the spectralnorm and so $\overline{\mathcal{O}(F)} = \bar{k}[X, Y]/(Y^2 - X^2(X-1))$.

The reduction $\bar{F}$ is a plane rational curve with one ordinary double point  .

It follows from (2.9.2) that F cannot be an affinoid subset of $\mathbb{P}(K)$.

(2.10) Pure coverings and reductions:

A rational domain Y in an affinoid space X is called pure if there exists an affine open U in $\bar{X}$ with $R^{-1}(U) = Y$. The canonical reduction of Y is then equal to $R|Y : Y \to U$, the restriction of $R : X \to \bar{X}$.

A covering (X_i) of an analytic space X is called pure if

a) (X_i) is an allowed affinoid covering of X.

b) each X_i meets only finitely many X_j's.

c) if $X_i \cap X_j \neq \emptyset$ then $X_i \cap X_j$ is pure in X_i (and X_j).

Given X and a pure covering $U = (X_i)$ of X we can form $\bar{X}_U$, the reduction of X with respect to U, as the algebraic variety over $\bar{k}$ formed by glueing the affine sets $\bar{X}_i$ over the $\overline{X_i \cap X_j} \hookrightarrow \bar{X}_i, \bar{X}_j$. If U is a finite covering then $\bar{X}_U$ is clearly a reduced algebraic variety over $\bar{k}$ of finite type. In general $\bar{X}_U$ is only locally of finite type over $\bar{k}$. As for affinoid spaces there is an associated surjective map $R : X \to \bar{X}_U$.

(2.10.1) Examples: Put $X = \{z \in k \mid |z| \leq 1\}$ and let $0 < |\pi| < 1$ for $\pi \in k$. The sets $X_1 = \{z \in k \mid |z| \leq |\pi|\}$ and $X_2 = \{z \in k \mid |\pi| \leq |z| \leq 1\}$ form a pure covering of X. The reduction is obtained by glueing $\bar{X}_1$ and $\bar{X}_2$ and is easily seen to be 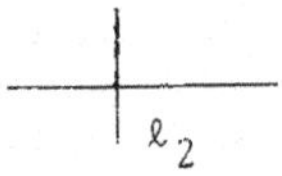 ℓ_1 with $\ell_1 = \mathbb{P}(\bar{k})$ and $\ell_2 \cong \bar{k}$.

ℓ_2

(2.10.2) $X = k$. Put $X_o = \{z \in k \mid |z| \leq 1\}$ and $X_n = \{z \in k \mid |\pi|^{-n+1} \leq |z| \leq |\pi|^{-n}\}$ for $n \geq 1$. This is a pure covering of k and has as reduction:

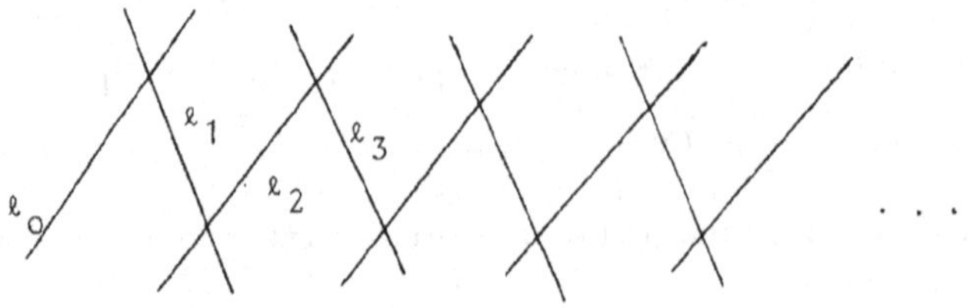

(2.10.3) The analytic space k^* has a pure covering $(X_n)_{n\in\mathbb{Z}}$ in which $X_n = \{z \in k^* \mid |\pi|^{n+1} \leq |z| \leq |\pi|^n\}$. The reduction is equal to

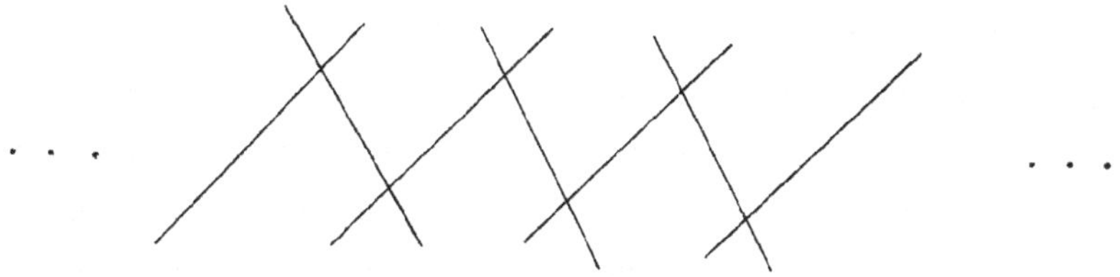

(2.10.4) $\mathbb{P}(k)$ has the pure covering $\{z \mid |z| \leq 1\}$, $\{z \mid |z| \geq 1\}$. The reduction is $\mathbb{P}(\bar{k})$ and the map $R : \mathbb{P}(k) \to \mathbb{P}(\bar{k})$ can be given by $\ulcorner x_0, x_1 \urcorner \mapsto \ulcorner \bar{x}_0, \bar{x}_1 \urcorner$ with $x_0, x_1 \in k$ chosen such that $\max(|x_0|, |x_1|) = 1$. This is the standard reduction considered in Chap. I.

(2.10.5) $\mathbb{P}(k)$ has $\{z \mid |z| \leq |\pi|\}$, $\{z \mid |\pi| \leq |z| \leq 1\}$, $\{z \mid |z| \geq 1\}$ as pure covering. The reduction has two components, each $\cong \mathbb{P}(\bar{k})$. The two components meet normally at one point. The reduction map is in fact the map corresponding to the finite subset $\{0, \pi, 1, \infty\}$ of $\mathbb{P}(k)$ considered in Chap. I.

(2.10.6) For a finite subset X of $\mathbb{P}(K)$ we have defined in Chap. I a tree T(X) and using $S = X^{(3)}$ we have a "reduction" $R_S : \mathbb{P}(K) \to Z \subset \mathbb{P}(\bar{K}) \times \ldots \times \mathbb{P}(\bar{K})$. We will now show that R_S is actually the reduction of $\mathbb{P}(K)$ with respect to a suitably chosen pure covering of $\mathbb{P}(K)$. The easiest way to do this is to observe that for any proper Zariski-open subset U in Z the inverse image $Y = R_S^{-1}(U)$ is affinoid in $\mathbb{P}(K)$. If U is moreover affine (in this case that U does not contain a complete component of Z) then the map $R_S : Y \to U$ is in fact the canonical reduction of Y. This is maybe difficult to verify, however if U contains at most 2 lines of Z (and is still affine) then the statement is easily seen to be true. Let $U_1, U_2, \ldots, U_n$ be a finite covering of Z by open affine subsets meeting at most two components of Z. Then $Y_i = R_S^{-1}(U_i)$ $(i = 1, \ldots, n)$ is a pure affinoid covering of $\mathbb{P}(K)$. The map $R_S : Y_i \to U_i$ is the canonical reduction and the $U_i = \bar{Y}_i$ are glued together in the proper way. Therefore $R_S : \mathbb{P}(K) \to Z$ is the reduction of $\mathbb{P}(K)$ with respect to the covering $\{U_i | i = 1, \ldots, n\}$.

Moreover as we know already $T(X)$ is the intersectiongraph of the components of Z. Reductions with respect to a compact set $X \subset \mathbb{P}(K)$ and more generally reductions of open subspaces Ω of $\mathbb{P}(K)$ will be treated in Chap. IV.

(2.11) The reductions of an affinoid in $\mathbb{P}(K)$:

A reduction $\bar{X}_U$ of a 1-dimensional analytic space X is called pre-stable if the only singularities of $\bar{X}_U$ are ordinary double points. The reduction is called stable if moreover every component Z of $\bar{X}_U$, which is isomorphic to $\mathbb{P}(\bar{K})$, meets the other components in at least 3 points.

Proposition: Any affinoid F in $\mathbb{P}(K)$ has a unique stable reduction $\bar{F}_U$ (U is not unique). Moreover the components of $\bar{F}_U$ are all is isomorphic to a Zariski-open subset of $\mathbb{P}(\bar{K})$ and the intersectiongraph of the components of $\bar{F}_U$ is a finite tree.

Proof: We may suppose that F is given as the complement of open disks $B_1, \ldots, B_s$ in K. For each disk B_i we take a point $b_i \in B_i$ and a point $c_i \in B_i^+ - B_i$. The finite set $X = \{b_1, c_1, \ldots, b_s, c_s\}$ induces a reduction $R : \mathbb{P}(K) \to Z$ (as defined in Chap. I or as in (2.10.6)). Put $Z_o = \{R(b_1), \ldots, R(b_s)\}$. Then $R^{-1}(Z - Z_o) = F$ and $R : F \to Z - Z_o$ is a prestable reduction (w.r.t. a suitable covering of F). Every component of $Z - Z_o$ is Zariski-open in $\mathbb{P}(\bar{K})$ and the intersection graph of $Z - Z_o$ is a tree.

If $Z - Z_o$ happens to have a component $\ell \cong \mathbb{P}(\bar{K})$ which meets the other components in less than three points, then we make the following covering $\mathcal{V}$ of $Z - Z_o$: a Zariski-open $V \subset Z - Z_o$ belongs to $\mathcal{V}$ if

$V \cap \ell = \emptyset$ or $V \supseteq \ell$, and V contains at most one complete component of $Z - Z_o$.

The pure covering $U = R^{-1}(\mathcal{V})$ of F gives a new reduction $R_1 : F \to \bar{F}_U$ of F. As one can easily show $\bar{F}_U$ is obtained from $Z - Z_o$ by contracting the line ℓ to a point. Continuing this process one arrives at a stable reduction of F with the required properties.

Let $R_2 : F \to \bar{F}_{U_2}$ be another stable reduction of F. The covering $U_1 \cap U_2 = U_3$ is also pure and gives a prestable reduction $R_3 : F \to \bar{F}_{U_3}$ of F. The maps R_1 and R_2 factor over R_3.
So we have morphisms $\phi_i : \bar{F}_{U_3} \to \bar{F}_{U_i}$ $(i = 1,2)$ with $R_i = \phi_i \circ R_3$ for $i = 1,2$. The maps ϕ_i contract every line ℓ of $\bar{F}_{U_3}$ which is isomorphic to $\mathbb{P}(\bar{K})$ and intersects the other components of $\bar{F}_{U_3}$ in less than 3 points, to a point. Take such a line ℓ in $\bar{F}_{U_3}$ and let $\bar{F}^*$ denote the result after contracting ℓ in $\bar{F}_{U_3}$. Then we have new maps $\phi_i^* : \bar{F}^* \to \bar{F}_{U_i}$ $(i = 1,2)$ and $R_3^* : F \to \bar{F}$ with $R_i = \phi_i^* \circ R_3^*$ $(i = 1,2)$. After finitely many steps of this type one obtains $\bar{F}_{U_1} \cong \bar{F}_{U_2}$ and the uniqueness in the proposition is proved.

Remark: In Chap. V we will prove that algebraic curves over k have a stable reduction.

(2.12.) The reductions of Ω/Γ for Schottky groups Γ:

We use the notation of (2.1). The covering pG, $p(B_i^+ - B_i)$ $(i = 1,\ldots, 2g)$ of Ω/Γ is pure and gives some reduction of Ω/Γ. But this reduction is not pre-stable (for $g \geq 2$) since the canonical reduction $\bar{G}$ of G consists of $2g$ lines ($\cong \bar{k}$) meeting "quasi-normally" at one point. We replace G by a pure covering $\mathcal{U} = (U_1,\ldots, U_n)$ such that $\bar{G}_U$ is stable. The covering $\mathcal{V} = \{U_1,\ldots, U_n, B_i^+ - B_i \ (i = 1,\ldots, 2g)\}$ is a pure covering of F (= the fundamental domain of Γ). The corresponding reduction $\bar{F}_{\mathcal{V}}$ contains exactly $2g$ lines $\cong \bar{K}$, they correspond to the points in $B_i^+ - B_i$ $(i = 1,\ldots, 2g)$. The other components of $\bar{F}_{\mathcal{V}}$ are isomorphic to $\mathbb{P}(\bar{K})$.

The covering $\{pV | V \in \mathcal{V}\}$ of Ω/Γ is pure and gives a pre-stable reduction $\overline{\Omega/\Gamma}$ which is obtained from $\bar{F}_{\mathcal{V}}$ by glueing the 2g affine components of $\bar{F}_{\mathcal{V}}$ pairwise to g projective lines over $\bar{K}$.

The reduction is in general not stable. To make it stable we have to show that for any line $\ell \cong \mathbb{P}(\bar{K})$, which meets the other components in less than three points, there is a new pure covering $\mathcal{W}$ of Ω/Γ such that the reduction with respect to $\mathcal{W}$ is equal to the original reduction in which ℓ is contracted to a point.

The existence of such a pure covering $\mathcal{W}$ follows at once from the next lemma.

(2.12.1) Lemma: Given an analytic space X and a pure covering $\mathcal{U}$ of X consisting of affinoid subsets of $\mathbb{P}(K)$ such that $\bar{X}$ has one of the following forms:

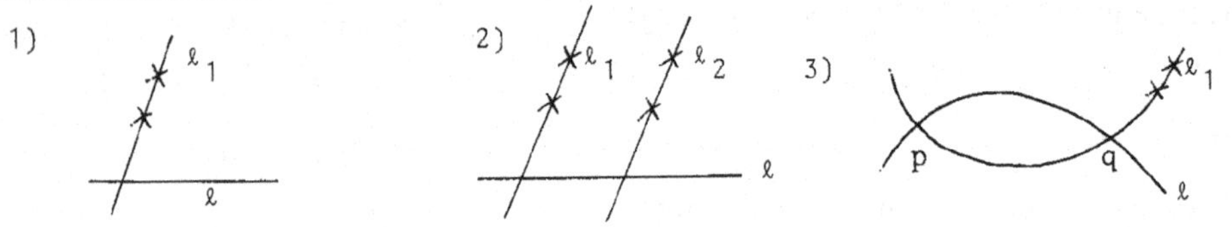

(The sign ✗ denotes a "missing" point. So $\ell \cong \mathbb{P}(\bar{K})$ and ℓ_1, ℓ_2 are proper open subsets in $\mathbb{P}(\bar{K})$).

Then X is affinoid and its canonical reduction $\bar{X}$ is

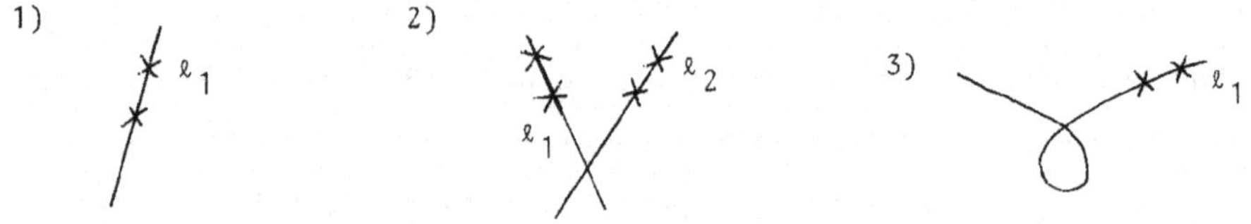

Proof: A direct proof seems to be rather difficult. We use a result that will be proved in Chap.IV, namely: X is affinoid and its ring $\overline{\mathcal{O}(X)}$ is equal to $\mathcal{O}(\bar{X}_U)$. In the cases 1), 2) one easily sees that the Max $(\mathcal{O}(\bar{X}_U))$ coincides with the picture. In case 3) $\mathcal{O}(\bar{X}_U)$ is equal to $\{f \in \mathcal{O}(\ell_1) | f(p) = f(q)\}$. So Max $(\mathcal{O}(\bar{X}_U))$ is obtained by identifying p and q on ℓ_1.

After this lemma it follows that Ω/Γ has a stable reduction and as in (2.11) this stable reduction is unique. Let $Z = \overline{\Omega/\Gamma}$ denote this stable reduction. We associate with Z the following graph:

a vertex for each component Z_i $(i = 1, \ldots, s)$ of Z

an edge for each double point of Z.

If the double point is an intersection of Z_i and Z_j $(i \neq j)$ then the edge connects the vertices Z_i and Z_j. If the double point is a self--intersection of Z_i, then the edge connects Z_i with itself.

The Betti-number of the graph is clearly g = the number of generators of Γ. The graph has the additional properties:

(1) the graph is connected.

(2) every vertex, which is not connected with itself, is beginpoint of at least 3 edges.

We summarize results in the following theorem:

(2.12.2) Theorem: *For a Schottky group Γ the analytic space Ω/Γ has a unique stable reduction Z. The components of Z are rational curves. The intersectiongraph has Betti-number g = the number of generators of Γ.*

(2.12.3) Remark: In Chap. I we have attached to $\Gamma (g \geq 2)$ the tree $T(\mathcal{L})$ of the limit points of Γ and the graph $T(\mathcal{L})/\Gamma$. This graph is isomorphic to the intersection graph of the stable reduction of Ω/Γ.

A natural proof of this statement will be given in Chap. IV after the introduction of the canonical reduction of Ω. Namely the canonical reduction $\bar{\Omega}$ has as intersection graph the tree $T(\mathcal{L})$. And $\bar{\Omega}/\Gamma$ is isomorphic to the stable reduction of Ω/Γ. So the intersectiongraph of the stable reduction of Ω/Γ is equal to $T(\mathcal{L})/\Gamma$.

(2.13) Explicit calculation of the stable reduction of Ω/Γ:

As usual we denote by g the number of generators of Γ. For $g = 1$ it is easily seen that (with the notations of (2.12)) the reduction $\bar{F}_{\mathfrak{V}}$ is equal to

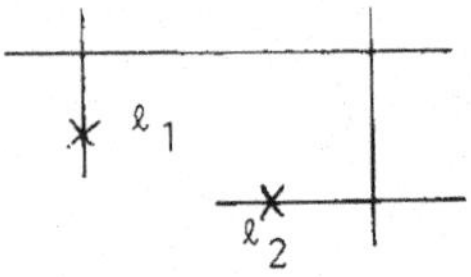

In the corresponding reduction of Ω/Γ the affine lines ℓ_1 and ℓ_2 are glued together to form a $\mathbb{P}(\bar{K})$. So one obtains

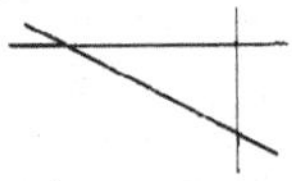

as pre-stable reduction of Ω/Γ . Using the lemma (2.12.1) twice one finds that the stable reduction of Ω/Γ is

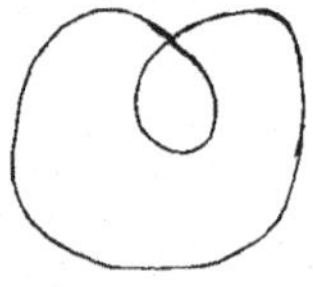

i.e. a rational curve with one double point.

It's graph is

For $g \geq 2$ the stable reduction is determined by the position of the 2g open disks $B_1, \ldots, B_{2g}$. Choose points $b_i \in B_i$ $(i = 1, \ldots, 2g)$, put $X = \{b_1, \ldots, b_{2g}\}$ and let $R : \mathbb{P}(K) \to Z$ be the reduction of the example (2.10.6) (or of Chap. I). The reduction $\bar{F}_{\mathfrak{V}}$ is obtained from Z by adding 2g affine lines to Z intersecting at the 2g points $R(b_1), \ldots, R(b_{2g})$. The corresponding reduction of Ω/Γ is obtained by connecting each pair $\{R(b_i), R(b_{i+g})\}$ $(i = 1, \ldots, g)$ by a projective line.

After contracting those g projective lines one obtains a reduction $Z/\sim$ of Ω/Γ, in which $\sim$ means the identification of the pairs of points $\{R(b_i, R(b_{i+g})\}$ $(i = 1, \ldots, g)$ in Z.

Possibly $Z/\sim$ is not yet stable. After contracting one by one projective lines in $Z/\sim$ one obtains the stable reduction of Ω/Γ. This process does not depend on the open disks $B_1, \ldots, B_{2g}$ since one can take for $b_1, \ldots, b_{2g}$ the 2g fixed points of the generators $\gamma_1, \ldots, \gamma_g$ of Γ.

We make the calculation for $g = 2$

Possibilities for Z are (dots denote the points $R(b_i)$).

1) 

2) 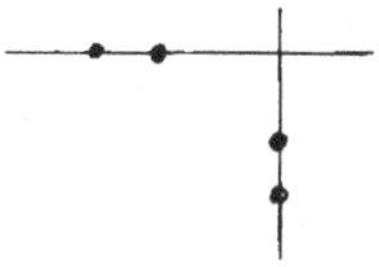

In case 1) the stable reduction is clearly 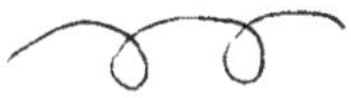 (a rational curve with two double points).

In case 2) there are two possibilities of identifying the dots pairwise. Namely:

a) 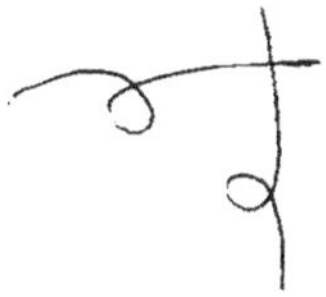

and b) 

So we have three possibilitics for the intersection praph of the stable reduction of Ω/Γ namely: 

A similar calculation gives for $g = 3$ all the possiblities of the intersection graph, namely:

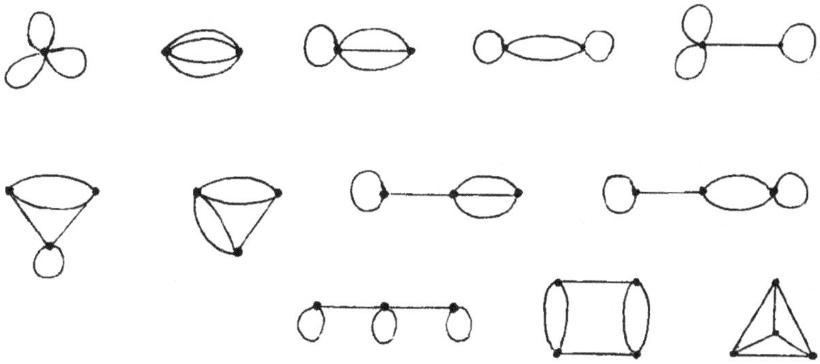

In principle one can now calculate all possibilities for the stable reductions of Ω/Γ and its intersectiongraph. As we will see later in Chap. IV every graph with the properties:

(1) the graph is finite and connected.

(2) every vertex, which is not connected with itself, is the begin-point of at least three edges.

(3) the Betti number is g.

is in fact the intersection graph of the stable reduction of Ω/Γ for some Schottky group Γ of rank g.

Chapter IV. Totally split curves and universal coverings.

Introduction:

The main result of this chapter (Thm. (3.10)) states that a totally split curve X over k (i.e, the stable reduction of X contains only rational curves over $\bar{k}$) can be parametrized by a Schottky group.

The proof goes along the following lines: First a universal covering $\hat{X} \to X$ (see definition (3.3)) of X is constructed. The analytic space $\hat{X}$ has a very nice reduction and is a "genus zero space" (see (2.1)). According to (2.4), X is isomorphic to an analytic subspace of $\mathbb{P}$. Using the notion of "angle" (1.9.) one shows that $\hat{X} \cong \mathbb{P} - \mathcal{L}$ where $\mathcal{L}$ is a compact and perfect subset of $\mathbb{P}$. The automorphism group Γ of the covering $\hat{X} \to X$ extends to a Schottky group on $\mathbb{P}$ with $\mathcal{L}$ as set of limit points. Then finally $\mathbb{P} - \mathcal{L}/\Gamma \cong X$.

§1 Analytic subspaces of $\mathbb{P}^1$

(1.1) A connected rational affinoid over k in $\mathbb{P}$ is given by a finite set of inequalities $|z - m_i| \geq r_i$, with $m_i \in \mathbb{P}(k)$ and $r_i \in |k^*|$. A rational affinoid over k in $\mathbb{P}$ is by definition a finite union of connected rational affinoides. If k is algebraically closed then we have shown in Ch. III that every affinoid subspace of $\mathbb{P}$ is a rational affinoid. In the sequel the field k is not necessarily algebraically closed and we will only work with affinoids in $\mathbb{P}$ which are rational over k (in the sense above). For simplicity we will drop the adjective "rational". Let $\mathcal{F}$ be a family of affinoids in $\mathbb{P}$ satisfying:

(1) if $X_1, X_2 \in \mathcal{F}$ then $X_1 \cup X_2 \in \mathcal{F}$

(2) if $X \subset Y$ and $Y \in \mathcal{F}$ then $Y \in \mathcal{F}$

(3) $\mathcal{F} \neq \emptyset$.

For any open set O in $\mathbb{P}$ we can define such a family $\mathcal{F}$ by : $X \in \mathcal{F}$ if and only if $X \subset O$. As we will see, there are more complicated examples. For a family $\mathcal{F}$ as above we construct an analytic space over k called $\Omega = \Omega(\mathcal{F})$, as follows:

(1) As a set $\Omega = \cup \{X | X \in \mathcal{F}\}$. A basis for the topology on Ω is the family $\mathcal{F}$.

(2) An allowed subset $Y \subsetneq \Omega$ for the Grothendieck topology means an affinoid subset (not necessarily rational over k) of some $X \in \mathcal{F}$. Also Ω is an allowed subset.

(3) A covering $\mathfrak{X}$ is allowed if any $X \in \mathcal{F}$ which is contained in $\cup \mathfrak{X}$ is already contained in a finite union of element of $\mathfrak{X}$.

(4) The structure sheaf $\mathcal{O}_\Omega$ is defined by $\mathcal{O}_\Omega(X) = \mathcal{O}(X)$ = the ususal set of holomorphic functions on X, if $X \subsetneq \Omega$, and $\mathcal{O}_\Omega(\Omega) = \varprojlim \{\mathcal{O}_\Omega(X) | X \in \mathcal{F}\}$.

So we have given Ω a Grothendieck topology and a structure sheaf. Locally Ω is affinoid. Hence Ω is an analytic space over k of dimension 1. In fact Ω is just $\varinjlim \{X | X \in \mathcal{F}\}$ in the category of analytic spaces over k. The space Ω is connected (i.e. $\mathcal{O}_\Omega(\Omega)$ has no idempotents $\neq 0, 1$) if and only if $\mathcal{F}$ has the property: for every affinoid $X \in \mathcal{F}$ there exists a connected affinoid $X' \in \mathcal{F}$ with $X \subset X'$.

From the description of the connected affinoids in $\mathbb{P}^1$ it follows that $\mathcal{F}$ is generated by a sequence $X_1 \subset X_2 \subset X_3 \subset \ldots$ of connected affinoids (rational over k) in $\mathbb{P}$. If this sequence is finite then Ω is just an affinoid in $\mathbb{P}$. If the sequence is infinite then Ω is called a <u>domain</u> in $\mathbb{P}$.

If Ω is not connected then each connected component of Ω corresponds with a subfamily $\mathcal{F}'$ of $\mathcal{F}$.

(1.2.) Examples:

(1) For any compact set $\mathcal{L} \subset \mathbb{P}(k)$, we define the family $\mathcal{F}$ by $X \in \mathcal{F}$ if $X \cap \mathcal{L} = \emptyset$. The corresponding space $\Omega = \mathbb{P} - \mathcal{L}$ is a domain in $\mathbb{P}$.

(2) Let X be a closed disk in $\mathbb{P}$ (center in $\mathbb{P}(k)$, radius in $|k^*|$), then $\Omega = \mathbb{P} - X$ corresponding to the family $\mathcal{F} = \{Y | Y \cap X = \emptyset\}$ is again a domain in $\mathbb{P}$.

(3) Suppose that k is not maximally complete. There exists a sequence of disks (centers in k, radii in $|k^*|$) $B_1 \supset B_2 \supset B_3 \supset \dots$ with $\cap B_n = \emptyset$. Define $\mathcal{F}$ by : $X \in \mathcal{F}$ if for some n, $X \cap B_n = \emptyset$. Then $\mathcal{F}$ is generated by the sequence of affinoids $\mathbb{P} - B_1 \subset \mathbb{P} - B_2 \subset \mathbb{P} - B_3 \subset \dots$. The corresponding space Ω is a domain. As a topological space Ω is equal to $\mathbb{P}$. However Ω and $\mathbb{P}$ have a different Grothendieck topology. It is this example which inspired as to the careful definition of a domain in (1.1).

(4) Suppose that k is maximally complete and algebraically closed. Let $\Omega = \Omega(\mathcal{F})$ be a domain in $\mathbb{P}$. Put $\Omega^* = \cup \{X | X \in \mathcal{F}\}$ and give the open subset $\Omega^*$the analytic structure defined by $\mathcal{F}^* = \{X \subset \Omega^* | X \text{ is affinoid}\}$. We will show that Ω and Ω^* are isomorphic. Let the sequence $X_1 \subset X_2 \subset X_3 \subset \dots$ generate $\mathcal{F}$ and let $X \in \mathcal{F}^*$. Then $X \subset \cup X_n$ and $X \cap (\bigcap_{n \geq 1} (\mathbb{P} - X_n)) = \emptyset$. Suppose that $X \cap (\mathbb{P} - X_n) \neq \emptyset$ for all $n \geq 1$. Then there a sequence of open disks $B_1 \supset B_2 \supset B_3 \supset \dots$ with

(i) B_n is one of the components of $\mathbb{P} - X_n$.

(ii) $X_n \cap B_n \neq \emptyset$.

Since $\cap B_n \neq \emptyset$ it follows that $\cap B_n$ is a disk which has points in common with X. This contradicts over hypotheses. So we have shown that $X \subset X_n$ for some n and hence $\mathcal{F} = \mathcal{F}^*$ and $\Omega \cong \Omega^*$.

We can draw the following consequence: If k is maximally complete and algebraically closed then every domain (or connected affinoid) in $\mathbb{P}$ has the form $\Omega = \mathbb{P} - X$, with the Grothendieck topology of all affinoid subsets of Ω, where the closed set X satisfies:

$X = \bigcap_{n=1}^{\infty} X_n$ and each X_n is a finite union of open disks.

A subset C of $\mathbb{P}$ is called <u>polynomially convex</u> if $C = \bigcap_{n\geq 1} C_n$ and each C_n is a finite union of closed disks. This notion plays a role in the theory of function algebras. One easily verifies that $\Omega = \mathbb{P} - C$ is a domain in $\mathbb{P}$.

(1.3) <u>A strange phenomenon for non-maximally complete fields:</u>

We return to example (1.2.3). We will show that there exist on Ω (defined by $\mathbb{P} - B_1 \subset \mathbb{P} - B_2 \subset \ldots$) non-constant bounded holomorphic functions. The construction of such a function requires some theory of Banach spaces.

Let ℓ^∞ denote the Banach space of all bounded sequences $\lambda = (\lambda(1), \lambda(2), \lambda(3), \ldots)$ of elements in k:

The norm on ℓ^∞ is given by $\|\lambda\| = \sup |\lambda(n)|$. In ℓ^∞ the sequences λ for which $\lim_{n\to\infty} \lambda(n)$ exists forms a closed subspace c of ℓ^∞.

A <u>Banach limit</u> (E, ϕ) consists of a closed subspace E of ℓ^∞ containing c and a bounded linear $\phi : E \to k$ such that $\phi(\lambda) = \lim (\lambda(n))$ for every $\lambda \in c$. We need the following result:

(1.3.1) <u>Lemma:</u>

a) <u>If</u> $\varepsilon > 0$ <u>and</u> E <u>is a subspace of</u> ℓ^∞, <u>containing</u> c, <u>of countable type, then there is a Banach limit</u> (E, ϕ) <u>with</u> $\|\phi\| \leq 1 + \varepsilon$.

b) <u>If</u> k <u>is maximally complete then there exists a Banach limit</u> (ℓ^∞, ϕ) <u>with</u> $\|\phi\| = 1$.

<u>Proof:</u> a) If E is of countable type (i.e. E is the closure of a subspace with a finite or countable base) then one can show that any bounded functional ϕ_1 on a subspace E_1 can be extended to a bounded functional ϕ on E. Moreover for $\varepsilon > 0$ there is a choice of ϕ with

$\|\phi\| \leq \|\phi_1\|(1+\varepsilon)$. Take $E_1 = c$ and $\phi_1(\lambda) = \lim \lambda(n)$, then a) follows.

b) If k is maximally complete then any bounded ϕ_1 defined on a subspace extends to a ϕ with the same norm. So statement b) follows.

(1.3.2) Proposition: $\varinjlim \mathcal{O}(\mathbb{P} - B_n)^o \neq k^o$. (In other words there are non-constant bounded holomorphic functions on Ω).

Proof: We suppose that B_n is the open disk with center b_n and radius $|\pi_n|$. Further we suppose $B_n \supsetneqq B_{n+1}$ and $\cap B_n = \emptyset$.
For $s \leq t$ we write

$$\frac{\pi_t}{z - b_t} = \sum_{j=1}^{\infty} \lambda_{s,j}(t) \left(\frac{\pi_s}{z - b_s}\right)^j \quad \text{with} \quad \lambda_{s,j}(t) \in k^o.$$

Let $\Lambda_{s,j} \in \ell^\infty$ be defined by $\Lambda_{s,j}(n) = \begin{cases} 0 & \text{if } n < s. \\ \lambda_{s,j}(n) & \text{if } n \geq s. \end{cases}$

Let E be a subspace of ℓ^∞, containing all $\Lambda_{s,j}$ and c, and of countable type. Let ϕ be a Banach limit on E.

We define $F_s = \sum_{j=1}^{\infty} \phi(\Lambda_{s,j}) \left(\frac{\pi_s}{z - b_s}\right)^j$. Since $\lim_{j\to\infty} \|\Lambda_{s,j}\| = 0$ the holomorphic function belongs to $\mathcal{O}(\mathbb{P} - B_s)$.

For any $u \in \mathbb{P}$ with $|u - b_s| \geq |\pi_s|$ the infinite series $\sum_{j=1}^{\infty} \left(\frac{\pi_s}{a - b_s}\right)^j \Lambda_{s,j}$ converges and lies in E since E is closed.

This infinite series is in fact equal to

$$(0, 0 - 0, \frac{\pi_s}{u - b_s}, \frac{\pi_{s+1}}{u - b_{s+1}}, \frac{\pi_{s+2}}{u - b_{s+2}}, \ldots).$$

The continuity of ϕ implies that

$$F_s(u) = \phi((0, \ldots, 0, \frac{\pi_s}{u - b_s}, \frac{\pi_{s+1}}{u - b_{s+1}}, \ldots)).$$

It follows that $F_s(a) = F_{s+1}(u)$ for $|u - b_s| \geq |\pi_s|$. So we have found a holomorphic function $F \in \varinjlim \mathcal{O}(\mathbb{P} - B_s)$.
Since $F(u) = \phi(0, \ldots, 0, \frac{\pi_n}{u - b_n}, \ldots)$ (for $n \gg 1$) it follows that

$|F(u)| \leq \|\phi\| \leq (1+\varepsilon)$ for all $u \in \mathbb{P}$. Clearly $F(\infty) = 0$. Moreover ϕ can be chosen such that for example $\phi(\Lambda_{1,1}) \neq 0$ (N. B. $\Lambda_{1,1} = (\pi_1, \pi_2, \pi_3, \ldots)$). Hence $F \neq 0$.

So F is a bounded holomorphic function on $\mathbb{P}^1$ and $F \notin k$. It follows also that $\varinjlim \mathcal{O}(\mathbb{P} - B_n)^o \neq k^o$.

(1.4) Theorem: Let Ω be a domain (or rational connected affinoid) in $\mathbb{P}$. Then Ω has arbitrarily fine pure coverings $\mathfrak{U}$ such that each $U \in \mathfrak{U}$ is a connected rational affinoid and $\bar{\Omega} = \bar{\Omega}_{\mathfrak{U}}$ satisfies:

(1) $\bar{\Omega}$ has at most countably many irreducible components.

(2) each component is $= \mathbb{P}^1_{\bar{k}} - V$ for some finite (or empty) subset V of $\mathbb{P}^1(\bar{k})$.

(3) every intersection of components is an ordinary double point.

(4) the point of intersection are rational over $\bar{k}$.

(5) the intersection graph of $\bar{\Omega}$ is a locally finite (connected) tree.

Proof: We follow closely the construction in Chap. I, §2 "The tree of a compact subset of $\mathbb{P}$". Let Ω be defined by a sequence of rational connected affinoids $X_1 \subset X_2 \subset X_3 \subset \ldots$ (the sequence may be finite or infinite). We may assume that $\infty \in X_1$ and we write
$\mathbb{P} - X_n = B_{1,n} \cup \ldots \cup B_{s(n),n}$. Each $B_{i,n}$ is an open disk with center in k and radius in $|k^*|$. The open disks are supposed to be disjoint. Further we may assume that the $X_1, X_2, X_3, \ldots$ are chosen such that each $B_{i,n}$ contains some $B_{j,n+1}$. Choose finite sets $A_1, A_2, A_3, \ldots$ in $\mathbb{P}(k)$ follows:

(1) A_1 contains ∞ and for each $i \leq s(1)$ a point in $B_{i,1}$ and a point in $B^+_{i,1} - B_{i,1}$. (As before B^+ denotes the smallest closed disk in k containing B).

(n) A_n (for $n > 1$) contains for each $i \leq s(n)$ such that $B_{i,n} \neq B_{j,n-1}$ for all j, a point in $B_{i,n}$ and a point in $B^+_{i,n} - B_{i,n}$.

Put $A = \bigcup_{n=1}^{\infty} A_n$ and $A^{(3)} = \{(a_0, a_1, a_\infty) \in A^3 | a_0, a_1, a_\infty$ are three different points$\}$. With respect to this A and $A^{(3)}$ we will construct a covering $\mathcal{U}$ of the required type. If one varies the defining sequence for Ω and the collection A of points in $\mathbb{P}(k)$ one obtains arbitrarily fine coverings of Ω. First a lemma.

(1.5) Lemma: For every $a = (a_0, a_1, a_\infty) \in A^{(3)}$ the set $R_a(A)$ is finite.

Proof: For an open disk B in $\mathbb{P}$ the image $R_a(B)$ is either a point or equal to $\mathbb{P}^1_{\bar{k}}$. If $R_a(B) = \mathbb{P}^1_{\bar{k}}$ then the three points a_0, a_1, a_∞ must belong to B. From the definition if A it follows that $R_a(B_{i,n})$ is a point for $n >> 0$. Hence $R_a(A)$ is finite.

(1.6) Continuation of the proof of (1.4)

As in Chap. I, §2 the set $T = A^{(3)}/\sim$ is a locally finite, connected, tree. Every edge of T corresponds to a reduction $R_{a,b} : \mathbb{P}^1_k \to \mathbb{P}^1_{\bar{k}} \times \mathbb{P}^1_{\bar{k}}$ with image $\ell_1 \cup \ell_2$, $\ell_i \simeq \mathbb{P}^1_{\bar{k}}$, intersecting in just one point p, and such that p is not contained in the finite set $R_{a,b}(A)$.

Let W denote $\bigcap_{n\geq 1} R_{a,b}(\mathbb{P} - X_n) \subseteq V$ and $U(a, b) = U = R^{-1}_{a,b}(\ell_1 \cup \ell_2 - W)$. Since $R_{a,b}(\mathbb{P} - X_n)$ is finite for $n >> 0$, it follows that $W = R_{a,b}(\mathbb{P} - X_m)$ for some m and $U \subseteq X_m$. The set U is a connected rational affinoid with canonical reduction $\ell_1 \cup \ell_2 - W$.

For a finite subtree T^o of T let $U(T^o)$ denote the union of the U's belonging to the edges of T^o. Clearly $U(T^o)$ is a connected rational affinoid and $\mathbb{P} - U(T^o)$ in equal to $R^{-1}(\bigcap_{n\geq 1} R(\mathbb{P} - X_n)) = R^{-1}(R(\mathbb{P} - X_m))$ for $m >> 0$, in which R denotes the reduction of $\mathbb{P}$ corresponding to the finite set T^o.

Now, we want to show that each X_n in contained in some $U(T^o)$. As before $\mathbb{P} - X_n = B_{1,n} \cup \ldots \cup B_{s(n),n}$ and choose $a_i, b_i \in A$ with $a_i \in B_{i,n}$, $b_i \in B^+_{i,n} - B_{i,n}$, We take T^o such that all $[(\infty, a_i, b_i)]$

belong to T^o. Then $\mathbb{P} - U(T^o) = D_1 \cup \ldots \cup D_r$ ($D_1, \ldots, D_r$ are disjoint open disks) and we have:

For each i, $1 \leq i \leq s(n)$, there is a $D_j \subset B_{i,n}$. Further any D_j has the form $R^{-1}(R(B_{\ell,m}))$ for some $m >> n$, $1 \leq \ell \leq s(m)$. So D_j contains some $B_{\ell,m}$. Hence any D_j is contained in some $B_{i,n}$. It follows that $U(T^o) \supseteq X_n$.

Let $\mathcal{U}$ denote the covering of Ω consisting of the U(a, b) where (a, b) is an edge of T. Then one finds that $\mathcal{U}$ is a pure covering of Ω. That $\bar{\Omega}$ has the required properties follows as in Chap. I, §2. In particular the intersection graph of $\bar{\Omega}$ is equal to the tree T.

(1.7) Remarks: The reduction $\bar{\Omega}$ of Ω constructed in (1.4) is far from unique. However in the most important case:

$\Omega = \mathbb{P} - \mathcal{L}$ in which $\mathcal{L}$ is a compact and perfect subset of $\mathbb{P}(k)$ there is a unique stable reduction of Ω. This stable reduction is obtained by taking $A = \mathcal{L}$ in the proof of (1.4). The tree of this reduction is equal to the tree of the compact set $\mathcal{L}$ introduced in Chap. I, §2.

More generally, let X be a compact subset of $\mathbb{P}(k)$ and let X_1 be the set of limit points of X. By taking A = X in the construction explained in (1.4) we find a reduction of $\Omega = \mathbb{P} - X_1$.

In the special case that Ω is an affinoid in $\mathbb{P}$ the then (1.4) gives a proof of a result that we obtained already in Chap. III (2.11).

In some other cases (e.g. $\Omega = k$ or $\Omega = k^*$ etc.) there is no canonical choice for the reduction and in particular Ω has no stable reduction.

(1.8) Definition of the angle

Choose two triples $a = (a_0, a_1, a_\infty)$, $b = (b_0, b_1, b_\infty)$ in $\mathbb{P}^1(k)^{(3)}$ and let $R_a, R_b: \mathbb{P}^1_k \to \mathbb{P}^1_{\bar{k}}$ and $R_{a,b} : \mathbb{P}^1_k \to \ell_1 \cup \ell_2 \subset \mathbb{P}^1_{\bar{k}} \times \mathbb{P}^1_{\bar{k}}$ be the corresponding reductions. We suppose that R_a and R_b are inequivalent.

We will define an invariant of the reduction $R_{a,b}$, and we will call this invariant the angle of R_a and R_b (or of $R_{a,b}$).

Let $X_1, X_2, Y_1, Y_2 \in \mathbb{P}^1_k$ be points with images x_1, x_2, y_1, y_2 (under $R_{a,b}$) such that $x_1, x_2 \in \ell_1 - \ell_2$ and $y_1, y_2 \in \ell_2 - \ell_1$. The cross-ratio $D(X_1, X_2, Y_1, Y_2) = \dfrac{(X_1 - X_2)(Y_1 - Y_2)}{(X_1 - Y_1)(X_2 - Y_2)}$ is an invariant under the action of PGL(2, k) and for the calculation of D we may assume that $R_{ab}(0)$ lies on $\ell_1 - \ell_2$ and $R_{ab}(1)$, $R_{ab}(\infty)$ are different points of $\ell_2 - \ell_1$.

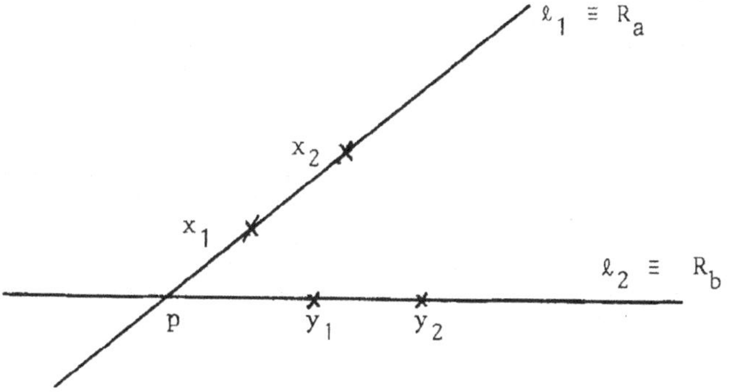

Then it is easily seen that $|D| = |X_1 - X_2| = |\rho|$ for some $\rho \in k^*$, $0 < |\rho| < 1$, independent of the choice of X_1, X_2, Y_1, Y_2. This constant will be called the "<u>absolute value of the angle</u>".

The image of D in $\{\lambda \in k \mid |\Lambda \leq |\rho|\} / \{\lambda \in k \mid |\lambda| < |\rho|\} \cong \bar{k}$ lies in $\bar{k}^*$ (if $x_1 \neq x_2$ and $y_1 \neq y_2$) and depends only on x_1, x_2, y_1, y_2.

We write $D(x_1, x_2, y_1, y_2)$ for this image. It depends on x_1, x_2, y_1, y_2 according to the formula:

$$D(\tilde{x}_1, \tilde{x}_2, \tilde{y}_1, \tilde{y}_2) = D(x_1, x_2, y_1, y_2) \quad \frac{\tilde{x}_1 - \tilde{x}_2}{x_1 - x_2} \cdot \frac{\tilde{y}_1 - \tilde{y}_2}{y_1 - y_2}.$$

In this formula we have identified $\ell_1 - \{p\}$ and $\ell_2 - \{p\}(\{p\} = \ell_1 \cap \ell_2)$ with $\bar{k}$.

We now define the angle of $R_{a,b}$ by:

(i) the absolute value $|\rho|$.

(ii) the map $D(x_1, x_2, y_1, y_2)$ which takes values in

$$\{\lambda \in k \mid |\lambda| \leq |\rho|\} / \{\lambda \in k \mid |\lambda| < |\rho|\}.$$

(1.9) The importance of the notion of angle is shown by the following:

Lemma: Let $R = R_{a,b} : \mathbb{P}^1_k \to \ell_1 \cup \ell_2 = Z$ be the reduction described in (1.8). An automorphism ϕ of Z (as algebraic veriety over $\bar{k}$) lifts to an automorphism Φ of $\mathbb{P}^1_k$ (i.e. $R\,\Phi = \phi R$) if and only if ϕ respects the angle of R.

Proof: We may assume that we are in the situation of the picture ,where $\rho \in k^*$, $0 < |\rho| < 1$. Then the angle is completely determined by $D(R(0), R(\rho), R(1), R(\infty)) = \rho$ modulo elements of absolute value $< |\rho|$.

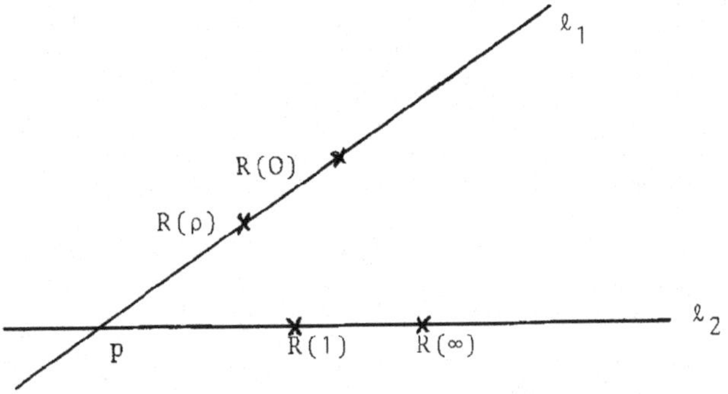

If an automorphism σ of $\mathbb{P}^1_k$ induces some automorphism $\bar{\sigma}$ of Z, then $\bar{\sigma}$ preserves the angle, since σ preserves D. Consider the map $\sigma_1 : z \to \rho/z$. Then $\bar{\sigma}_1$ is an automorphism of Z that interchanges ℓ_1 and ℓ_2. The map $\sigma_2 : z \mapsto \frac{z}{cz + d}$, with $|c| \leq 1$, $|d| = 1$, induces an automorphism $\bar{\sigma}_2$ of Z such that $\bar{\sigma}_2$ restricted to ℓ_2 has the form $(z \mapsto \frac{z}{\bar{c}z + \bar{d}}) \in PGL(2, \bar{k})$. It is the general automorphism of ℓ_2 which leaves p fixed.Next consider $\sigma_3 : z \mapsto z + b$ where $|b| \leq |\rho|$. Then σ_3 induces an automorphism $\bar{\sigma}_3$ of Z such that $\bar{\sigma}_3 = \mathrm{id}$ on ℓ_2 and $\bar{\sigma}_3 = (u \mapsto \frac{1}{Bu + 1})$, where u denotes the residue of ρ/z, parametrizing ℓ_1 and where B is the residue of b/ρ in $\bar{k}$.

Any automorphsims of Z which respects the angle has the form $\bar{\sigma}_3\bar{\sigma}_2\bar{\sigma}_1$ or $\bar{\sigma}_3\bar{\sigma}_2$, and as a consequence lifts ot an automorphism of $\mathbb{P}^1_k$.

(1.10) Lemma: Let $R_1 : \mathbb{P}^1_k \to Z_1$ and $R_2 : \mathbb{P}^1_k \to Z_2$ denote two reductions of the type described in (1.8). Then an isomorphism $\phi : z_1 \to z_2$ lifts to an isomorphism ϕ of $\mathbb{P}^1_k$ (i.e. $\phi R_1 = R_2 \phi$) if and only if ϕ preserves the angles.

Proof: Let Z_1 have components ℓ_1, ℓ_2. Take points X_1, X_2, X_3, Y_1, Y_2, $Y_3 \in \mathbb{P}^1_k$ such that $R_1(X_1)$, $R_1(X_2)$ are different points of $\ell_1 - \{\ell_1 \cap \ell_2\}$ and $R_1(X_3)$ lies on $\ell_2 - \{\ell_1 \cap \ell_2\}$, and such that $R_2(Y_i) = \phi R_1(X_i)$. Let σ be the automorphism of $\mathbb{P}^1_k$ given by $\sigma(X_i) = Y_i (i = 1, 2, 3)$. Then σ induces at least a rational "map" $\bar{\sigma} : Z_1 \to Z_2$. But, since $\bar{\sigma}\ (R_1(X_i)) = R_2(Y_i)\ (i = 1, 2, 3)$, one finds that $\bar{\sigma}$ is an isomorphism. Further $\bar{\sigma}$ and ϕ agree on ℓ_1 and $\phi^{-1}\bar{\sigma}$ is an automorphism of Z_1 which preserves the angle of Z_1. According to (1.9) $\phi^{-1}\bar{\sigma} = \bar{\sigma}_1$ where σ_1 is an automorphism of $\mathbb{P}^1_k$. Hence ϕ lifts. Finally, given an isomorphism Φ such that $\phi R_1 = R_2 \Phi$ holds for some isomorphism ϕ. Then clearly ϕ preserves the angles.

(1.11) Lemma and definition: Let $\Omega \subset \mathbb{P}^1_k$ be a domain of the form $R^{-1}(Z - W)$ where

a) $R : \mathbb{P}^1_k \to Z$ is a reduction of the type described in (1.8). (So Z has two components ℓ_1, ℓ_2).

b) W is a finite $\bar{k}$-rational subset of Z not containing $\ell_1 \cap \ell_2$. Then "the angle of Ω" is defined as the angle of R. This is independent of the embedding of Ω in $\mathbb{P}^1_k$.

Proof: Let $s : \Omega \to \mathbb{P}^1_k$ be another embedding of Ω. Let σ be the automorphism of $\mathbb{P}^1_k$ which sends X_1, X_2, X_3 to $s(X_1)$, $s(X_2)$, $s(X_3)$ where X_1, X_2, $X_3 \in \Omega$ are chosen as in the proof of (1.10). Then instead of s we may consider the embedding $t = \sigma^{-1}s$ which satisfies $t(X_i) = X_i (i=1,2,3)$;

$t(\Omega) = \Omega$ and as a consequence t is an isomorphism of Ω.
The automorphism $\bar{t}$ of $\bar{\Omega} = R(\Omega)$, (we may assume that W has at least one point on each line of Z and $R : \Omega \to \bar{\Omega}$ is then the canonical reduction of Ω; so $\bar{t}$ exists), is the identity on ℓ_1 and has two fixed points on ℓ_2.
So if we show that $\bar{t} = \mathrm{id}$, then the embedding t (and also s) gives the same angle as the given embedding.

The affinoid algebra of Ω has the form $A = k < T_1, T_2, S_1, S_2 > /_I$ where I is the ideal generated by $\{T_1T_2 - \rho,\ Q_1(T_1)S_1 - 1,\ Q_2(T_2)S_2 - 1\}$, and Q_1, Q_2 are monic polynomials with coefficients in k^o such that their reductions $P_1, P_2 \in \bar{k}[T_1, T_2]$ satisfy:

$\bar{A} = \bar{k}[T_1, T_2, S_1\ S_2] / (T_1T_2,\ S_1P_1(T_1) - 1,\ S_2P_2(T_2) - 1) =$

the affine ring of $\bar{\Omega}$. Let ϕ be the automorphism of A, corresponding to t. Then ϕ the $\bar{k}$-automorphism of $\bar{A}$ must have the form $\bar{\phi}(T_1, = T_1)$,

$\bar{\phi}(T_2) = \dfrac{T_2}{\bar{c}T_2 + (1 - \bar{c})}$ where $c \in k$ satisfies $|c| \leq 1$.

This follows from "$\bar{t} = \mathrm{id}$ on ℓ_1, $\bar{t}$ fixed two points of ℓ_2".

Hence $\phi(T_1) = T_1 + \varepsilon$; $\phi(T_2) = \dfrac{T_2}{cT_2 + (1 - c)} + \delta$ where $\varepsilon, \delta \in A$ have spectral norms < 1.

The function $f = \dfrac{\phi(T_1)}{T_1} = \dfrac{T_2}{\phi(T_2)}$ lies in A. Choose $\delta \in k^*$ such that $\lambda f \in A^o$ and $\overline{(\lambda f)}$ is non-zero element of $\bar{A}$. Then $\overline{(\lambda f)}$ is non-zero on some open, non-empty, subset of $\ell_1 \cup \ell_2$. From the explicit form of $\phi(T_1)$ and $\phi(T_2)$ it follows that we have $\lambda = 1$. So $|f(x)| \leq 1$ for all $x \in \Omega$; similary $|(f(x))^{-1}| \leq 1$ for all $x \in \Omega$. Further $|f(x) - 1| < 1$ holds for all $x \in \Omega$ with $R(x) \in \ell_1$.

As a consequence we find $\left|\dfrac{T_2}{\phi(T_2)} - 1\right| < 1$ holds if $|\rho| < |T_2| < 1$.

It follows that $|c| < 1$ and that $\bar{t} = \mathrm{id}$ on $\bar{\Omega}$.

(1.12) Remarks: For a domain Ω we found a reduction $R : \Omega \to \bar{\Omega}$ (a unique one if $\Omega = \mathbb{P}^1_k$ - {perfect compact subset of $\mathbb{P}^1(k)$}) in (1.4). But according to (1.11) we find moreover for any two interesting lines in $\bar{\Omega}$ an angle which is independent of the embedding of Ω in $\mathbb{P}^1_k$. In particular any automorphism Φ of Ω which preserves the reduction R, gives an automorphism ϕ of $\bar{\Omega}$ which preserves all the angles of $\bar{\Omega}$.

However the converse of the statement fails for more complicated $\bar{\Omega}$'s. Example:

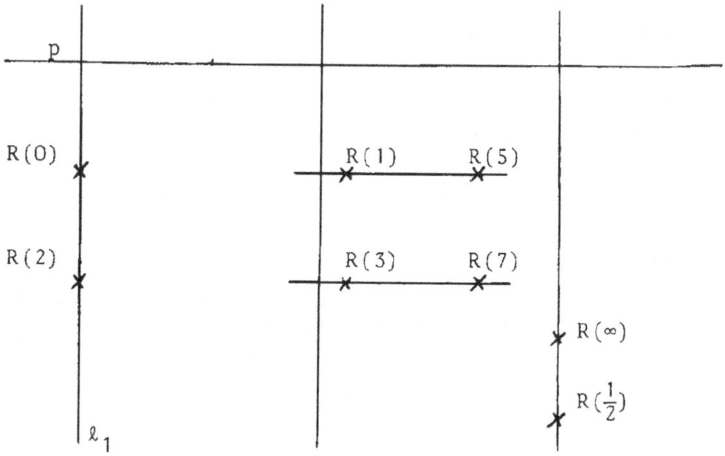

Consider the reduction $R : \mathbb{P}^1_k \to Z$, where $k = \mathbb{Q}_2$ and R has the properties shown in the picture. Let ϕ denote the automorphism of Z given by:

(1) ϕ is the identity on all lines but ℓ_1.

(2) ϕ acts on ℓ_1 by $\phi(p) = p$, $\phi(R(0)) = R(2)$, $\phi(R(2)) = R(0)$.

Then ϕ preserves all possible angles, but as one easily verifies ϕ cannot be lifted to an automorphism of $\mathbb{P}^1_k$.

Problem: What invariants should one add to $\bar{\Omega}$ such that every automorphism of Ω which respects the invariants, does lift to an automorphism of Ω?

§2 Genus zero spaces

(2.1) In this section we suppose that the field k is either discrete or stable. This condition is equivalent to the following: For every finite field extension $\ell \supset k$ the ring ℓ^o is a finitely generated k^o--module. As explained in Ch. III (1.20) the condition k implies the following statement:
If $A \to B$ is a finite map between k-affinoid algebras then also $A^o \to B^o$ is finite.

We consider in this section analytic spaces Ω over k with the properties:

(1) Ω is purely 1-dimensional and reduced.

(2) Ω has a pure covering $\mathcal{U} = (U_i)$ such that the spectral norm on each U_i has values in $|k|$.

We are interested in the reduction $\bar{\Omega} = \bar{\Omega}_{\mathcal{U}}$ of Ω and in the Čech-complex $\check{C}$ of $\mathcal{O}^o$ with respect to $\mathcal{U}$. As in Ch. III (1.14) defined by

$$0 \to \Pi\, \mathcal{O}(U_{i_o})^o \xrightarrow{d^o} \prod_{i_o<i_1} \mathcal{O}(U_{i_o} \cap U_{i_1})^o \xrightarrow{d^1} \prod_{i_o<i_1<i_2} \mathcal{O}(U_{i_o} \cap U_{i_1} \cap U_{i_2})^o \to \dots$$

The complex $\check{C}$ forms the link between Ω and $\bar{\Omega}$. The space Ω is called a <u>genus zero space</u> is Ω and $\mathcal{U}$ satisfy:

(a) The reduction $\bar{\Omega}_{\mathcal{U}} = \bar{\Omega}$ has at most countably many components; each component is isomorph to $\mathbb{P}^1_{\bar{k}} - V$, where V is some finite subset (possibly empty) of $\mathbb{P}^1(\bar{k})$; every intersection of components is an ordinary double point (i.e. normal crossing); the points of intersection are rational over $\bar{k}$.

(b) The intersectiongraph of $\bar{\Omega}$ is a locally finite tree.

(c) For every $U \in \mathcal{U}$, $\mathcal{O}(U)$ has no zero-divisors (and the spectralnorm on $\mathcal{O}(U)$ has values in $|k|$).

As we have seen in §1, every domain or connected rational affinoid in $\mathbb{P}$ is a genus zero space. Our aim is to prove the converse. So we will show that "genus zero-space" and "subspace of $\mathbb{P}^1$" are one and the same thing.

(2.2) Let Ω (as in (2.1)) satisfy:

(1) Ω is purely 1-dimensional and reduced.

(2) Ω has a pure covering $\mathcal{U} = (U_i)$ such that the spectralnorm on each U_i has values in $|k|$.

That $\check{C}$ is indeed the link between Ω and $\bar{\Omega} = \bar{\Omega}_{\mathcal{U}}$ is shown in the following rather technical theorem.

<u>Theorem: Suppose that $\bar{\Omega}$ has finitely many components. Then</u>

(1) $H^i(\check{C}) = 0$ <u>for</u> $i \geq 2$.

(2) $H^1(\check{C})$ <u>is a finitely generated</u> k^o<u>-module.</u>

(3) <u>If</u> $\bar{\Omega}$ <u>is complete and connected then</u> $H^o(\check{C}) = k^o$ <u>and</u> $H^1(\check{C})$ <u>is a finitely generated free</u> k^o<u>-module. Moreover</u> Ω <u>is a complete algebraic curve and</u> Ω <u>and</u> $\bar{\Omega}$ <u>have the same arithmetic genus.</u>
(i.e. $\dim_k H^1(\Omega, \mathcal{O}_\Omega) = \dim_{\bar{k}} H^1(\bar{\Omega}, \mathcal{O}_{\bar{\Omega}})$).

(4) <u>Suppose that</u> Ω <u>is irreducible and that</u> $\bar{\Omega}$ <u>is not complete. Then</u> Ω <u>is affinoid. If moreover</u> $H^1(\bar{\Omega}, \mathcal{O}_{\bar{\Omega}}) = 0$ <u>then</u> $\mathcal{O}(\Omega)^o \otimes \bar{k} = \mathcal{O}(\bar{\Omega})$ <u>and the canonical reduction on</u> Ω <u>is</u> Spec $(\mathcal{O}(\bar{\Omega}))$.

(2.3) <u>Proof of the theorem:</u> <u>(Compare [43] Thm. 2.8)</u>

The case where the field k has a discrete valuation is fairly easy. In the general case we replace k^o be a convenient "discrete" subring $R = k_1^o + \pi k^o$ in which k_1 is a suitably chosen discrete subfield of k and $0 < |\pi| < 1$. (Compare Ch. III (1.20)). Put $Z = \bar{\Omega}$ and let $\mathcal{J}$ be a finite covering of Z by Zariski-open affine subsets satisfying:

1) if $T_1, T_2 \in \mathcal{T}$ then $T_1 \cap T_2 \in \mathcal{T}$;

2) $R^{-1}(T)$ is affinoid with canonical reduction T.

Any $\mathcal{O}(R^{-1}T)^o$ is presented as $k^o\langle X_1, \ldots, X_n\rangle/(f_1, \ldots, f_s)$; the ring R is supposed to contain the coefficients of $f_1, \ldots, f_s$ and $\mathcal{O}(R^{-1}T)^o$ is given an orthonormal base over k^o, which is a subset of $\{X_1^{\alpha_1} \ldots X_n^{\alpha_n}\}$. For $T_1 \subset T_2$ and $T_2 = k^o\langle X_1, \ldots, X_n\rangle/(f_1, \ldots, f_s)$ and $T_1 = k^o\langle Y_1, \ldots, Y_m\rangle/(g_1, \ldots, g_t)$ the restriction map $\mathcal{O}(T_2)^o \to \mathcal{O}(T_1)^o$ lifts to a map $\phi : k^o\langle X_1, \ldots, X_n\rangle \to k^o\langle Y_1, \ldots, Y_m\rangle$. We have $\phi(f_i) = \Sigma h_{ij} g_j$ for certain elements $h_{ij} \in k^o\langle Y_1, \ldots, Y_m\rangle$. We assume that R contains all the coefficients of the $\phi(X_i)$ and the h_{ij}. We arrive at the following: the Čech-complex $\check{C}_\Omega = C$ of $\mathcal{O}^o$ with respect to the covering $\{R^{-1}T\}$ is $0 \to C^o \to C^1 \to C^2 \to \ldots \to C^n \to 0$. The k^o-modules C^i have orthonormal bases such that the coboundary maps d^i have coordinates in R with respect to those bases. Further we know that $H^i(C \otimes \bar{k}) = H^i(Z, \mathcal{O}_Z)$ are finite-dimensional over $\bar{k}$. We need the following:

(2.3.1) Lemma: Let $E_o \overset{d^o}{\to} E_1 \overset{d^1}{\to} E_2$ be a complex of complete k^o-modules with orthonormal bases. Suppose that d^o and d^1 have coordinates in R with respect to those bases. If $H^1(E \otimes \bar{k})$ is finite-dimensional then $H^1(E)$ is a finitely generated k^o-module.

Proof: We write e^o and e^1 for the maps in the complex $E_o \otimes \bar{k} \to E_1 \otimes \bar{k} \to E_2 \otimes \bar{k}$.

(i) Take a subset A of the basis of E_o such that $\{e^o(a \otimes 1) | a \in A\}$ is a basis of im e^o. We may replace E_o by the complete submodule generated by A. In other words, we may assume that e^o is injective.

(ii) Take a subset B of the basis of E_1 such that $\{b \otimes 1 | b \in B\} \cup \{e^o(a \otimes 1) | a \in A\}$ is a basis of $E_1 \otimes \bar{k}$. Using the method of lemma (1.20) Chap. III one sees that $B \cup \{d^o(a) | a \in A\}$ is a basis of E_1. We replace now E_o by 0 and E_1 by the complete submodule generated by B. So we have reduced the lemma to the case $E_o = 0$.

(iii) Choose a subset C of the basis of E_1 such that $\{e^1(c \otimes 1) | c \in C\}$ is a base of im e^1. Choose a subset D of the base of E_2 such that $\{e^1(c \otimes 1) | c \in C\} \cup \{d \otimes 1 | d \in D\}$ is a basis of $E_2 \otimes \bar{k}$. As before, it follows that $\{d^1(c) | c \in C\} \cup D$ is a basis of E_2. Let F be the complete submodule of E_2 generated by D. We replace now E_2 by E_2/F. In other words we have to consider the special case of the lemma where $E_o = 0$, d^1 surjective and $\dim \ker e^1 < \infty$.

(iv) Let C' be the basis of E_1. For any $c' \in C' - C$ there is a unique convergent sum $\sum_{c \in C} \lambda(c', c)c$ $(\lambda(c', c) \in k^o)$ with $d^1(c')$ equal to $d^1(\sum_{c \in C} \lambda(c', c)c)$. Since $\dim \ker e^1 = n < \infty$, it follows that $\#(C' - C) = n$ and $\ker d^1 \cong (k^o)^n$.

We conclude from the lemma that our complex C has finitely generated k^o-modules $H^i(C)$ $(i \neq 0)$ as cohomology groups. As usual there are exact sequences

$$0 \to H^i(C) \otimes \bar{k} \to H^i(C \otimes \bar{k}) \to \mathrm{Tor}_1^{k^o}(H^{i+1}(C), \bar{k}) \to 0.$$

For $i > n$ one has $H^i(C) = 0$, and one easily finds then $H^i(C) = 0$ for $i \geq 2$.

<u>Proof of part (3) of Thm (2.2).</u> If Z is complete and connnected then $H^o(C \otimes \bar{k}) = H^o(Z, \mathcal{O}_Z) = \bar{k}$. It follows that $H^o(C)$ is finitely generated; $H^o(C) \otimes \bar{k}$ has dimension ≤ 1. Since $k^o \subset H^o(C)$ one obtains $H^o(C) = k^o$. Moreover in our complex $C^o \to C^1 \to C^2 \to \dots$ the term $\ker d^1/\mathrm{im}\, d^o$ has no torsion, because the injective map $C^o/_{k^o} \to C^1$ remains injective after $- \otimes \bar{k}$. So $H^1(C^\bullet)$ is a finitely generated free k^o-module. It follows that $H^1(\Omega, \mathcal{O}_\Omega) = H^1(C \otimes k)$ and $H^1(\bar{\Omega}, \mathcal{O}_{\bar{\Omega}}) = H^1(C \otimes \bar{k})$ have the same dimension. Finally as in Chap. III (2.2) one finds that Ω is a complete algebraic curve.

Proof of Thm. (2.2) Part (4): Since $H^1(\check{C}_\Omega)$ is finitely generated there is an exact sequence $0 \to M \to (k^o)^s \to H^1(\check{C}_\Omega) \to 0$ (with s minimal) Hence Tor $(H^1(\check{C}_\Omega), \bar{k}) = M \otimes \bar{k}$ has dimension $\leq s$. As a consequence the injective map $\mathcal{O}(\Omega)^o \otimes \bar{k} \to \mathcal{O}(\bar{\Omega})$ has a finite dimensional cokernel and there exists $F \in \mathcal{O}(\Omega)^o$ such that $\bar{k}[\bar{F}] \to \mathcal{O}(\bar{\Omega})$ is finite and $\bar{F}$ is zero on the complete components of $\bar{\Omega}$.

By induction on the number of complete components of $\bar{\Omega}$ we will show that Ω is affinoid. It is easily seen that we have only to consider the cases

(0) $\bar{\Omega}$ has no complete components.

(1) $\bar{\Omega}$ has exactly one complete component.

We will need the following result:

(2.3.2) Lemma: Let $\phi : X \to Y$ be a finite morphism between k-analytic-spaces. If Y is affinoid then so is X.

Proof: By definition, ϕ is finite if there exists a finite affinoid covering $\{Y_1, \ldots, Y_n\}$ of Y such that $X_i = \phi^{-1}(Y_i)$ is affinoid and $\mathcal{O}(Y_i) \to \mathcal{O}(X_i)$ is finite morphism.

Since Y is affinoid we can refine this covering to a covering $(Y_i^*)_{i=1}^m$ given as follows:

$s_1, \ldots, s_m$ are elements in $\mathcal{O}(Y)$ generating the unit ideal and $Y_i^* = \{y \in Y \mid |s_i(y)| \geq |s_j(y)| \text{ for all } j\}$. Consider the coherent sheaf $\mathcal{S}$ on Y defined by $\mathcal{S}(Y_i^*) = \mathcal{O}(X_i)$. There exists a finitely generated $\mathcal{O}(Y)$-module M such that $\mathcal{S}(U) \cong M \otimes_{\mathcal{O}(Y)} \mathcal{O}(U)$ for every affinoid U in Y. Since $0 \to \mathcal{O}(X) \to \otimes\, \mathcal{O}(X_i) \to \bigoplus_{i<j} \mathcal{O}(X_i \cap X_j)$ is exact it follows that $M = \mathcal{O}(X)$ and $\mathcal{O}(X)$, being a finite extension of $\mathcal{O}(Y)$, is an affinoid algebra. Further $\mathcal{O}(X_i) \cong \mathcal{O}(X) \otimes_{\mathcal{O}(Y)} \mathcal{O}(Y_i^*) \cong \mathcal{O}(X) \langle \frac{s_1}{s_i}, \ldots, \frac{s_m}{s_i} \rangle$. Hence is an affinoid space.

We continue now the proof of (2.2) part (4).

<u>Case (0):</u> Let $\mathcal{T}$ be a covering of Z by small enough open affine sets. For $T \in \mathcal{T}$ the set $R^{-1}T \subset \Omega$ is affinoid and has canonical reduction T. Further $\mathcal{O}(R^{-1}T)^o \otimes \bar{k} = \mathcal{O}(T)$ is a localisation of $\mathcal{O}(Z)$. Since $\bar{k}[\bar{F}] \to \mathcal{O}(Z)$ is finite, one finds that a suitable localisation of $\bar{k}[\bar{F}]$ maps finitely in $\mathcal{O}(T)$. Hence $\Omega \to Sp(k\langle F\rangle)$ is a finite map and according to (2.3.2) the set Ω is affinoid.

<u>Case (1):</u> Let L be the complete of $Z = \bar{\Omega}$ and let $\{p_1, \ldots, p_a\}$ be the points where L intersects the other components of Z. Put $L^* = L - \{p_1, \ldots, p_a\}$ and put $\Omega_1 = R^{-1}(L^*)$. There exists $F \in \mathcal{O}(\Omega)^o$ such that $\bar{k}[\bar{F}] \to \mathcal{O}(Z)$ is finite and $\bar{F}$ is zero on L. We may assume that $L = \{z \in Z \mid \bar{F}(z) = 0\}$. Since Ω is irreducible, the function F is not identical zero on Ω_1. Let the spectral norm of F on Ω_1 be $|\pi|$, with $\pi \in k$, $0 < |\pi| < 1$. Put $t = F/\pi$ and let $\bar{t}$ be its image in $\mathcal{O}(L^*) = \mathcal{O}(\Omega_1)^o \otimes \bar{k}$. Then $\bar{t}$, considered as rational function on L has poles in $p_1, \ldots, p_a$ and zeros in $n_1, \ldots, n_b$. Put $\Omega_2 = R^{-1}(Z - \{n_1, \ldots, n_b\})$.

The map $\phi : \Omega \to Y = Sp(k\langle F\rangle) = Y_1 \cup Y_2$, where $Y_1 = \{y \in Y \mid |F(y)| \leq |\pi|\}$ and $Y_2 = \{y \in Y \mid |F(y)| \geq |\pi|\}$ satisfies $\phi^{-1}(Y_i) = \Omega_i$ $(i = 1, 2)$.

The map $\Omega_1 \to Y_1$ is finite since $\mathcal{O}(Y_1)^o \otimes \bar{k} = \bar{k}[t] \to \mathcal{O}(L^*) = \mathcal{O}(\Omega_1)^o \otimes \bar{k}$ is finite. ALso $\Omega_2 \to Y_2$ is finite, since

$\mathcal{O}(Y_2)^o \otimes \bar{k} = \bar{k}[\frac{1}{t}, \bar{F}] \to \mathcal{O}(Z - \{n_1, \ldots, n_b\}) = \mathcal{O}(\Omega_2)^o \otimes \bar{k}$ is finite.

The lemma (2.3.2) implies that Ω is affinoid.

Finally if $H^1(\bar{\Omega}, \mathcal{O}_{\bar{\Omega}}) = 0$ then $H^1(\check{C}_\Omega) = 0$ and $\mathcal{O}(\Omega)^o \otimes \bar{k} \to \mathcal{O}(\bar{\Omega})$ is an isomorphism.

(2.4) Theorem: Let Ω be genus zero space. Then Ω is isomorphic to a domain in $\mathbb{P}$ or to a connected rational affinoid in $\mathbb{P}$.

Proof: We will prove the statement in a number of steps.

(a) If $\bar{\Omega}$ has finitely many components and every component is complete then by (2.2) part (3) it follows that $\Omega \simeq \mathbb{P}$ and $H^i(\check{C}_\Omega) = 0$ for $i \geq 1$ and $H^o(\check{C}_\Omega) = k^o$.

(b) Suppose that $\bar{\Omega}$ has finitely many components, but not all components are complete. By (2.2) part (4) we know that $H^i(\check{C}_\Omega) = 0$ for $i \geq 1$ and Ω is affinoid. We have still to see that Ω is isomorphic to a connected rational domain in $\mathbb{P}$.

Let ℓ be a non-complete component; let $b_1, \ldots, b_n$ the points where ℓ intersects the other components of $\bar{\Omega}$. Put $\ell^* = \ell - \{b_1, \ldots, b_n\}$. Then $R^{-1}(\ell^*)$ is according to (2.2) affinoid and on easily verifies that $R^{-1}(\ell^*) \cong \mathbb{P} - (B_1 \cup \ldots \cup B_n \cup C_1 \cup \ldots \cup C_m)$ where the B_i and C_j are open disjoint disks with radii 1. The points $b_1, \ldots, b_n$ correspond to $B_1, \ldots, B_n$. Let U_ℓ denote $\mathbb{P} - (B_1 \cup \ldots \cup B_n)$. We glue U_ℓ to Ω by identifying $R^{-1}(\ell^*)$ with $\mathbb{P} - (B_1 \cup \ldots \cup B_n \cup C_1 \cup \ldots \cup C_m)$ in U_ℓ.

This construction we do for every non-complete component of $\bar{\Omega}$. The space so obtained will be called Ω^+. The reduction $\overline{\Omega^+}$ is obtained from $\bar{\Omega}$ by completing every non-complete component of $\bar{\Omega}$.

According to (a), $\Omega^+ \simeq \mathbb{P}$ and $\Omega \subset \Omega^+$ is clearly a connected rational domain in $\mathbb{P}$.

(c) Suppose that $\bar{\Omega}$ has infinitely many components. Let T denote the tree of $\bar{\Omega}$; t_o a vertex of T and put $T_n = \{t \in T \mid \text{distance}(t, t_o) \leq n\}$. Put $\Omega_n = R^{-1}(\bar{\Omega} - \bigcup_{t \notin T_n} t)$. Then $\bar{\Omega}_n$ has finitely many components and according to (b) the space Ω_n is isomorphic to a connected rational affinoid in $\mathbb{P}$.

The inclusion $\Omega_n \subset \Omega_{n+1}$ extends in a natural way to an idenfication $\Omega_n^+ = \Omega_{n+1}^+$. Choose three different points $q_o, q_1, q_\infty \in \Omega_1$ and let $F_n : \Omega_n^+ \xrightarrow{\sim} \mathbb{P}$ be the unique isomorphism with $F_n(q_i) = i$ $(i = 0, 1, \infty)$. Then $F_{n+1}|\Omega_n = F_n|\Omega_n$ for all n. So we have found a morphism $F : \Omega \to \mathbb{P}$ with $F|\Omega_n = F_n$. Let $\mathcal{F}$ be the family of affinoid subspaces of $\mathbb{P}$ generated by the $\{F(\Omega_n) \mid n \geq 1\}$. Then clearly $\Omega \cong \Omega(\mathcal{F})$. So we have shown that Ω is isomorphic to a domain in $\mathbb{P}$.

(2.5) *Corollary: Let* Ω *be a genus zero space. The following conditions are eqüivalent.*

(1) $\Omega \cong \mathbb{P} - X$ *for some compact subset* X.

(2) $\mathcal{O}(\Omega)^o = k^o$

(3) a) *every component of* $\bar{\Omega}$ *is a complete curve* (i.e $= \mathbb{P}^1_{\bar{k}}$).

b) *for every end of the tree of* $\bar{\Omega}$ (*i.e. a subtree of the form*

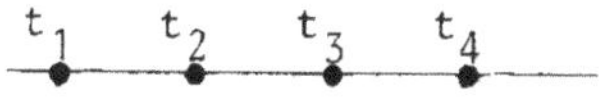

... ,

the product of the absolute values of the angles (t_1, t_2), (t_2, t_3), (t_3, t_n), ... *is zero.*

Proof: The only interesting case is the case where $\bar{\Omega}$ has infinitely many components. According to (2.4) we may regard Ω as a domain in $\mathbb{P}$. Let

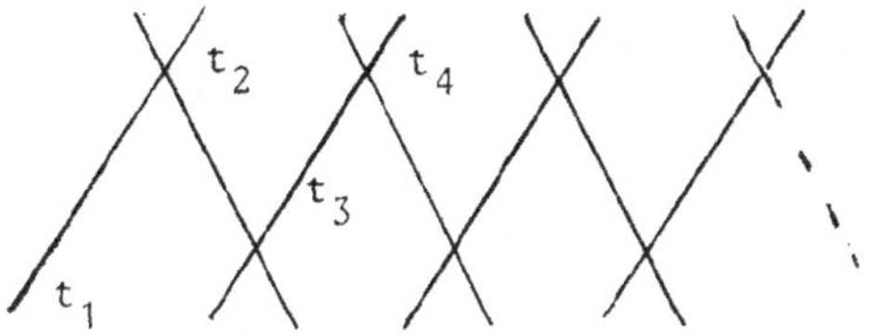

be an end of the tree of $\bar{\Omega}$.

Put $\{X_n\} = t_n \cap t_{n+1}$ and let $R_n : \mathbb{P} \to t_n$ denote the standard reduction. The set $R_n^{-1}(\{x_n\}) = B_n$ is an open disk with radius r_n. According to §1 we have $r_{n+1} = r_n \cdot$(the absolute value of (t_n, t_{n+1})). So the condition in (3) b) is equivalent to $\lim r_n = 0$. And one easily verifies now that (3) and (1) are equivalent.

If $\bar{\Omega}$ contains a component which is non-complete, then $\mathbb{P} - \Omega$ contains some open disk. In that case clearly $\mathcal{O}(\Omega)^o \neq k^o$.

If (3) b) is not satisfies for Ω, then we will also construct a bounded non-constant function. If k is maximally complete then $\bigcap_{n=1}^{\infty} B_n \neq \emptyset$ and $\mathbb{P} - \Omega$ contains some closed disk. So in this case again $\mathcal{O}(\Omega)^o \neq k^o$. If $\cap B_n = \emptyset$ then we have to do some more work. It is clear that any $F \in \varprojlim \mathcal{O}(\mathbb{P} - B_n)^o$ defines a bounded holomorphic function on Ω. So we have to see that $\varprojlim \mathcal{O}(\mathbb{P} - B_n)^o$ contains a non-constant function. But that is shown in (1.3.2).

So we have shown that (2) implies (3).

Finally we will show (1) $\Rightarrow$ (2). Let Ω be given as a union $\cup \Omega_n$, with $\Omega_n \subset \Omega_{n+1}$ and each Ω_n is a connected rational domain in $\mathbb{P}$. Suppose that $\infty \in \Omega_1$ and put $\mathcal{O}(\Omega_n)_+^o = \{f \in \mathcal{O}(\Omega_n)^o \mid f(\infty) = 0\}$. From the compactness of $\mathbb{P} - \Omega$ and CH. III (1.8.3) it follows that for every $\varepsilon > 0$ there is an integer n such that

$\mathrm{im}\,(\mathcal{O}(\Omega_n)_+^o \to \mathcal{O}(\Omega_1)_+^o) \subseteq \{f \in \mathcal{O}(\Omega_1)_+^o \mid \|f\| \leq \varepsilon\}$.

An element $F \in \varprojlim \mathcal{O}(\Omega_n)_+^o = \mathcal{O}(\Omega)_+^o$ maps onto zero in $\mathcal{O}(\Omega_1)_+^o$. Hence $\mathcal{O}(\Omega)_+^o = 0$ and $\mathcal{O}(\Omega)^o = k^o$.

(2.6) Remarks: Let Ω be a genus zero space. If $\bar{\Omega}$ has finitely many components then we know by (2.2) that $H^1(\check{C}_\Omega) = 0$.

If $\bar{\Omega}$ has infinitely many components then this statement is in general false. However one can show the following:

(2.6.1) <u>If either</u> (a) $\mathbb{P} - \Omega$ <u>is compact, or</u>

(b) <u>the field</u> k <u>is maximally complete</u>

<u>then</u> $H^i(C_\Omega) = 0$ <u>for</u> $i \geq 1$.

We will give the outline of the <u>proof:</u>

Write Ω as union of connected rational affinoids Ω_n (with $\Omega_n \subset \Omega_{n+1}$ for all n). Since for each Ω_n the Cech-complex $\check{C}_\Omega/\Omega_n$ has trivial cohomology one can see that $H^i(\check{C}_\Omega) = H^i(\check{C})$ where $\check{C}$ is the Čech-complex of $\mathcal{O}^o$ with respect to the covering $\{\Omega_n\}$ of Ω.

This Čech-complex is essentially equal to

$0 \to \Pi\mathcal{O}(\Omega_n)^o \overset{\alpha}{\to} \Pi\mathcal{O}(\Omega_n)^o \to 0$ where α is given by $(h_n) \mapsto (h_{n+1} - h_n)_n$. Hence $H^i(\check{C}) = 0$ for $i \geq 2$ and $H^1(\check{C})$ = the cokernel of α. We suppose that $\infty \in \Omega$, and write as before $\mathcal{O}(\Omega_n)^o_+ = \{f \in \mathcal{O}(\Omega_n)^o | f(\infty) = 0\}$. Then $H^1(\check{C})$ is also equal to the cokernel of

$$\alpha : \Pi\mathcal{O}(\Omega_n)^o_+ \to \Pi\mathcal{O}(\Omega_n)^o_+.$$

In case (a) we know that for every $\varepsilon > 0$ there exists an $n \geq 1$ with $\text{im}\,(\mathcal{O}(\Omega_n)^o_+ \to \mathcal{O}(\Omega_1)^o_+) \subseteq \{f \in \mathcal{O}(\Omega_1)^o_+ | \|f\|_{\Omega_1} \leq \varepsilon\}$. Let now $(f_n) \in \Pi\mathcal{O}(\Omega_n)^o_+$. Then $\lim \|f_n\|_{\Omega_1} = 0$ and $h_1 = \sum_1^\infty f_n$ converges and lies in $\mathcal{O}(\Omega_1)^o_+$.

Similarly $h_N = \sum_{n \geq N} f_n \in \mathcal{O}(\Omega_N)^o_+$. Then clearly $f_n = -h_{n+1} + h_n$ for all $n \geq 1$. So α is surjective and $H^1(\check{C}) = 0$.

In case (b) we have to use a variant ot the proof of (1.3.2). Let $(f_n) \in \Pi\mathcal{O}(\Omega_n)^o$ be given.

Let $\phi : \ell^\infty \to k$ be a linear map with $\|\phi\| = 1$ satisfying $\phi(\lambda) = \lim_{n\to\infty} \lambda(n)$ if $\lambda = (\lambda(1), \lambda(2), \ldots,$ in ℓ^∞ is a convergent sequence (see (1.3)).

Define the function h_1 on Ω_1 as follows:

$h_1(p) = \phi(f_1(p), f_1(p) + f_2(p), \ldots, \sum_{i=1}^n f_i(p), \ldots)$.

One easily sees that $|h_1(p)| \leq 1$ for all p.

Every f in $\mathcal{O}(\Omega_1)^o$ can be decomposed as (see Ch. III (1.8.3))

$f = c_o + \sum_i \sum_{\ell \geq 1} c_{i,\ell} \left(\frac{\pi_i}{z - a_i}\right)^\ell$.

This decomposition can be made for every f_n (coefficients $c^n_{i,\ell}$, c^n_o).
Let $\tilde{c}_o$ be the sequence $(c^1_o, c^1_o + c^2_o, c^1_o + c^2_o + c^3_o, \ldots)$
and similary $\tilde{c}_{i,\ell} = (c^1_{i,\ell}, c^1_{i\ell} + c^2_{i\ell}, \ldots)$.
Then one easily verifies $h_1 = \phi(\tilde{c}_o) + \sum_i \sum_{\ell \geq 1} \phi(\tilde{c}_{i,\ell})(\frac{\pi_i}{z - a_i})^\ell$.

Hence $h_1 \in \mathcal{O}(\Omega_1)^o$.

In a similar way one defines h_N for all $N \geq 1$.
Then $(h_{n+1} - h_n)(p) = \phi(f_n(p), f_n(p), f_n(p), \ldots,) = f_n(p)$.
This shows that α is surjective and $H^1(\check{C}) = 0$.

(2.6.2) <u>If k is not maximally complete and $\mathbb{P} - \Omega$ is not compact then in general $H^1(\check{C}) \neq 0$. However $H^1(\check{C})$ is a vector space over $\bar{k}$.</u> (i.e. <u>for $\xi \in H^1(\check{C})$ and $\pi \in k^o$ with $|\pi| < 1$, the element $\pi\xi = 0$</u>).

<u>Proof:</u> We can reproduce the proof of (2.6.1) since the Banach-limit ϕ is only needed on a subspace of countable type (see (1.3)). However $\|\phi\|$ is in general > 1. For a given $\pi \in k^o$, $0 < |\pi| < 1$, we can construct a ϕ with $\|\phi\| \leq |\pi|^{-1}$. This means that $\pi\xi = 0$.

To show that in general $H^1(\check{C}) \neq 0$ we look at a typical case, namely Ω defined by a sequence $\mathbb{P} - B_1 \subset \mathbb{P} - B_2 \subset \ldots$ and $\cap B_n = \emptyset$. Let B_n have center b_n and radius $|\pi_n| < 1$.

If $H^1(\check{C}) = 0$ then $\alpha : \Pi\mathcal{O}(\mathbb{P} - B_n)^o_+ \to \Pi\mathcal{O}(\mathbb{P} - B_n)^o_+$ is surjective. We have a surjective map $\mathcal{O}(\mathbb{P} - B_n)^o_+ \to \pi_n k^o$ given by $f \mapsto (\lim_{z\to\infty} zf(z))$. It follows that also $\alpha : \Pi(\pi_n k^o) \to \Pi(\pi_n k^o)$ is surjective. Consider the exact sequence of complexes

$0 \to \Pi(\pi_n k^o) \to \Pi(k^o) \to \Pi(k^o/\pi_n k^o) \to 0$. The exact sequence of cohomology is: $0 \to \{\lambda \in k^o \mid |\lambda| \leq \inf |\pi_n|\} \to k^o \overset{\beta}{\to} \varprojlim (k^o/\pi_n k^o) \to H^1 = 0$.

However β is not surjective since the sequence $(b_1, b_2, b_3, \ldots) \in \varprojlim (k^o/ {}_n k^o)$ does not lie in the image of β. From this contradiction the statement $H^1(\check{C}) \neq 0$ follows.

§3 The universal covering of a totally split curve

(3.1) Any algebraic curve X over k is in this section supposed to be connected, complete and non-singular. X is called <u>totally split over</u> k if X has a pure (finite) covering $\mathfrak{U}$ such that:

(1) for every $U \in \mathfrak{U}$, the spectral norm on $\mathcal{O}(U)$ has values in $|k|$.

(2) $\bar{X}_U$ has components $\ell_1, \dots, \ell_s$ such that

a) each ℓ_i is isomorphic to $\mathbb{P}^1_{\bar{k}}$.

b) every intersection is a normal crossing.

c) every point of intersection is $\bar{k}$-rational.

Let G denote the intersection graph of $\bar{X} = \bar{X}_{\mathfrak{U}}$, i.e. the vertices of G are the components of $\bar{X}$ and the edges of G correspond to the points of intersection on $\bar{X}$. The graph is connected since X is connected. Let $\hat{G} \overset{\pi}{\to} G$ denote the universal covering of G. Then $\hat{G}$ is a connected locally finite tree. The group Γ of automorphisms of $\pi : \hat{G} \to G$, acts freely on $\hat{G}$, is isomorphic to the fundamentalgroup of G, is a free finitely generated group and satisfies $\hat{G}/\Gamma \overset{\sim}{\to} G$. We are going to lift the above to a mapping of k-analytic spaces $\hat{X} \to X$.

(3.2) <u>Construction of $\hat{X}$:</u> We assume for convenience that $\hat{G} \neq G$ (otherwise $X = \mathbb{P}^1$). For any subgraph H of G we define a Zariski open subset $\bar{X}(H)$ of $\bar{X}$ by $\bar{X} - \bar{X}(H)$ = the union of the lines corresponding with vertices not in H and the points corresponding to the edges not in H. Let $R : X \to \bar{X}$ denote the given reduction, then $X(H) = R^{-1}(\bar{X}(H))$. If H happens to be a connected subtree of G then X(H) is isomorphic to a connected rational affinoid in $\mathbb{P}^1_k$. In particular if H has the form •——• or • then X(H) is a connected rational affinoid. We can replace the given pure covering $\mathfrak{U}$ of X by the covering $\mathfrak{U}^* = \{X(H) \mid$ H isomorphic to •——• $\}$. Then $\mathfrak{U}^*$ has also the properties 1) and 2) of (3.1) and moreover the new intersection graph G^* has no endpoints.

In the sequel we will assume $G = G^*$ and $\mathfrak{U} = \mathfrak{U}^*$.

Let e be an edge of $\hat{G}$ with vertices a and b. Define $U(e) = X(\{\pi(a), \pi(b); \pi(e)\})$. For a vertex a of $\hat{G}$ we define $U(a) = X(\{\pi(a)\})$. Then $\hat{X}$ is the k-analytic space obtained by glueing the $\{U(e) \mid e \text{ edge of } \hat{G}\}$ according to:

(i) $U(e) \cap U(f) = \emptyset$ if the edges e and f have no common vertex.

(ii) If e and f have the common vertex a then U(e) and U(f) are glued with respect to the inclusion maps $U(a) \to U(e)$, $U(a) \to U(f)$.

It is clear that $\hat{X}$ is a k-analytic space with $\{U(e)\}$ as a pure covering. The reduction $\hat{R} : \hat{X} \to \bar{\hat{X}}$ obtained from this covering has as intersection graph the tree $\hat{G}$. The obvious map $a : \hat{X} \to X$ is surjective, lifts the map $\hat{G} \overset{\pi}{\to} G$, and has the property:

For every connected subtree H of G, $u^{-1}(X(H)) = \dot{\bigcup}_i Z_i$ a disjoint union of affinoid subspace and $u : Z_i \overset{\sim}{\to} X(H)$ holds for all i.

The group Γ lifts in an obvious way to a group of automorphisms of $\hat{X}$. This group will also be denoted by Γ. Further X is the categorical quotient of $\hat{X}$ with respect to Γ.

(3.3) We will give a precise meaning to the statement: $\hat{X} \to X$ is the universal covering of X.

<u>Definition:</u> a morphism $\phi : Y \to X$ of k-analytic spaces is called a <u>covering</u> if X has an allowed (affinoid) covering $(X_i)_{i \in I}$ and each $\phi^{-1}(X_i)$ has a disjoint covering $\{Y_{i,j}\}_{j \in I_j}$ such that

(i) $\{Y_{i,j}\}_{i,j}$ is an allowed covering of Y.

(ii) $\phi : Y_{i,j} \overset{\sim}{\to} X_i$ holds for all i and j.

<u>Definition:</u> A covering $\phi : Y \to X$ is called the <u>universal covering</u> if

(i) Y and X are connected.

(ii) For every covering $\phi' : Y' \to X$, with Y' connected, and every point $y_o' \in (\phi')^{-1}\phi(y_o)$ with $y_o \in Y)$, there exists a unique morphism $f : Y \to Y'$ with $\phi' f = \phi$ and $f(y_o) = y_o'$.

As usual the universal covering is unique if it exists.

(3.4) Proposition: Let $\phi : Y \to X$ be a connected covering of a domain in $\mathbb{P}^1_k$. Then ϕ is an isomorphism.

Proof: By (1.4) there exists a pure covering $\mathfrak{U}$ of X such that the reduction $R : \bar{X} \to \bar{X} = \bar{X}_{\mathfrak{U}}$ has the properties stated in (1.4) and moreover $\phi^{-1}(U_i) = \cup Y_{ij}$ (every $U_i \in \mathfrak{U}$) where the Y_{ij} are disjoint affinoid subsets of Y, $\{Y_{ij}\}$ is an allowed covering of Y and $Y_{ij} \tilde{\to} U_i$ is an isomorphism for all i and j. The covering $\mathcal{Y} = (Y_{ij})$ of Y is pure and we have a commutative diagramm. The map $\bar{\phi}$ is with respect to the Zariski-topology a covering. The induced map $\bar{\bar{\phi}}$: (Graph of $\bar{Y}$) $\to$ (Graph of $\bar{X}$) is again a connected covering. Since the graph of $\bar{X}$ is a tree, it follows that $\bar{\bar{\phi}}$ is an isomorphism. Then it easily follows that ϕ is bijective and hence is an isomorphism.

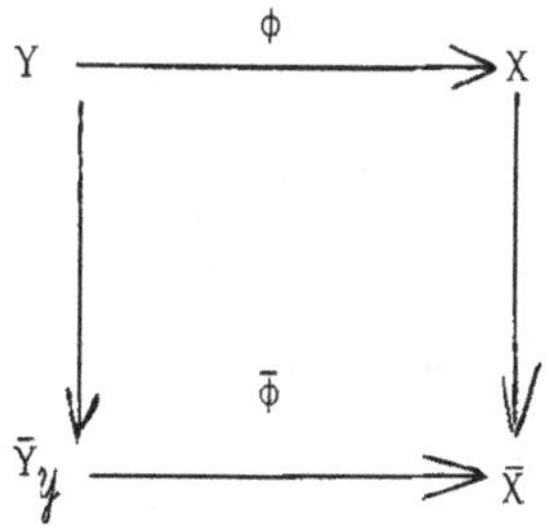

(3.5) Proposition: $u : \hat{X} \to X$ is the universal covering of X.

Proof: Let $\phi : Y \to X$ be a covering of X. Then the fibre product $Y \times_X \hat{X}$ (defined in the obvious way) is a covering of $\hat{X}$. This covering is trivial according to (3.4) and it follows that we have a unique morphism $f : \hat{X} \to Y$ with $\phi f = u$ and $f(y_o) = y_o'$.

(3.6) Remarks: (1) As usual it follows that every connected covering of X has ths form $\hat{X}/_H$ where H is a subgroup of Γ.

(2) Further we note that the definition of covering does not correspond to "covering for algebraic varieties". As an axample: $\phi : k^* \to k^*$, given by $z \mapsto z^n$, ($n > 1$, prime to the characteristic), is in the algebraic sense a covering, but it is not a covering according to (3.3).

(3) The statement (3.5) implies that $\hat{X}$ does not depend on our construction in (3.2) and is also independent of the choice of $\mathcal{U}$ in (3,1).

(3.7) Since $\hat{X}$ is a genus zero space we can embed $\hat{X}$ as a domain in $\mathbb{P}^1_k$. For the absolute values of the angles in $\hat{X}$ we have only finitely many possiblities, namely the angles of the edges in G or X. According to (2.5) we find that $\hat{X} \simeq \mathbb{P}^1_k - \mathcal{L}$ where $\mathcal{L}$ is a compact subset of $\mathbb{P}^1_k$. Further we can lift the automorphism group Γ of $\hat{X}$ to a group of automorphisms of $\mathbb{P}^1_k$ which has $\mathcal{L}$ as invariant set. This follows from:

(3.8) Lemma: If $\mathcal{L}$ is a compact and perfect subset of $\mathbb{P}$ and $\Omega = \mathbb{P} - \mathcal{L}$ then $\mathrm{Aut}(\Omega) = \{\sigma \in PGL(2, k) \mid \sigma(\Omega) = \Omega\}$.

Proof: Assume that $\infty \in \Omega$ and that σ is an automorphism of Ω with $\sigma(\infty) = \omega \in \Omega$. Let F be a meromorphic function on $\mathbb{P}$ of the form $F(z) = \frac{az + b}{cz + d}$ with $F(\infty) = \omega$ and $F - \sigma$ is bounded on some neighbourhood of ∞ in Ω. Then $F - \sigma$ is bounded on all of Ω since F and σ are already bounded outside any neighbourhood of $\infty \in \Omega$. Using (2.5) one finds that $F - \sigma$ is constant. Hence σ extends to an automorphism of $\mathbb{P}$ with $\sigma(\Omega) = \Omega$.

(3.9) The group Γ extended to a group of automorphisms of $\mathbb{P}$ is also denoted by Γ. Since Γ has a compact set $\mathcal{L}$ as invariant set it follows that Γ is discontinuous. Further Γ is finitely generated and free. So Γ is a Schottky group. Finally it is easily seen that $\mathcal{L}$ is the set of limit points of Γ.

We summarize our results in

(3.10) *Theorem:* *Let* X *be an algebraic curve, which is totally split over* k. *Then there exists a Schottky group* Γ *in* PGL(2, k), *unique up to conjugation, such that* $X \approx \Omega/\Gamma$ *where* $\Omega = \mathbb{P} - \mathcal{L}$ *and where* $\mathcal{L}$ *is the set of limit points of* Γ.

Proof: The uniqueness of Γ, up to conjugation in PGL(2, k), follows from the uniqueness of $\hat{X}$ and lemma (3.8).

Chapter V. Analytic reductions of algebraic curves

Introduction: As in chapter IV, the field k is supposed to be stable or discrete. For an algebraic curve X (complete, irreducible and non--singular) we have seen that X carries the structure of an analytic space over k. In particular X has analytic reductions $\bar{X}_{\mathfrak{U}}$)w.r.t. a pure covering $\mathfrak{U}$ of X).

In contrast to analytic reductions there are also algebraic reductions of X, defined as follows:

Let $Y \to \mathrm{Spec}\,(k^o)$ be a proper flat map of schemes (where $k^o = \{\lambda \in k \mid |\lambda| \leq 1\}$ is the valuationsring of k) such that $X \cong Y \times_{k^o} k$. We call Y a model of X over k^o. Then $Y \times_{k^o} \bar{k}$ ($\bar{k}$ the residue field of k) is called an algebraic reduction of X. An algebraic reduction of X is not necessarily reduced, but it is connected, purely 1-dimensional and complete over $\bar{k}$.

A reduction Z (algebraic of analytic) of X is called:

(a) pre - stable, if Z is reduced and the only singularities of Z are ordinary double points.

(b) stable, if Z is pre-stable and every non-singular rational component of Z has at least three points in common with the other components.

We will show in §1 that algebraic reductions and analytic reductions of X are practically one and the same thing. The central point in this chapter is:

Main Theorem: Let X be an algebraic curve over k (non-singular, complete, irreducible) of genus $g \geq 1$. Then there is a finite field-extension k' of k and a pure covering $\mathfrak{U}$ of $X' = X \times_k k'$ such that the analytic reduction $\overline{X'}_{\mathfrak{U}}$ is stable.

Unfortunately we have no complete proof yet. Instead we will give a proof in the following cases:

(0.1) the valuation of k is discrete.

(0.2) the characteristic of $\bar{k}$ is zero.

(0.3) the characteristic of $\bar{k}$ is $p \neq 0$ and X admits a simple covering $X \to \mathbb{P}^1_k$ with degree $< p$.

(0.4) X is locally isomorphic to $\mathbb{P}^1_k$.

§1. Fields with a discrete valuation:

In this section we assume that the valuation of k is discrete. Then according to Deligne-Mumford [5] Corollary (2.7) the following holds:

(1.1) Stable reduction for curves: Let X be an algebraic curve of genus ≥ 2 over k. Then there exists a finite extension $k' \supset k$ such that $X' = X \times_k k'$ has a stable algebraic reduction.

The proof of this statement is highly complicated. First one translates the stability condition for X to a stability condition for the Jacobean variety J of X (prop. (2.3) in [5]). The stability condition for J means that the Nêron minimal model $\mathcal{J}$ of J satisfies: the algebraic group $\mathcal{J} \otimes_{k^o} \bar{k}$ has no unipotent radical. The existence of $\mathcal{J}$ is proved by Nêron and Raynaud. The existence of a finite field extension $k' \supset k$ such that $\mathcal{J} \otimes_k k'$ is stable is proved by Grothendieck in [18]. The main difficulties in the proofs arise in case char $\bar{k} \neq 0$.

The following lemma shows that (1.1) implies part (0.1) of the "main result".

(1.2) Lemma: Let $Y \to \mathrm{Spec}\,(k^o)$ be a model of X such that $Y \times_{k^o} \bar{k}$ is reduced. Then X has a pure covering $\mathfrak{U}$ such that $\bar{X}_{\mathfrak{U}} \cong Y \times_{k^o} \bar{k}$.

Proof: Let $Y_1, \ldots, Y_n$ denote an affine covering of Y. Then $Y_i = \mathrm{Spec}(B_i)$ where B_i is some finitely generated flat k^o-algebra. So B_i has the form $k^o[T_1, \ldots, T_n]/I$. Let π generate the maximal ideal of k^o and let $\hat{B}_i$ denote the completion of B_i with respect to the π-adic topology. Then $\hat{B}_i = \varprojlim B_i/\pi^s B_i = k^o\langle T_1, \ldots, T_n\rangle/(I)$, where $k^o\langle T_1, \ldots, T_n\rangle$ denotes the ring of restricted power series, i.e. all power series $\Sigma a_\alpha T^\alpha$ with $a_\alpha \in k^o$ and $\lim a_\alpha = 0$.

Further (I) denotes the ideal $\bigcap_{s\geq 1}(I + \pi^s k^o\langle T_1, \ldots, T_n\rangle)$. Since $k^o\langle T_1, \ldots, T_n\rangle$ is noetherian one finds that (I) is infact the ideal in $k^o\langle T_1, \ldots, T_n\rangle$ generated by I.

Further $\hat{B}_i \otimes \bar{k} = B_i \otimes_{k^o} \bar{k}$ is reduced and it follows that $A_i = \hat{B}_i \otimes_{k^o} k$ is reduced affinoid algebra over k. For any affinoid algebra A over k we denote as usual by A^o the subring of A consisting of all elements with spectral norm ≤ 1. If A is reduced then the spectral norm is the unique power-multiplicative norm for which A is complete. The subring $\hat{B}_i$ of A_i induces a power-multiplicative norm for which A_i is complete, since $\hat{B}_i \otimes \bar{k}$ has no nilpotent elements ($\neq 0$). It follows that $A_i^o = \hat{B}_i$.

We have a natural map $\mathrm{Sp}(A_i) \to Y_i \times_{k^o} k$, induced by $B_i \otimes_{k^o} k \to A_i$, which identifies $\mathrm{Sp}(A_i)$ with an affinoid subset of the k-affine variety $Y_i \times_{k^o} k$. The given glueing of the pieces $\{Y_i\}$ induces a glueing of the $\mathrm{Sp}(A_i)$ to $T = \cup\, \mathrm{Sp}(A_i)$ and we have a natural embedding $T \to Y \times_{k^o} k = X$. The k-analytic space T has already a pure affinoid covering $\{\mathrm{Sp}(A_i)\}$ and the reduction w.r.t this covering is equal to

$\bar{T} = \cup\, \mathrm{Spec}(\bar{A}_i) = Y \times_{k^o} \bar{k}$.

The proof of the lemma will be complete if we have shown $T = X$. If $T \neq X$ then $\mathcal{O}(T)$ contains non-constant holomorphic functions. Let $f \in \mathcal{O}(T)$ and $f \neq 0$; after multiplying f by a constant we may assume $\|f\| = 1$, where $\|\ \|$ denotes the spectral norm on T. The induced function $\bar{f}$ on $\bar{T}$ is constant since $\bar{T}$ is complete.

Hence $f = \lambda_1 + \pi f_1$ with $\lambda_1 \in k^o$ and $\|f_1\| \leq 1$. Repeating this argument one finds $f = \lambda_s + \pi^s f_1$ where $\lambda_s \in k^o$ and $\|f_s\| \leq 1$, for all $s \geq 1$.

Hence f is constant and T = X.

(1.3) Remarks:

a) In more geometric terms the proof of (1.2) reads: Lets $\mathcal{Y}$ be the formal completion of Y with respect to the closed subscheme $Y \times_{k^o} \bar{k}$. Then $\mathcal{Y} \times_{k^o} k$ is the k-analytic space X provided with a pure covering induced by an affine covering of Y (or of $Y \times_{k^o} \bar{k}$). This pure covering satisfies $\bar{X}_{\mathfrak{U}} \cong Y \times_{k^o} \bar{k}$.

b) If the model $Y \to \operatorname{Spec}(k^o)$ has an non-reduced $Y \times_{k^o} \bar{k}$, then the model still induces a pure covering U of X. But now one can only assert the existence of a canonical surjective morphism $\bar{X}_{\mathfrak{U}} \to (Y \times_{k^o} \bar{k})$ red.

(1.5) Corollary: The "Main Theorem" holds for fields with a discrete valuation.

(1.5) Lemma (1.2) has the following converse:

Lemma: Let $\mathfrak{U}$ be a pure covering of X. Suppose that the valuation is discrete and that the spectral norms on all $U \in \mathfrak{U}$ have values in $|k|$. Then there is a model $Y \to \operatorname{Spec}(k^o)$ with $Y \times_{k^o} \bar{k} \cong \bar{X}_{\mathfrak{U}}$.

Proof: Let $\mathfrak{U} = \{U_1, \ldots, U_n\}$ and let A_i be the affinoid algebra of U_i. Then we can associate with $\mathfrak{U}$ a formal scheme of finite type over k^o, namely $\cup \operatorname{Spf}(A_i^o) = \mathcal{Y} \to \operatorname{Spf}(k^o)$. This formal scheme is proper and flat over k^o and one can even show that $\mathcal{Y}$ is projective.

According to E. G. A. III (5.1.8) it follows that $\mathcal{Y}$ is the formal completion of some $Y \to \operatorname{Spec}(k^o)$ proper and flat, along the closed subscheme $Y \times_{k^o} \bar{k}$. This Y satisfies the assertions of the lemma.

§2 Generalities on analytic reductions

In this section we compare pre-stable and stable reductions. For later use we include two formulas for the cohomology on a complete reduced curve Z over $\bar{k}$, the residue field of k. Let $\eta : \tilde{Z} = C_1 \cup \ldots \cup C_s \to Z$ denote the normalisation of Z. The components C_i of $\tilde{Z}$ are complete, nonsingular, irreducible curves of genus g_i. We are interested in the cohomology groups of $\mathcal{O}_Z$ and $\bar{k}_Z$ = the constant sheaf on Z with stalk $\bar{k}$.

(2.1) Proposition:

(1) $\dim H^0(Z, \mathcal{O}_Z) = \dim H^0(Z, \bar{k}_Z)$ = the number of connected components of Z.

(2) $\dim H^1(Z, \mathcal{O}_Z) - \dim H^0(Z, \mathcal{O}_Z) = \sum_{i=1}^{s}(g_i - 1) + \sum_{p\in Z} \dim \tilde{\mathcal{O}}_{Z,p}/\mathcal{O}_{Z,p}$, in which $\tilde{\mathcal{O}}_{Z,p}$ denotes the normalisation of $\mathcal{O}_{Z,p}$.

(3) $\dim H^1(Z, \bar{k}_Z) - \dim H^0(Z, \bar{k}_Z) = -s + \sum_{p\in Z}(n_p - 1)$, where n_p denotes the number of irreducible components of Z passing through p.

(4) $\dim H^1(Z, \bar{k}_Z) \leq \dim H^1(Z, \mathcal{O}_Z)$. Equality holds if and only if every irreducible component of Z is a non-singular rational curve and Z has only ordinary multiple points as singularities.

Proof: On Z we have two exact sequences of sheaves, namely $0 \to \mathcal{O}_Z \to \eta_*\mathcal{O}_{\tilde{Z}} \to A \to 0$ and $0 \to \bar{k}_Z \to \eta_*\bar{k}_{\tilde{Z}} \to B \to 0$. The sheaves A and B are given by their stalks:

$A_p \cong \tilde{\mathcal{O}}_{Z,p}/\mathcal{O}_{Z,p}$ and $B_p \cong (\bar{k})^{n_p - 1}$. Then (1), (2) and (3) follow easily from $H^1(Z, A) = H^1(Z, B) = 0$.

By an ordinairy multiple point p we mean the following: "every branch through p is non-singular and the tangents at the branches have different directions". Statement (4) follows easily from $n_p - 1 \leq \dim \tilde{\mathcal{O}}_{Z,p}/\mathcal{O}_{Z,p}$ and equality holds if and only if p is an ordinary multiple point and the branches at p come from different components of Z.

We remark that the proposition remains valid for non-complete Z if one replaces Z by the union of its complete pomponents.

(2.2) Theorem: Let X be a 1-dimensional non-singular k-analytic space which admits a pre-stable reduction having finitely many components. Then:

(1) X has a unique stable reduction Z.

(2) Every pre-stable reduction of X is obtained from Z by blowing up points.

(3) For every prestable reduction $X \overset{R_1}{\to} Z_1$ and point p of Z_1 there is a prestable reduction $X \overset{R_2}{\to} Z_2$ and a morphism $\phi : Z_2 \to Z_1$ such that

a) $\phi R_2 = R_1$

b) $\phi : Z_2 - \phi^{-1}(p) \to Z_1 - \{p\}$ is an isomorphism.

c) $\phi^{-1}(p) \cong \mathbb{P}^1_{\bar{k}}$.

(We say that Z_2 is obtained from Z_1 by blowing up the point p).

Remark: In the algebraic case (i.e models of X over the valuation ring k^o of k) a prestable reduction Z_1 of X corresponds with a model Y of X over k^o and p is a point of $Z_1 = Y \times \bar{k} \subset Y$. The new prestable reduction Z_2 of X is in fact obtained by blowing up the point p on Y (in other words a quadratic transform). Contracting a $\mathbb{P}^1_{\bar{k}}$ in Z_1 corresponds to "blowing down". Statement (2) of the theorem corresponds to the factorisation of a birational map into quadratic transforms.

Our theorem is however not purely algebraic, since we work with more general valued fields k and since we are working with analytic reductions instead of algebraic reductions.

Proof: Using Ch. IV Thm. (2.2) we can contract in a given prestable reduction $X \to Z_1$ any component $L \cong \mathbb{P}^1_{\bar{k}}$ which meets the other components in at most 2 points. Repeating this proces we find a stable reduction of X. (Compare Ch. III lemma (2.12.1)).

If $R_i : X \to Z_i$ ($i = 1, 2$) are two stable reductions of X with respect to pure coverings $\mathcal{U}_i$($i = 1$,) then the covering $\mathcal{U}_3 = \mathcal{U}_1 \cap \mathcal{U}_2$ of X is also pure. The corresponding reduction $R_3 : X \to Z_3$ is easily seen to be the map $R_1 \times R_2 : X \to Z_1 \times Z_2$ with image $\cong Z_3$.

So R_3 is also a prestable reduction and we have a commutative diagram, in which ϕ_i are morphisms of algebraic varieties over $\bar{k}$.

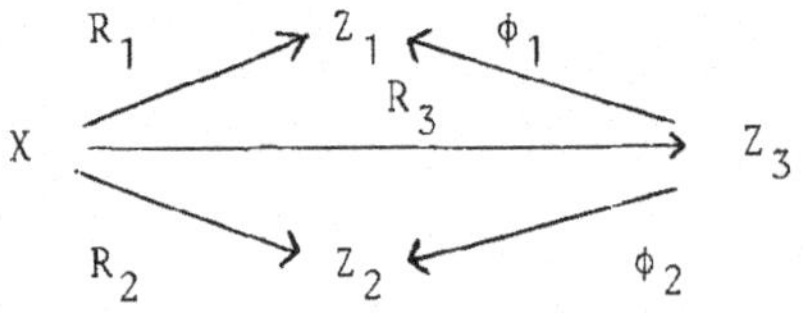

Every component L of Z_3 which is $\cong \mathbb{P}^1_{\bar{k}}$ and meets at most in two points the other components of Z_3 in mapped to one point under ϕ_1 and ϕ_2. After contracting this component of Z_3 we find a reduction Z_3' and maps $\phi_i' : Z_3' \to Z_1$. Continuing this proces one obtains $Z_1 \cong Z_2$. This proves (1) and it proves that every prestable reduction can be contracted to the stable reduction. As "contracting" and "blowing up a point" are each others inverses we have shown statement (2).

(3) The algebraic version of the statement is rather easy to verify. The analytic statement however is somewhat complicated. Since the result is local, we may suppose that $X = \operatorname{Sp} A$, where A is affinoid with canonical reduction $\bar{X} = Z_1$. If p is an ordinary point of Z_1 one can easily verify the statement. We concentrate on the difficult case where p is an ordinary double point, i.e. $\hat{\mathcal{O}}_{\bar{X},p} = \bar{k}[[s,t]]/(st)$.

First we consider two special cases.

1) $\bar{X}$ has two components C_1, C_2, intersecting at p. The components C_1 and C_2 are rational.

2) $\bar{X}$ is irreducible, rational and p is a double point.

Case 1): We can glue two affinoids (subsets of $\mathbb{P}^1_k$) to X such that the resulting analytic space Y has reduction $\bar{Y} = \mathbb{P}^1_{\bar{k}} \cup \mathbb{P}^1_{\bar{k}}$, where the two lines intersect normally at p. According to Chap. IV (2.2) we find $Y \cong \mathbb{P}^1_k$. And we know already that Y has a pure covering $\mathfrak{U}$ with reduction.

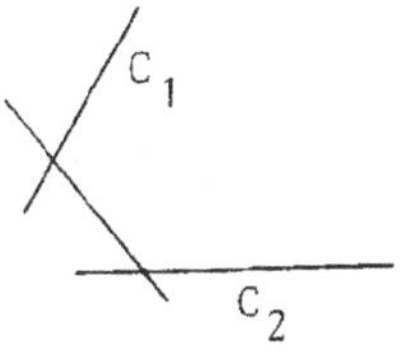

So this case is settled.

Case 2): Again we may suppose that $\bar{X}$ is a complete rational curve with as only singulary the double point p. According to IV (2.2) X is an elliptic curve over k and $\bar{X}$ is its unique stable reduction. Using the equation of X one can find explicitely a reduction $\bar{X}_{\mathfrak{U}}$ such that every component is a non-singular rational curve (and all intersection are normal coverings). (Compare the explicit calculations in §3).

After contracting superfluous $\mathbb{P}^1_{\bar{k}}$'s in $\bar{X}_{\mathfrak{U}}$ we arrive at a reduction

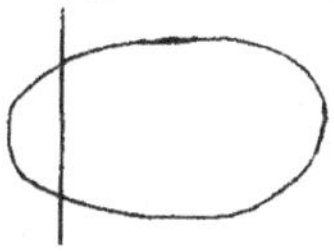

So this case is settled.

The general case: $\bar{X}$ is a finite unramified extension of Z where Z is a Zariski-open subset of 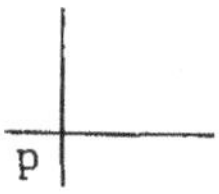

or 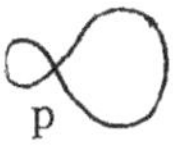

(All curves rational).

There is an affinoid Y and a finite map $\phi : Y \to X$ such that $\bar{\phi} : Z \cong \bar{Y} \leftarrow \bar{X}$ is finite and unramified.

Let $\mathfrak{U} = (U_1, \ldots, U_n)$ be a pure covering of Y such that the reduction $\bar{Y}_{\mathfrak{U}}$ has the required property, (i.e $\bar{Y}_{\mathfrak{U}}$ is  or 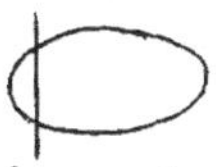).

Then $\mathfrak{V} = (\phi^{-1}U_1, \ldots, \phi^{-1}U_n)$ is a pure covering of X and we claim that the reduction $\bar{X}_{\mathfrak{V}}$ is a finite, unramified extension of $\bar{Y}_{\mathfrak{U}}$. Then $\bar{X}$ is prestable and every component of $\bar{X}_{\mathfrak{V}}$ is non-singular.

After contracting (if necessary) some $\mathbb{P}^1_{\bar{k}}$'s in $\bar{X}_{\mathfrak{V}}$ one obtains the required reduction of X.

Let X = Sp A and Y = Sp B then $\bar{B} \to \bar{A}$ has the properties:
$\bar{A}$ is a free (finitely generated) $\bar{B}$-module with discriminant 1. So also A^o is a free B^o-discriminant 1.

Let U be an affinoid part of Y, then $\mathcal{O}_Y(U)^o \to A^o \otimes_{B^o} \mathcal{O}_Y(U)^o$ has again this property. Hence $A^o \otimes_{B^o} \mathcal{O}_Y(U)^o$ is integrally closed and must be the same as $\mathcal{O}_X(\phi^{-1}U)^o$. So we find that also $\overline{\mathcal{O}_Y(U)} \to \overline{\mathcal{O}_X(\phi^{-1}U)}$ is free and unramified. It follows that $\bar{X}_{\mathfrak{V}} \to \bar{Y}_{\mathfrak{U}}$ is finite and unramified.

(2.3) <u>Remark:</u> Let X be a complete non-singular, irreducible curve over k which has a stable reduction.

For any prestable reduction Z of X such that every component of Z is non-singular, one can form $n_Z = \dim H^1(Z, \bar{k}_Z)$. From (2.1) and (2.2) it follows that n_Z is independent of the choice of Z.

However if one allows a component of Z to have an ordinary double point, then this is no longer true. The essential example is:

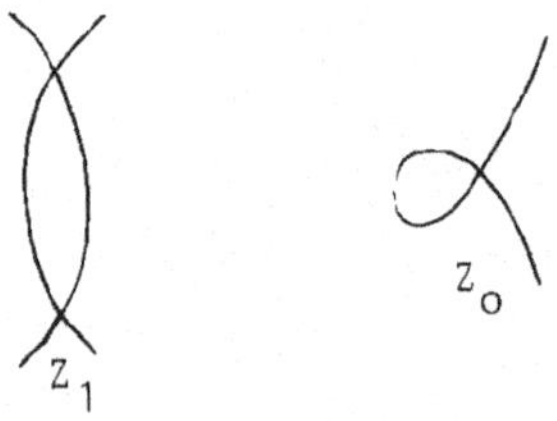

In which Z_1 and Z_o are two reduction of X and Z_o is obtained from Z_1 by contracting a component. $\dim H^1(Z_1, \bar{k}_{Z_1}) = 1$ and $\dim H^1(Z_o, \bar{k}_{Z_o}) = 0$. This inconvenience can be overcome by taking instead of the Zariski--topology, the ètale-topology. So we can formulate the result above as: $n_X = \dim H^1_{\acute{e}t}(Z, \bar{k}_Z)$ is independent of the chosen prestable reduction Z of X. Moreover $\dim H^1_{\acute{e}t}(Z, \bar{k}_Z) = \dim H^1(Z, \bar{k}_Z) + \tau$ where τ is the number of double points of Z lying on exactly one component of Z. Further, again by (2.1), we have $0 \leq n_X \leq g$ = genus of X.

And $n_X = g$ if and only if X is totally split (i.e. every component of Z is rational).

The number n_X can also be defined directly for X. Let $\bar{k}_X$ denote the constant sheaf on X with its Grothendieck topology. Then one can show that $n_X = \dim H^1_{an}(X, \bar{k}_X)$, where "an" means that X is given the Grothendieck topology as k-analytic space.

§3 Hyperelliptic curves

In this section we assume that the characteristic of $\bar{k}$ is $\neq 2$. Let a hyperelliptic curve $\phi : X \to \mathbb{P}^1$ (degree ϕ is 2) be given. We will show that the "Main Theorem" holds in this case. In fact we will assume that k is algebraically closed; this suffices to prove the general case.

In order to state the result more precisely, we recall that any finite subset $S \subseteq \mathbb{P}^1_k = \mathbb{P}$ (with at least 3 points) induces a unique reduction $R = R_S : \mathbb{P} \to Z = (\overline{\mathbb{P}, S})$. The components of Z are lines (i.e. $\cong \mathbb{P}^1_{\bar{k}}$); two lines intersect in at most one point, every intersection is an ordinary double point; the intersection graph of the components of Z forms a finite tree. Every component L of Z gives a partition of S in at least 3 parts.

Let $\mathfrak{U}$ denote the pure covering of $\mathbb{P}$ given by the sets $R^{-1}(T)$, where T is a Zariski-open subset of Z which meets at most 2 components of Z. Then $\bar{\mathbb{P}}_{\mathfrak{U}} = (\overline{\mathbb{P}, S})$.

(3.1) <u>Theorem:</u> <u>Let</u> $\phi : X \to \mathbb{P}$ <u>be a hyperelliptic curve and let</u> $S \subset \mathbb{P}$ <u>denote set of ramification points of</u> ϕ. <u>Then</u>:

(1) <u>Let</u> $\mathfrak{U}$ <u>denote the pure covering of</u> $\mathbb{P}$ <u>corresponding to</u> S. <u>Then</u> $\phi^{-1}(\mathfrak{U})$ <u>is a pure affinoid covering of X with prestable reduction</u> $\bar{X}$.

(2) The following properties are equivalent:

(a) X has a totally split reduction (i.e. X has a prestable reduction Z such that every component of Z is $\cong \mathbb{P}^1_{\bar{k}}$).

(b) X is totally split (i.e. every component of $\bar{X}$ is $\cong \mathbb{P}^1_{\bar{k}}$).

(c) Every line L in the reduction $\overline{(\mathbb{P}^1_k, S)}$ gives a partition $\{S_i, \ldots, S_a\}$ of S such that for at most 2 indices i, S_i has an odd number of elements.

Proof: (1) For $U_1, U_2 \in \mathcal{U}$ there exists $h \in \mathcal{O}(U_1)$ with spectral norm ≤ 1, such that $U_1 \cap U_2 = \{x \in U_1 \mid |h(x)| = 1\}$. Then $\phi^{-1}(U_1) \cap \phi^{-1}(U_2) = \{z \in \phi^{-1}(U_1) \mid |h(\phi(z))| = 1\}$ and $h \circ \phi \in \mathcal{O}(\phi^{-1}U_1)$ has again spectral norm ≤ 1. As a consequence, $\phi^{-1}(\mathcal{U})$ is a pure covering of X. The map ϕ induces a morphism $\bar{\phi} : \bar{X} \to Z = \overline{(\mathbb{P}, S)} = \ell_1 \cup \ldots \cup \ell_a$ where $\ell_1, \ldots, \ell_a$ are the components of Z. We will show that X is a prestable reduction by analysing the map $\bar{\phi}$.

(a) Let ℓ be a component of Z; after deleting $R(S) \cap \ell$ and the points where ℓ intersects other components of Z we have ℓ^*, Zariski-open in $\ell \cong \mathbb{P}^1_{\bar{k}}$. Let $U = R^{-1}(\ell^*)$. For a suitable coordinate t on $\mathbb{P}$ we have $U = \mathbb{P} - (B_1 \cup \ldots \cup B_s)$ where $B_i = \{t \in \mathbb{P} \mid |t - c_i| < 1\}$ and $c_1, \ldots, c_s \in k^o$ have mutual distance 1. For convenience we work with $U^* = \{t \in \mathbb{P} \mid |t| \leq 1 \text{ and all } |t - c_i| \geq 1\}$. Then $\phi^{-1}(U^*) \cong \{(y, t) \mid t \in U^* \text{ and } y^2 = (t - \alpha_1) \ldots (t - \alpha_{2g+2})\}$ where $S = \{\alpha_1, \ldots, \alpha_{2g+2}\}$. Let n_j be the number of α_b's contained in B_j and suppose that $n_1, \ldots, n_i$ are odd an $n_{i+1}, \ldots, n_s$ are even. Since $(t - \alpha_1) \ldots (t - \alpha_{2g+2}) = (t - c_1) \ldots (t - c_i) f^2$ for some $f \in \mathcal{O}(U^*)$, we may also write $\phi^{-1}(U^*) \cong \{(z, t) \mid t \in U^* \text{ and } z^2 = (t - c_i) \ldots (t - c_i)\}$.

$W = \{(z, t) \in \bar{k} \times \bar{k} \mid t \neq \bar{c}_1, \ldots, \bar{c}_i \text{ and } z^2 = (t - \bar{c}_1) \ldots (t - \bar{c}_i)\}$.

This set W is an open dense part of $\bar{\phi}^{-1}(\ell)$. It follows at once that:

(i) $\bar{\phi} : \bar{\phi}^{-1}(\ell^*) \to \ell^*$ is unramified.

(ii) if i = 0 then $\bar{\phi}^{-1}(\ell^*)$ has two components, both $\cong \ell^*$.

(iii) if $i \neq 0$, then i is even, since the number of elements of S is $2g + 2$, where g = genus of X.

(iv) if $i \neq 0$, then $\bar{\phi}^{-1}(\ell)$ is an irreducible curve of genus $\frac{i}{2} - 1$, and $\bar{\phi} : \bar{\phi}^{-1}(\ell) \to \ell$ is only ramified in the images of $c_1, \ldots c_i$.

(b) Let $p \in Z$ be the intersection of ℓ_1, ℓ_2, components of Z. Let $T \subset Z$ be $\ell_1 \cup \ell_2$ where $R(S)$ and the intersection with other components is deleted. Then $U = R^{-1}(T)$ can be written, for a suitable coordinate t on $\mathbb{P}$ as $U^* \cup \{t \in \mathbb{P} \,|\, |t| > 1\}$, where

$U^* = \{t \in \mathbb{P} \,|\, |\rho| \leq |t| \leq 1\} - (B_1 \cup \ldots \cup B_a \cup C_1 \cup \ldots \cup C_b)$

with $B_1, \ldots, B_a$ open spheres with radii 1; $b_1, \ldots, b_a$ their centers have absolute value 1 and mutual distance 1; $C_1, \ldots, C_b$ are open spheres with centers $c_1, \ldots, c_b$, all with absolute value $|\rho|$ and mutual distances $|\rho|$; all radii $|\rho|$.

As before we prefer to work with U^*; the affinoid algebra of U^* is

$\mathcal{O}(U^*) = k\langle T, S, V\rangle / (TS - \rho, U(T - b_1)\ldots(T - b_a)(S - \frac{\rho}{c_1})\ldots(S - \frac{\rho}{c_b}) - 1)$.

The affinoid algebra of $\phi^{-1}(U^*)$ is the normalisation of the ring

$\mathcal{O}(U^*)[Y]/(Y^2 - (t - \alpha_1) \ldots (t - \alpha_{2g+2}))$.

Let $n_1, \ldots, n_a$, $m_1, \ldots, m_b$ be the number of elements of S that lie in $B_1, \ldots, B_a$, $C_1, \ldots, C_b$. Suppose that $n_1, \ldots, n_i$ are odd; $n_{i+1}, \ldots, n_a$ are even; $m_1, \ldots, m_j$ are odd; $m_{j+1}, \ldots, m_b$ are even. Then

$(t - \alpha_1)\ldots(t - \alpha_{2g+2}) = (t - b_1)\ldots(t - b_i)(s - \frac{\rho}{c_1})\ldots(s - \frac{\rho}{c_j})s^{-j}f^2$

for some $f \in \mathcal{O}(U^*)$.

If j is even then the affinoid algebra of $\phi^{-1}(U^*)$ is $\mathcal{O}(U^*)[z]$, where

$z^2 = (t - b_1)\ldots(t - b_i)(s - \frac{\rho}{c_1})\ldots(s - \frac{\rho}{c_j})$.

The reduction of this algebra is a localisation of the ring

$A = \bar{k}[T, S, Z]/(TS, Z^2 - (T - \bar{b}_1)\ldots(T - \bar{b}_j)(S - \frac{\bar{\rho}}{c_1})\ldots(S - \frac{\bar{\rho}}{c_j}))$.

Since $\bar{b}_1, \ldots, \bar{b}_i, \frac{\bar{\rho}}{c_1}, \ldots, \frac{\bar{\rho}}{c_j}$ is $\neq 0$, it follows that $\bar{\phi}^{-1}(p)$ consists of two points, both unramified over p. The completion of the local rings of the two points are both $= \bar{k}[[T, S]]/TS$.

If j is odd then the affinoid algebra of $\phi^{-1}(U^*)$ is $\mathcal{O}(U^*)[Z]$ where $Z^2 = (t - b_1) \dots (t - b_i)(s - \frac{\rho}{c_1}) \dots (s - \frac{\rho}{c_j})t$.
One finds that $\bar{\phi}^{-1}(p)$ consists of one point, ramified over p.

The completion of the local ring at that point is
$\cong \bar{k}[\![T, S, Z]\!]/(TS, Z^2 - T, ZS) \cong \bar{k}\ [\![S,Z]\!]/_{ZS}$.

We draw the conclusion that the singularity at $q \in \bar{X}$ with $\bar{\phi}(q) = p$ is an ordinairy double point.

(c) Let $p \in R(S)$, then p lies on precisely one component ℓ of Z. After a good choice of the coordinate t we have
$U = \{t \in \mathbb{P} \,|\, |t| \leq 1;\ |t - c_i| \geq 1$ for $i = i, \dots, s\}$, where $c_1, \dots, c_s$ have absolute value 1 and mutual distances 1. The equation for X becomes $y^2 = t(t - \alpha_2) \dots (t - \alpha_{2g+2})$ where $\alpha_2, \dots, \alpha_{2g+2}$ lie in the spheres $\bigcup_i \{t \in \mathbb{P} \,|\, |t - c_i| < 1\}$. Further the point p on ℓ corresponds with the point $\bar{t} = 0$ of $\overline{\mathcal{O}(U)} = \bar{k}[t]_{\text{localized}}$. As before $\mathcal{O}(\phi^{-1}U) = \mathcal{O}(U)[y]$ and $\bar{\phi}^{-1}(p)$ consists of one point q ramified over p.
The completion of the local ring at q is
$\hat{\mathcal{O}}_{\bar{X},q} = \bar{k}[\![y, t]\!]/(y^2 - t(t - \bar{\alpha}_2)\dots(t - \bar{\alpha}_{2g+2})) \cong \bar{k}[\![y]\!]$. Hence q is a regular point of $\bar{X}$.

(d) The equivalence of (2b) and (2a) follows from part (a) of the proof of (1). The equivalence of (2a) and (2b) follows from (2.5) and (2.6).

(3.2) Examples: We can now easily calculate the possibilities for curves of genus 1 and 2 (char $\bar{k} \neq 2$; x = image of a point of S; $\boxed{i}$ means that the curve has genus i).

the reduction $(\mathbb{P}, \bar{S})$	the reduction $\bar{X}$ of (3.1)	the stable reduction
genus 1:		
		the same
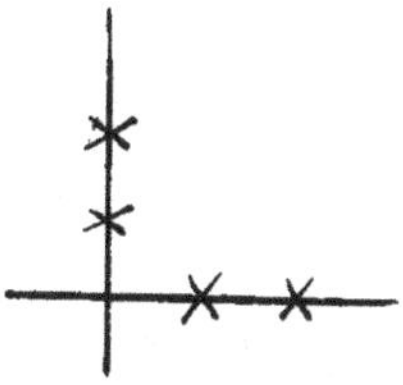		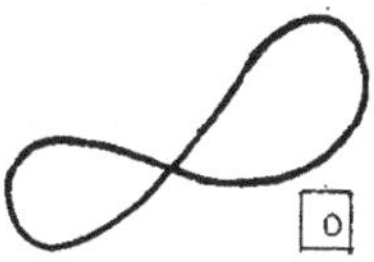 totally split

the reduction $(\overline{\mathbb{P}}, \overline{S})$	the reduction $\tilde{X}$ of (3.1)	the stable reduction
genus 2: 	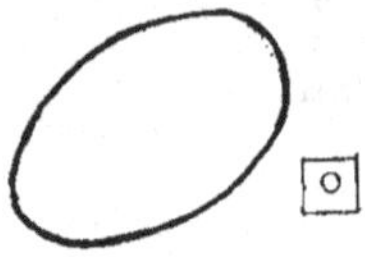	the same
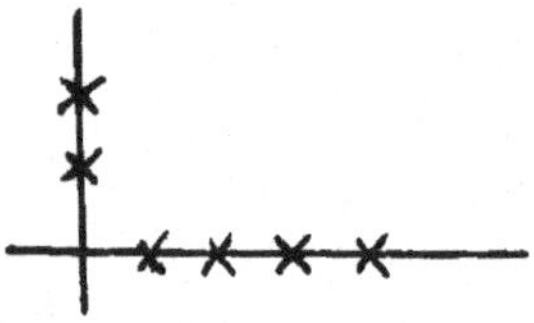	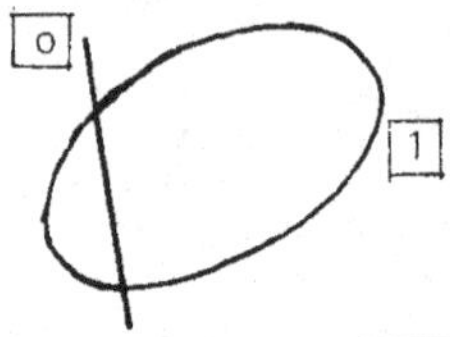	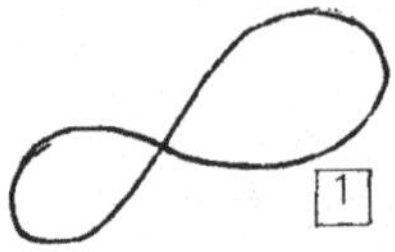
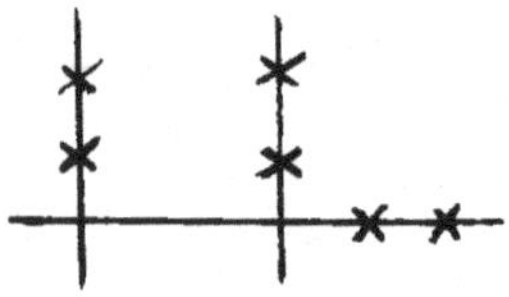	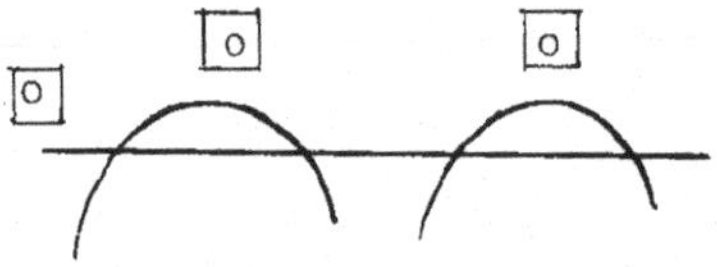	totally split 
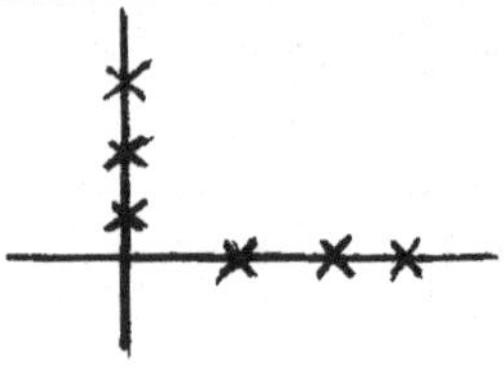	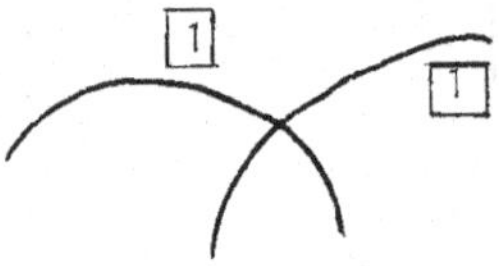	the same
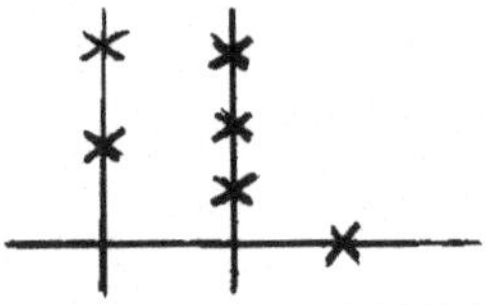	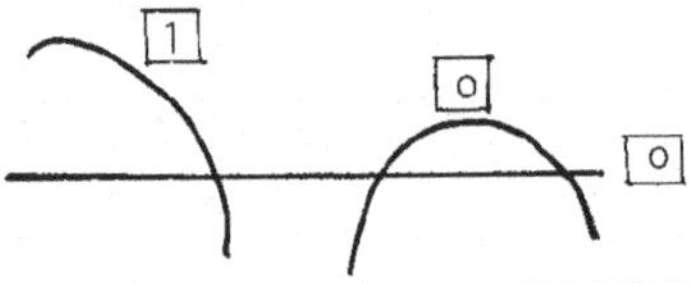	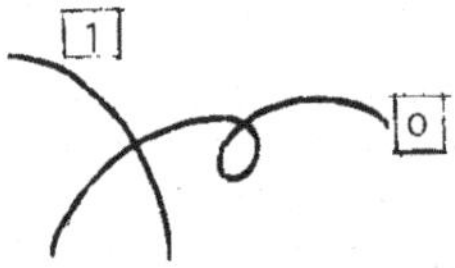
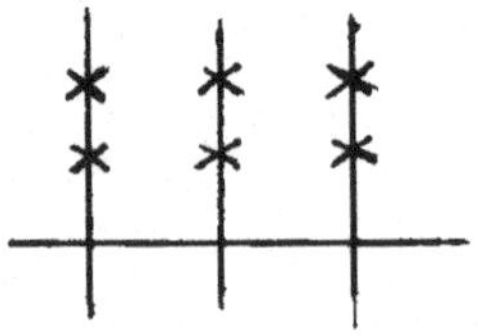	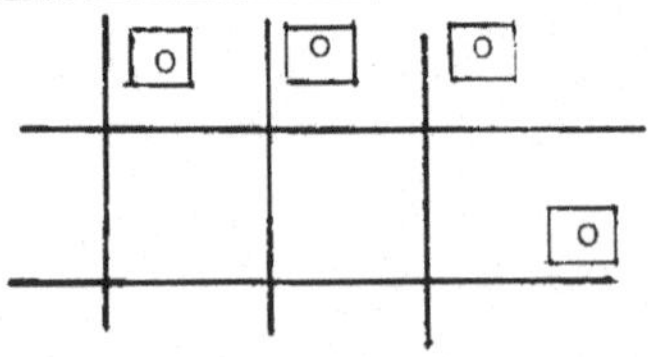	totally split 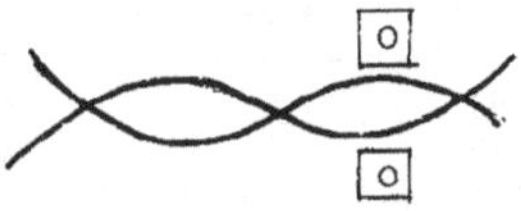
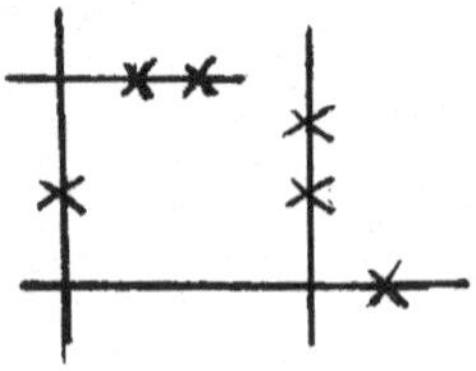	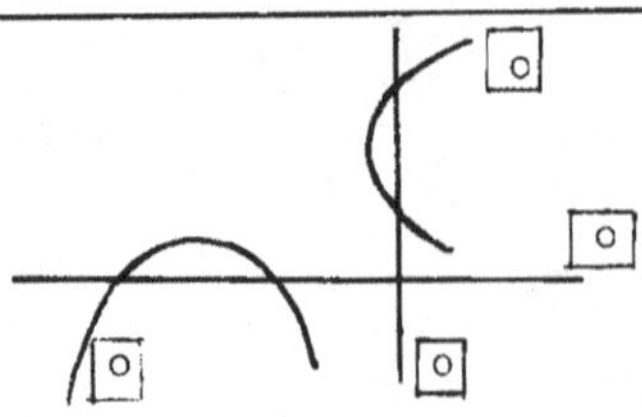	totally split 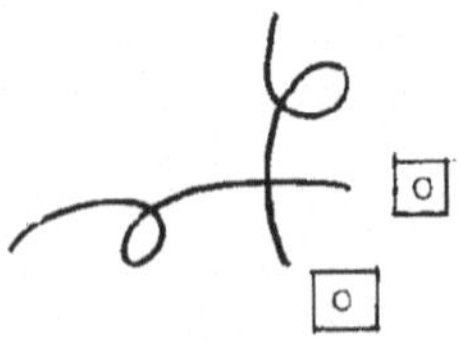

(3.3) Another combinatorical formula equivalent with $\bar{X}$ totally split.

We consider $(\overline{\mathbb{P}, S})$. A line L of $(\overline{\mathbb{P}, S})$ gives a partition $\{S_1, \ldots, S_a\}$ of L. This is called an even partition if each S_i has an even number of elements. A double point p of $(\overline{\mathbb{P}, S})$ given a partition $\{S_1, S_2\}$ of S. This partition is called even if S_1 and S_2 have an even number of elements.

(3.3.1) Corollary: The number $h = \#\{p \mid p$ double point with even partition$\} - \#\{L \mid L$ line with even partition$\}$ satisfies: $0 \leq h \leq g$. Further $h = g$ if and only if $\bar{X}$ is totally split.

Proof: The number h does not change if we refine the reduction $(\overline{\mathbb{P}, S})$ to $(\overline{\mathbb{P}, T})$, where T is some finite set containing S. We may take T such that the corresponding reduction $\bar{\bar{X}}$ of X satisfies: $\bar{\bar{X}}$ is prestable and every component of $\bar{\bar{X}}$ is non-singular.

Using the covering $\bar{\bar{X}} \to (\overline{\mathbb{P}, T})$ and the fact that $(\overline{\mathbb{P}, T})$ is a tree, one easily finds that $h = \dim H^1(\bar{\bar{X}}, \bar{k}_{\bar{\bar{X}}})$. Using (2.1) the corollary now follows.

§4 Tame coverings of $\mathbb{P}$.

In this section we generalize (3.1) to the case of simple coverings $\phi : X \to \mathbb{P}$ of degree n (i.e. for every $p \in \mathbb{P}^1_k$ the set $\phi^{-1}(p)$ consists of at least (n - 1) element. According to W. Fulton [6] for $n \geq g + 1$ (g = genus of X) the curve X admits a simple covering of degree n.

(4.1) Theorem: Let X be a non-singular complete irreducible curve (of genus $g \geq 1$) and let $\phi : X \to \mathbb{P}$ be a simple covering of degree n. Assume that either (a) char $\bar{k} = 0$, or (b) $n <$ char $\bar{k} = p \neq 0$.
Let $S \subset \mathbb{P}$ be the set of ramification points of ϕ and let $\mathcal{U}$ denote the pure covering of $\mathbb{P}$ corresponding to S.

Then $\phi^{-1}(\mathcal{U})$ is a pure covering of X with a prestable reduction.

Contrary to the proof of (3.1) we can not make a very explicit proof. since we have not much information on the equation defining X. In the proof of (4.1) we need a technical theorem (Compare with similar results in S. Bosch [4]) which is interesting in itself.

(4.2) Theorem: Let X be affinoid and reduced with canonical reduction $R : X \to \bar{X}$ and let $p \in \bar{X}$.

(1) $R^{-1}(p)$ has a canonical structure of k-analytic space and $\mathcal{O}(R^{-1}p)$ has a canonical residue class ring $\overline{\mathcal{O}(R^{-1}p)}$.

(2) There exists a morphism of $\bar{k}$-algebras $\phi : \hat{\mathcal{O}}_{\bar{X},p} \to \overline{\mathcal{O}(R^{-1}p)}$; ϕ depends functorically on X and p.

(3) If X is purely one-dimensional then ϕ is an isomorphism.

Remark: More generally, if X is purely d-dimensional and satisfies some weak additional property, then ϕ can also be shown to be an isomorphism. However, we will only need the one-dimensional case in the sequel.

Proof: (1) Choose $f_1, \ldots, f_n \in \mathcal{O}(X)^o$ such that their images $\bar{f}_1, \ldots, \bar{f}_n \in \mathcal{O}(\bar{X})$ have only p as common zero. For ρ, with $0 < \rho < 1$ and $\rho \in |k^*|$ we define: $U_\rho = \{x \in X \mid |f_i(x)| \leq \rho \text{ for } i = 1, \ldots, n\}$. This is an affinoid space and $R^{-1}(p) = \cup U_\rho$. As a consequence this introduces a k-holomorphic structure on $R^{-1}(p)$. For any $g \in \mathcal{O}(X)^o$ such that $\bar{g} \in \mathcal{O}(\bar{X})$ has p as zero, we find $\bar{g}^s = \Sigma \bar{c}_i \bar{f}_i$ with $s \geq 1$; $c_i \in \mathcal{O}(X)^o$. Hence $g^s = \Sigma c_i f_i + h$ where $|h(x)| < 1$ for all $x \in X$. For $\rho < 1$, close enough to 1 and $x \in U_\rho$ one finds $|g(x)| \leq \sqrt[s]{\rho}$. We draw from this the conclusion that the holomorphic structure on $R^{-1}(p)$ does not depend on the choice of $f_1, \ldots, f_n$. We consider the subset I of $\mathcal{O}(R^{-1}p)^o$ consisting of the f with, for every $N \geq 1$ and for $\rho < 1$ close enough to 1, $\|f\|_{U_\rho} \leq \rho^N$. One easily verifies that I is an ideal and does not depend on the choice of $f_1, \ldots, f_n$. $\overline{\mathcal{O}(R^{-1}p)}$ denotes $\mathcal{O}(R^{-1}p)^o/_I$. If the valuation of k is discrete then $\overline{\mathcal{O}(R^{-1}p)}$ is simply $\mathcal{O}(R^{-1}p)^o \otimes \bar{k}$.

(2) The definition of $\phi_{X,p} = \phi : \hat{\mathcal{O}}_{\bar{X},p} \to \overline{\mathcal{O}(R^{-1}p)}$.

Let $\phi_1 : \mathcal{O}(X)^o[\![T_1, \ldots, T_n]\!] \to \mathcal{O}(R^{-1}p)^o$ be given by the formula $\phi_1(\Sigma c_\alpha T^\alpha) = \Sigma c_\alpha f_1^{\alpha_1} \ldots f_n^{\alpha_n}$. The right-hand-side converges on each U_ρ with respect to the spectral norm. Since X and hence each U_ρ is reduced, $\mathcal{O}(U_\rho)$ is complete with respect to the spectral norm. Hence the expression belongs to $\mathcal{O}(R^{-1}p)^o = \varprojlim \mathcal{O}(U_\rho)^o$.

The kernel of ϕ_1 contains the ideal generated by $T_1 - f_1, \ldots, T_n - f_n$.
One finds a morphism:

$$\phi_2 : \hat{\mathcal{O}}_{\bar{X},p} = \mathcal{O}(\bar{X})[\![T_1, \ldots, T_n]\!]/(T_1 - \bar{f}_1, \ldots, T_n - \bar{f}_n) \to \overline{\mathcal{O}(R^{-1}p)}$$

This map is independent of the choice of $f_1, \ldots, f_n$. Indeed let $g_1, \ldots, g_m \in \mathcal{O}(X)^{o}$ be such that p is the only zero $\bar{g}_1, \ldots, \bar{g}_m$. Then with respect to $\{g_1, \ldots, g_m\}$ one finds a morphism ϕ_3 and with respect to $\{f_i, \ldots, g_n, g_1, \ldots, g_m\}$ one finds a morphism ϕ_4.

$$\phi_4 : \hat{\mathcal{O}}_{\bar{X},p} = \overline{\mathcal{O}(X)}[\![T_1, \ldots, T_n, S_1, \ldots, S_m]\!]/(T_1 - \bar{f}_1, \ldots, S_m - \bar{g}_m) \to \overline{\mathcal{O}(R^{-1}p)}$$

Since ϕ_2, ϕ_3, ϕ_4 all make the diagram

$$\begin{array}{ccccc}
\overline{\mathcal{O}(X)} & \longrightarrow & \hat{\mathcal{O}}_{\bar{X},p} & \xrightarrow{\phi_\alpha} & \overline{\mathcal{O}(R^{-1}p)} \\
\uparrow & & & & \uparrow \\
\mathcal{O}(X)^{o} & & \longrightarrow & & \overline{\mathcal{O}(R^{-1}p)}
\end{array}$$

commutative, one finds $\phi_2 = \phi_3 = \phi_4$. This implies that ϕ depends functionically on X and p.

(3) If $X = Sp(k\langle T_1, \ldots, T_m\rangle) \overset{R}{\to} \bar{X} = \bar{k}^n \ni p = (0, \ldots, 0)$ then clearly $\mathcal{O}(R_p^{-1})^{o} = k^{o}[\![T_1, \ldots, T_m]\!]$ and $\overline{\mathcal{O}(R^{-1}p)} = \bar{k}[\![T_1, \ldots, T_n]\!]$. So ϕ as defined in (2) is in this case clearly an isomorphism.

Let now X = Sp (A) be reduced and purely 1-dimensional. Then $\bar{X}$ = Sp $(\bar{A})$ is also reduced and purely 1-dimensional. Let $p \in \bar{X}$. There exists a morphism $\tau : \bar{k}[T] \to \bar{A}$ having the properties: τ is finite and injective; im τ contains no zero-divisors $\neq 0$; τ is separable. The induced map $\mu : \bar{X} \to \bar{k}$ sends p to $0 \in \bar{k}$ as we may assume.
Let n be the rank of $\bar{A}$ over $\bar{k}[T]$. Then $\bar{A}$ is a free module over $\bar{k}[T]$ on n generators.

The map τ can be lifted to a morphism $f : k\langle T\rangle \to A$. Both f and $f^o : k^o\langle T\rangle \to A^o$ have the properties: finite, injective, separable and free of rank n.

Let $e_1,\ldots, e_n \in A^o$ be chosen such that $\bar{e}_1,\ldots, \bar{e}_n$ is a free basis of $\bar{A}$ over $\bar{k}[T]$. Then $e_1,\ldots, e_n$ is a free basis of A^o over $k^o\langle T\rangle$. We can form the discriminant $D = D(e_1,\ldots, e_n) = \det (Tr(e_ie_j)) \in k^o\langle T\rangle$. Clearly its residue $\bar{D} \in \bar{k}[T]$ is equal to $D(\bar{e}_1,\ldots, \bar{e}_n) \neq 0$.

There are elements $e_1^*,\ldots, e_n^* \in A^o$ with $Tr(e_ie_j^*) = D\delta_{ij}$. Put $U_\rho = \{x \in X \mid |f(T)(x)| \leq \rho\}$ where $0 < \rho < 1$ and let $U = \cup U_\rho =$ $= \{ x \in X \mid |f(T)(x)| < 1\}$. Then $U = R^{-1}p_1 \cup \ldots \cup R^{-1}p_s$ where $\{p_1,\ldots, p_s\} = \mu^{-1}(0)$ and $p = p_1$.

We have $\mathcal{O}(U_\rho) = A^o \otimes k\langle T/\sigma\rangle$ where $\sigma \in k^*$ satisfies $|\sigma| = \rho$ and of course $\mathcal{O}(U_\rho)^o \supseteq A^o \otimes k^o\langle T/\sigma\rangle$.

For $w \in \mathcal{O}(U_\rho)^o$ one has $Dw = \sum_i Tr(we_i)e_i^* \in A^o \otimes k^o\langle T/\sigma\rangle$.
So $D\mathcal{O}(U_\rho)^o \subseteq A^o \otimes k^o\langle T/\sigma\rangle \cap A^o \otimes Dk\langle T/\sigma\rangle$.

The image of D in $\bar{k}[[T]]$ is a power series ($\neq 0$) of some order e: an easy calculation shows that $k^o\langle T/\sigma\rangle \cup Dk\langle T/\sigma\rangle \subseteq \sigma^{-e}Dk^o\langle T/\sigma\rangle$ for $|\sigma| = \rho$ close to 1. It follows that $\mathcal{O}(U_\rho)^o \subseteq A^o \otimes \sigma^{-e}k^o\langle T/\sigma\rangle$. Taking the limit for $\rho = |\sigma| \to 1$ one finds
$\mathcal{O}(R^{-1}p_1)^o \oplus \ldots \oplus \mathcal{O}(R^{-1}p_s)^o = \mathcal{O}(U)^o = \varprojlim \mathcal{O}(U_\rho)^o \subseteq A^o \otimes \varprojlim \sigma^{-e}k^o\langle T/\sigma\rangle$
Since $\varprojlim \sigma^{-e}k^o\langle T/\sigma\rangle = k^o[[T]]$ one obtains:
$\mathcal{O}(U)^o = A^o \otimes k^o[[T]]$. An inspection of the proof above shows that the canonical ideal $I \subset \mathcal{O}(U)^o$ is $A^o \otimes k^{oo}[[T]]$. It follows that

$\oplus\, \overline{\mathcal{O}(R^{-1}p_i)} = \bar{A} \otimes \bar{k}[[T]] = \oplus\, \hat{\mathcal{O}}_{\bar{X},p_i}$. Hence all the mappings

$\phi_{\bar{X},p_i} : \hat{\mathcal{O}}_{\bar{X},p_i} \to \overline{\mathcal{O}(R^{-1}p_i)}$ are isomorphisms.

most an ordinairy double point.

Case (c): This seems to be the most complicated case. Let $U = \phi^{-1}(R_o^{-1}p)$ then $U \not\cong R_o^{-1}(p) \cong \{t \in \mathbb{P}^1 | \rho < |T| < 1\}$ is unramified of degree n. That $\overline{\delta(U)}$ has only ordinary double points will follows from the following lemma.

Lemma (4.3): Suppose $n < \text{char } \bar{k} = p \neq 0$ or $\text{char } \bar{k} = 0$. Let U be a connected k-analytic space and let $\phi : U \to \{t \in \mathbb{P} | \rho < |t| < 1\}$ be a finite, unramified map of degree n. Then $U \cong \{s \in \mathbb{P} | \mu < |s| < 1\}$ and $\phi(s) = s^n$ (and $\mu^n = \rho$).

Proof: Let $\psi : V \to \{t \in \mathbb{P} | |t| = 1\}$ be connected and unramified of degree n. Then $V = \{s \in \mathbb{P} | |s| = 1\}$ and $\psi(s) = s^n$. This follows as in case (b). We can extend ψ to a morphism $\tilde{\psi} : \{s \in \mathbb{P} | |s| \leq 1\} \to \{t \in \mathbb{P} | |t| \leq 1\}$ which is only ramified in $s = 0$ of order n.

Let $\rho < \rho_1 < \rho_2 < 1$ and put $U_{\rho_1,\rho_2} = \phi^{-1}(\{t \in \mathbb{P} | \rho_1 \leq |t| \leq \rho_2\})$ and put $U_{\rho_i} = \phi^{-1}(\{t \in \mathbb{P} | |t| = \rho_i\})$. The unramified covering $U_{\rho_1} \to \{t \in \mathbb{P} | |t| = \rho_1\}$ can be extended to a covering of $\{t \in \mathbb{P} | |t| \leq \rho_1\}$ only ramified above $t = 0$. Further $U_{\rho_2} \to \{t \in \mathbb{P} | |t| = \rho_2\}$ can be extended to a covering of $\{t \in \mathbb{P} | |t| \geq \rho_2\}$ only ramified above $t = \infty$. Glueing this to $U_{\rho_1\rho_2} \to \{t \in \mathbb{P} | \rho_1 \leq |t| \leq \rho_2\}$ we obtain a covering $Z \to \mathbb{P}$ only ramified in $t = 0$ and $t = \infty$.

Then $Z = Z_1 \cup \ldots \cup Z_s$; each $Z_i \cong \mathbb{P}$ and $Z_i \to \mathbb{P}$ is a map of degree n_i only ramified above $t = 0$ and $t = \infty$ (or unramified if $n_i = 1$). It follows that $U_{\rho_1\rho_2}$ has components $V_1, \ldots, V_s$ and

$$V_i \cong \{t \in \mathbb{P} | \rho_1^{\frac{1}{n_i}} \leq |t| \leq \rho_2^{\frac{1}{n_i}}\} \not\cong \{t \in \mathbb{P} | \rho_1 \leq |t| \leq \rho_2\} \text{ is given}$$

by $t \to t^{n_i}$.

Proof of (4.1): Let $\bar{X}$ be the reduction of the statement in (4.1); then ϕ induces a morphism $\bar{\phi} : \bar{X} \to \overline{(\mathbb{P}, S)}$ and we have a commutative diagram.

$$\begin{array}{ccc} X & \xrightarrow{\ \phi\ } & \mathbb{P}^1 \\ {\scriptstyle R}\downarrow & & \downarrow{\scriptstyle R^o} \\ \bar{X} & \xrightarrow[\ \bar{\phi}\ ]{} & \overline{(\mathbb{P}^1, S)} \end{array}$$

The map $\bar{\phi}$ is finite and separable (since its degree $n < \text{char } \bar{k} = p$ or char $\bar{k} = 0$). For $p \in \overline{(\mathbb{P}, S)}$ we are going to calculate the fibre $\bar{\phi}^{-1}(p)$.

There are 3 cases to consider:

a) p is an ordinairy point of $\overline{(\mathbb{P}, S)}$.

b) p is the image of a point of S.

c) p lies on two components of $\overline{(\mathbb{P}, S)}$.

Case (a) and (b): Let ℓ be the component of $\overline{(\mathbb{P}, S)}$ on which p lies and let $\ell^* = \ell$ - {the other points of type b) or c) on ℓ}. Let $U = R_o^{-1}(\ell^*)$ and $V = \phi^{-1}(U)$. Take elements $e_1, \ldots, e_n \in \mathcal{O}(V)^o$ such that their images in $\mathcal{O}(\bar{\phi}^{-1}(\ell^*))$ form a free base of this $\mathcal{O}(\ell^*)$-module. Then the discriminant $D(e_1, \ldots, e_n) \in \mathcal{O}(U)^o$ has spectral norm 1 and also $\{e_1, \ldots, e_n\}$ is a free base of $\mathcal{O}(V)^o$ as $\mathcal{O}(U)^o$-module. But $D(e_1, \ldots, e_n)$ is also the discriminant of $\mathcal{O}(V)^o$ over $\mathcal{O}(U)^o$. Since $\phi : X \to \mathbb{P}^1$ is a simple covering it follows that $D(e_1, \ldots, e_n)$ has at most a simple zero in U. It follows that $D(\bar{e}_1, \ldots, \bar{e}_n)$ has also at most a simple zero of order 1 at p.

Then $\mathcal{O}(\bar{\phi}^{-1}p)^o = \mathcal{O}(V)^o \otimes (R_o^{-1}p)^o$ is still a free-module over $\mathcal{O}(R_o^{-1}p)^o$ with base $\{e_1, \ldots, e_n\}$ and also

$$\overline{\mathcal{O}(\bar{\phi}^{-1}p)} = \bigoplus_{q \in \bar{\phi}^{-1}(p)} \hat{\mathcal{O}}_{\bar{X},q} \text{ is a free } \overline{\mathcal{O}(R_o^{-1}p_o)} = \bar{k}[[T]]\text{- module}$$

with the same base.

Since $D(\bar{e}_1, \ldots, \bar{e}_n)$ has at most a zero of order 1, it follows that at most one of the $\hat{\mathcal{O}}_{\bar{X},q}$ can have a singularity and the this singularity is at

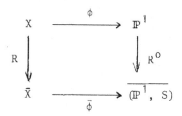

For other constants ρ_1^*, ρ_2^* with $\rho < \rho_1^* < \rho_1 < \rho_2 < \rho_2^* < 1$ one finds a similar description of $U_{\rho_1^*,\rho_2^*}$. Hence the number of components of $U_{\rho_1\rho_2}$ (and the numbers $n_1,\ldots, n_s$) are independent of the choice of ρ_1 and ρ_2. We have assumed that U is connected, hence s = 1 and we get

$$\begin{array}{ccc} U_{\rho_1\rho_2} & \xrightarrow{\sim\alpha} & \{t \in \mathbb{P} \mid \rho_1^{\frac{1}{n}} \le |t| \le \rho_2^{\frac{1}{n}}\} \\ \downarrow\phi & \swarrow\psi & \\ \{t \in \mathbb{P} \mid \rho_1 \le |t| \le \rho_2\} & & \end{array}$$

where $\psi(t) = t^n$; the diagram is commutative.

In this set up the isomorphism α is unique up to an nth-root of unity. Take now sequences ρ_1^n, ρ_2^n with $\lim \rho_1^n = \rho$, $\lim \rho_2^n = 1$ and such that $\rho < \ldots < \rho_1^3 < \rho_1^2 < \rho_1^1 < \rho_1^0 < \rho_2^0 < \rho_2^1 < \rho_2^2 < \rho_2^3 < \ldots < 1$, then using a coherent choice of the isomorphisms

$\alpha_n : U_{\rho_1^s\rho_2^s} \tilde{\to} \{t \in \mathbb{P} \mid \sqrt[n]{\rho_1^s} \le |t| \le \sqrt[n]{\rho_2^s}\}$ we can glue them to an isomorphism

$$\begin{array}{ccc} \alpha : U & \xrightarrow{\simeq} & \{t \in \mathbb{P} \mid \sqrt[n]{\rho} < |t| < 1\} \\ \phi\downarrow & \swarrow\psi & \\ \{t \in \mathbb{P} \mid \rho < |t| < 1\} & & \end{array}$$

where the diagram is commutative and ψ is defined by $\psi(t) = t^n$. This proves the lemma and finishes the proof of (4.1).

§5 Curves locally isomorphic to $\mathbb{P}$.

A k-analytic space X is said to be locally isomorphic to $\mathbb{P}$ if X admits an affinoid covering $\mathfrak{U}$ such that each $U \in \mathfrak{U}$ is isomorphic to some affinoid subset of $\mathbb{P}_k^1$.

A Mumford curve (i.e. a curve of the type Ω/Γ, where Γ is a Schottky group) has certainly this property. This explains our interest in curves of this type. We will show the following.

(5.1) Theorem: Let X be a curve over k, locally isomorphic to $\mathbb{P}$. Then X has a stable reduction. Moreover this stable reduction $\bar{X}$ of X is totally split (i.e. all the components of $\bar{X}$ are rational).

Remark: In view of Chap. IV (3.10) we have now established an analytic criterium for a curve to be a Mumford curve, namely: the curve must be locally isomorphic to $\mathbb{P}$.

(5.2) Lemma: Let X be a projective algebraic variety over k. Then X has arbitrarily fine, pure coverings.

Proof: Let $D \geq 0$ be a very ample divisor. As usual $\mathcal{L}(D) = \{f | f$ is a rational function X and $(f) \geq -D\}$. Let $f_o, \ldots, f_n \in \mathcal{L}(D)$ generate $\mathcal{L}(D)$ then $\phi : X \to \mathbb{P}_k^n$, given by $x \to ((f_o(x), \ldots, f_n(x))$ is an embedding of X.

$\mathbb{P}_k^n$ has a standard pure covering, namely $\mathcal{X} = \{X_o, \ldots, X_n\}$ where $X_i = \{((\lambda_o, \ldots, \lambda_n)) \in \mathbb{P}_k^n | |\lambda_i| \geq |\lambda_j|$ for all $j\}$ $(i = 0, \ldots, n)$.

Let $\mathfrak{U} = \{U_1, \ldots, U_s\}$ be another affinoid coverings of X. Then $U_i \cap U_j$ is a finite union of rational subsets of X_j. After refining $\mathfrak{U}$ we may suppose that each $U_i \cap X_j$ is a rational subset of X_j. Hence there are elements $f_{ijk} \in \mathcal{O}(X_j)$ $(k = 1, \ldots, m_{ij})$, generating the unit ideal and such that

$U_i \cap X_j = \{x \in X_j | |f_{ij1}(x)| \geq |f_{ijk}(x)|$ for all $k\}$.

One easily sees that the rational functions (with poles outside X_j) are dense in $\mathcal{O}(X_j)$. So we may suppose that all f_{ijk} are rational functions on X.

Let $E \geq D$ be a divisor such that $f_{ijk} \in \mathcal{L}(E)$ for all i, j, k.
The map $\psi : X \to \mathbb{P}_k^m$ given by
$\psi(x) = ((f_o(x), \ldots, f_n(x), f_{ijk}(x), \ldots))$, is again an embedding. The standard pure covering on $\mathbb{P}_k^m$ induced a pure covering X which is finer than $\mathfrak{U}$.

Proof of (5.1): From (5.2) it follows that X contains a pure covering $\mathfrak{U}$ such that every $U \in \mathfrak{U}$ is an affinoid subset of $\mathbb{P}$.

Let F be any affinoid subset of $\mathbb{P}$. If F is connected (and k is algebraically closed) then $F \cong \mathbb{P} - (B_1 \cup \ldots \cup B_s)$ where the B_i are open disks. Let S be a finite subset of $\mathbb{P}$ containing the centers of the B_i, some points on the "boundaries" of the B_i and ∞. Then the reduction $R : \mathbb{P} \to \overline{(\mathbb{P}, S)}$ has the property: there exists a Zariski-open $Z \subset \overline{(\mathbb{P}, S)}$ with $R^{-1}(Z) = F$. Hence F has a prestable reduction. According to §2, F has also a stable reduction.

Replace now every $U \in \mathfrak{U}$ be a finite pure covering of U which induces the stable reduction of U. Let $\mathfrak{U}^*$ denote the new covering of X obtained in this way. Then $\bar{X}_{\mathfrak{U}^*}$ is a pre-stable reduction of $\bar{X}$. Using §2, then (5.1) now follows.

(5.3) Analytic construction of Mumford curves by glueing $\mathbb{P}^1_k$'s.

We will use following terminology: A ring area $F \subset \mathbb{P}$ is an affinoid subset defined by $F = \{t \in \mathbb{P} \mid \rho_1 \leq |t - a_1| \leq \rho_2\}$ with $\rho_1 < \rho_2$. The two components of $\mathbb{P} - F$ are called $F^+ = \{t \in \mathbb{P} \mid |t - a| > \rho_2$ and $F^- = \{t \in \mathbb{P} \mid |t - a| < \rho_1\}$. The two "boundary components" of F are $\partial^+F = \{t \in \mathbb{P} \mid |t - a| = \rho_2\}$ and $\partial^-F = \{t \in \mathbb{P} \mid |t - a| = \rho_1\}$.

The data for construction of a Mumford curve can be given as follows:

1) $X_1, \ldots, X_s$ copies of $\mathbb{P}^1_k$.

2) ring areas $F_{i,1}, \ldots, F_{i,n_i}$ on X_i such that for $a \neq b$:
$$(F_{ia} \cup F^-_{ia}) \cap (F_{ib} \cup F^-_{ib}) = \emptyset.$$

3) a map from pairs to pairs: $(i, j) \to (i^*, j^*)$ such that $(i, j) \neq (i^*, j^*)$ and $(i^{**}, j^{**}) = (i, j)$.

4) for every pair (i, j) an isomorphism $F_{ij} \to F_{i^*j^*}$ which interchanges the boundary components.

The affinoid sets $X_i - (F_{i1}^- \cup \ldots \cup F_{in_1}^-)$ $(i = 1, \ldots, s)$ are glued together according to 3) and 4); the result is a k-analytic space X. It is clear that X has a stable reduction which is totally split. According to Chap. IV, (2.2), X is a complete algebraic curve (possibly not connected). If X happens to be connected then X is in fact a Mumford curve.

On the other hand, every Mumford curve can be obtained in this way. We sketch the proof of this statement. Let X be the Mumford curve and let $R_o : X \to Z$ be the stable reduction. This stable reduction wil be refined to a reduction $R : X \to \bar{X}$, by removing some configurations in Z by "refinements" (Compare §2). The refinements that we want are shown in the pictures:

refined to

refined to

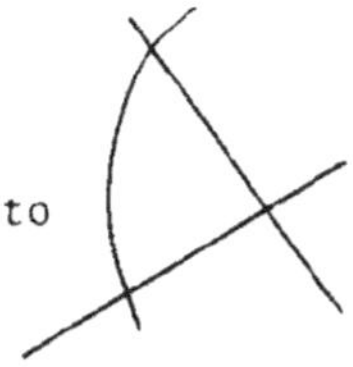

refined to ...

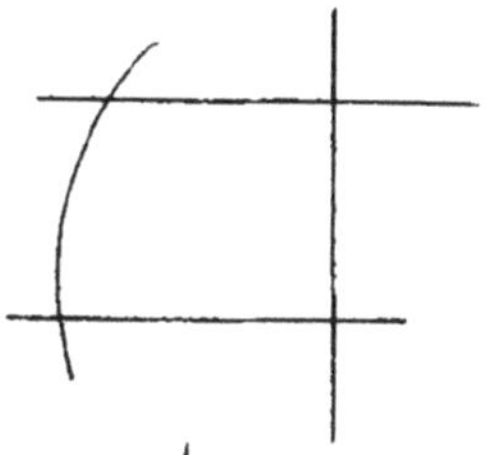

and

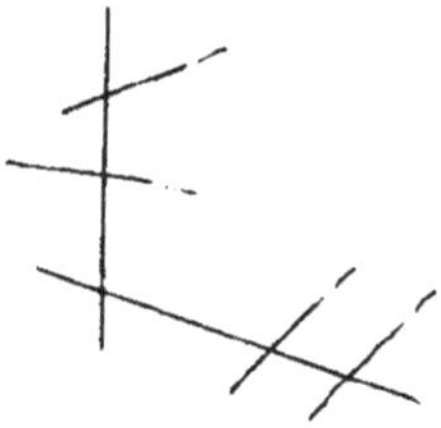

refined to ...

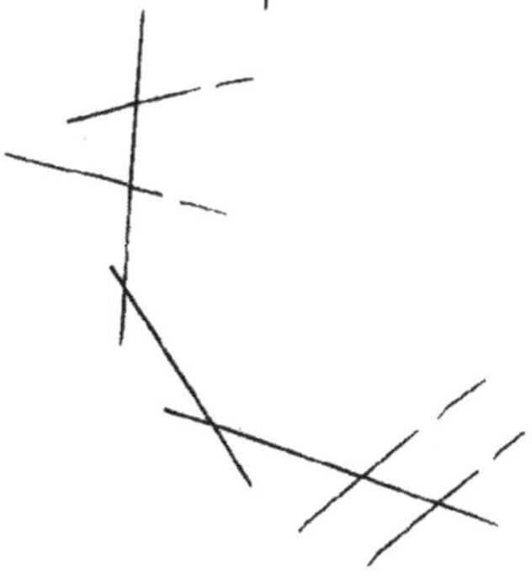

Let ℓ be a component of $\bar{X}$ which meets more than two other component $\ell_1, \ldots, \ell_s$. Then each ℓ_i meets only two components. Let $\tilde{\ell}$ be the Zariski-open subset: $\bar{X} - \cup$ {all components $\neq \ell_1, \ell_2, \ldots, \ell_s, \ell$} and let $U(\ell)$ be $R^{-1}(\tilde{\ell})$. Then $U(\ell) \cong \mathbb{P}^1$ - {s open disks}.

Let ℓ be a component of $\bar{X}$ which meets only two other components ℓ_1, ℓ_2. If also ℓ_i meets only two other components then we define a ringarea $U(\ell_i, \ell) = R^{-1}(\bar{X} - \{\text{all components} \neq \ell, \ell_i\})$.

In this manner we find a covering of X by affinoid subsets of $\mathbb{P}^1$, glued in the manner described above.

e.g. the last example of a totally split genus 2 curve (§3), has also the prestable reduction.

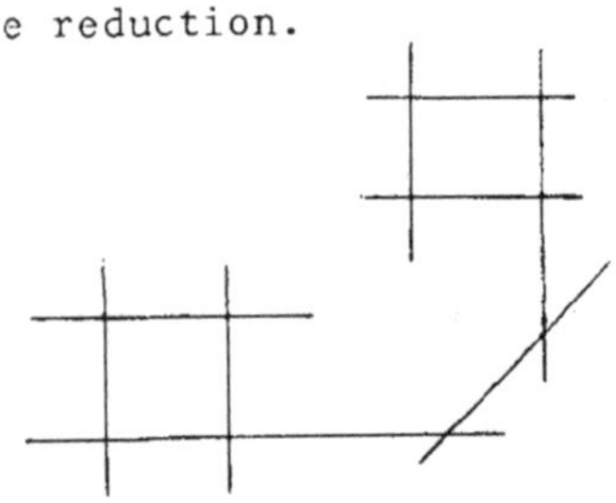

Hence it can be obtained by glueing 4 copies of $\mathbb{P}^1_k$.

Chapter VI. Jacobian Varieties

The explicit construction of the automorphic forms for a p-adic Schottky group Γ makes it possible to give a direct analytic construction of the Jacobian variety $\mathfrak{J}(S)$ for a Mumford curve S and the canonical mapping of the curve into its Jacobian. This mulitplicative theory for the Jacobian has appartently not been considered in the complex case. This construction has been given independently by Manin-Drinfeld [27] and Myers [29], see also Manin [26]. However the functional aspect of this approach became completely clear only after it was proved in [9] that any automorphic form is a product of the basic forms $\Theta(a, b; z)$.

In §1 we show that the group $\bar{\mathfrak{D}}$ of divisor classes of degree zero can be identified with the residue classes of the algebraic torus $C(\Gamma) = \mathrm{Hom}(\Gamma, K^*)$ modulo a lattice L.

In §2 we give a proof that the period matrix associated to Γ is symmetric and positive definit and that the lattice L is discrete in $C(\Gamma)$ which is not obvious if the ground field is not discrete.

§3 contains the fundamentally new results of this chapter. We consider the theta function $\vartheta(c; u)$ on the algebraic torus $(K^*)^g$ associated to the period matrix for Γ and show that the divisor of $\vartheta(c; u(z))$ is of degree g if $\vartheta(c; u(z))$ does not vanish identically in the variable z. This is obtained through the method explained in Chap. II, §3, which allows the determining of the number of zeroes of an analytic function on the fundamental domain F by its behaviour at the boundary of F. As a quite formal consequence of this result we obtain as in the complex case Riemann's vanishing theorem, see [41], Chap. 4.10, which gives the equation for the hypersurface $\phi(S^{g-1}) \subset \mathfrak{J}(S)$.

As the curve S has genus g it is known that the vectorspace of analytic differential forms on S has dimension g.

In §4 we verify this result with elementary methods of p-adic function theory and give a basis for the analytic differential forms on S.

§1 Divisor classes and automorphy factors

Throughout this chapter we use the notations that have been introduced at the beginning of §2 and §3 of Chap. II.

(1.1) If f(z) is a nontrivial automorphic form, then the automorphy factor $c = c_f$ is determined by f(z) and is a group homomorphism $c : \Gamma \to K^*$. Because if $\alpha, \beta \in \Gamma$, then

$$f(z) = c(\alpha\beta) \cdot f(\alpha\beta z)$$
$$f(\beta z) = c(\alpha) \cdot f(\alpha\beta z)$$
$$f(z) = c(\beta) \cdot f(\beta z).$$

(1.2) If $a \in \Omega$ is a zero (resp. pole) of f(z) of order k, then also αa is a zero (resp. pole) of order k and thus all the point of Γa are zeroes (resp. poles) of f(z) of order k. Therefore $\text{ord}_s f(z)$ is well-defined for any non-trivial automorphic form f(z) and any point $s \in S = \Omega/\Gamma$.

Let

$$\mathfrak{a}(f) := \sum_{\substack{s\in S \\ \text{ord}_s f(z)>0}} \text{ord}_s f(z) \cdot s$$

be the divisor of zeroes of f(z) and

$$\mathfrak{b}(f) := \sum_{\substack{s\in S \\ \text{ord}_s f(z)<0}} -\text{ord}_s f(z) \cdot s$$

be the divisor of poles f(z).

Denote div f = $\mathfrak{a}(f) - \mathfrak{b}(f)$ and call it the divisor of f(z).

If $f(z) = a_0 \cdot \Theta(a_1, b_1; z) \cdot \ldots \cdot \Theta(a_r, b_r; z)$ and if $a_i\Gamma \neq b_j\Gamma$ for all i, j, then

$$\mathfrak{a}(f) = \bar{a}_1 + \bar{a}_2 + \ldots + \bar{a}_r$$
$$\mathfrak{b}(f) = \bar{b}_1 + \bar{b}_2 + \ldots + \bar{b}_r$$
$$\text{div } f = \sum_{i=1}^{r} \bar{a}_i - \bar{b}_i$$

where $\bar{a}_r = a_i\Gamma$, $\bar{b}_i = b_i\Gamma$.

If for example $b_1 = \alpha a_1$, $\alpha \in \Gamma$, we use the facts

$$\Theta(a_1, \alpha a_1; z) = \Theta(b_2, \alpha b_2; z)$$
$$\Theta(b_2, \alpha b_2; z) \cdot \Theta(a_2, b_2; z) = \Theta(a_2, \alpha b_2; z)$$

to see that

$$f(z) = a_0 \cdot \Theta(a_2, \alpha b_2; z) \cdot \Theta(a_3, b_3; z) \ldots \Theta(a_r, b_r; z)$$

if $r \geq 2$.

This shows that any automorphic form which has zeroes and poles has a representation

$$f(z) = a_0 \sum_{i=1}^{r} \Theta(a_i, b_i; z)$$

such that $a_i \neq b_j$ for all i, j.

If however f(z) has no zeroes and poles then

$f(z) = a_0 \cdot \Theta(a, \alpha a; z)$ with some $\alpha \in \Gamma$ and

$\mathfrak{a}(f) = \mathfrak{b}(f) = \text{div } f = 0$.

Let D_0 denote the group of all divisors on S of degree 0.

Two divisors $\mathfrak{a}_1$, $\mathfrak{b}_2$ are called linearly equivalent if

$$\mathfrak{a}_1 - \mathfrak{b}_2$$

is the divisor of some Γ-invariant meromorphic function h(z). The group of classes of divisors of degree 0 is denoted by $\bar{D}_0$.

(1.3) If $f_1(z)$, $f_2(z)$ are non-trivial automorphic forms with the same automorphy factor c, then

$$\frac{f_1(z)}{f_2(z)}$$

is a Γ-invariant function on Ω and thus div $f_1(z)$ is linearly equivalent to div $f_2(z)$.

Let now C^* denote the subgroup of $C = \mathrm{Hom}\ (\Gamma, K^*)$ consisting of those group homomorphism

$$c : \Gamma \to K^*$$

for which there exists a non-trivial automorphic form f(z) with automorphy factor c.

We will show in (5.2) that in fact $C = C^*$.

For $c \in C^*$ denote by $\lambda(c)$ the divisor class of div f(z), if f(z) is automorphic with factor c. Then the mapping

$$\lambda : C^* \to \bar{D}_o$$

is a group homomorphism.

<u>Proposition:</u> λ ist onto and the kernel of λ is a subgroup L of rank g.

<u>Proof:</u> 1) Let $\sum_{i=1}^{r} \bar{a}_i - \bar{b}_i$ be a divisor of degree 0 on S, then let c be the factor of the automorphic form $f(z) = \prod_{i=1}^{r} \Theta(a_i, b_i; z)$.

Then c is the class of $\sum_{i=1}^{r} \bar{a}_i - \bar{b}_i$.

2) $\lambda(c)$ = zero class if and only if there is an automorphic form f(z) with factor c such that div f(z) is linearly equivalent to the divisor 0. This means there is a Γ-invariant function h(z) such that

$$\mathrm{div}\ f(z) = \mathrm{div}\ h(z).$$

Then div $\frac{f(z)}{h(z)} = 0$ and $\frac{f(z)}{h(z)}$ is also automorphic with factor c.

But now $\frac{f(z)}{h(z)}$ has no zeroes and poles and thus

$$\frac{f(z)}{h(z)} = a_0 \cdot u_1(z)^{n_1} \cdot u_2(z)^{n_2} \cdot \ldots \cdot u_g(z)^{n_g}$$

$$= a_0 \cdot \quad u_\alpha(z) \text{ with } \alpha = \gamma_1^{n_1} \ldots \gamma_g^{n_g}$$

with a constant a_0.

Now if $u_\alpha(z) \not\equiv$ const, at least for one $\gamma \in \Gamma$ we must have

$$|c(\gamma)| \neq 1.$$

Otherwise we would get

$$|f(z)| = |f(\gamma z)|$$

for all $\gamma \in \Gamma$. Then $|f(z)|$ would be bounded on Ω and therefore $f(z)$ would be a constant.

But $u_\alpha(z)$ is a constant if and only if $\alpha \in [\Gamma, \Gamma]$.
This shows that the mapping

$$\alpha \to \text{automorphy factor of } u_\alpha(z)$$

induces a group isomorphism

$$\bar{\Gamma} = \Gamma/[\Gamma, \Gamma] \to L.$$

As $\bar{\Gamma}$ is free of rank g, so is L.

(1.4) We identify C with $(K^*)^g$ through

$$c \to (c(\gamma_1), \ldots, c(\gamma_g)).$$

Proposition: L is discrete in C; that is: there is a $\varepsilon > 0$ such that

$$L \cap \{(c_1, \ldots, c_g) \in (K^*)^g : 1 - \varepsilon \leq |c_i| \leq 1 + \varepsilon\} = \{(1, \ldots, 1)\}$$

Proof: In the course of the proof for the above proposition, we have seen that

$$L \cap \{(c_1, \ldots, c_g) \in C : |c_i| = 1\} = \{(1, \ldots, 1)\}.$$

If the matrix elements of all the transformations $\gamma \in \Gamma$ are lying in some discrete subfield K_o of K, then we obtain the statement of the proposition automatically. In the non-discrete case one has to produce a proof.

Let $\rho > 1$ be a real number such that

$$\rho r_i < |m_i - m_j|$$

for $j \neq i$, $i \leq j \leq 2g$, $1 \leq i \leq 2g$.

Let $(c_1, \ldots, c_g) \in L$ such that $\rho^{-1/2} < |c_i| < \rho^{1/2}$. We will show that then $(c_1, \ldots, c_g) = (1, \ldots, 1)$.

Let $f(z)$ be an automorphic form without poles and zeroes and with factor $(c_1, \ldots, c_g)$.

Let

$$f(z) = f(\infty) + \sum_{i=1}^{2g} f_i(z)$$

where $f_i(z)$ are analytic function on $\mathbb{P} - B_i$ such that $f_i(\infty) = 0$.

We have

$$\|f\|_F = \max\left(|f(\infty)|, \sup_{i=1}^{2g} \|f_i\|_F\right).$$

As $|f(z)|$ is not constant on F, there is an index $i \leq 2g$

$$\|f\|_F = \|f_i\|_F.$$

Assume that $i = g + 1$.

We get

$$f_i(z) = \sum_{\nu=1}^{\infty} f_{i\nu} \frac{1}{(z - m_i)^\nu}.$$

As $|f_i(z)| \leq \|f\|_F$ we get

$$|f_{i\nu}| \frac{1}{r_i^{\nu}} \leq \|f\|_F.$$

If $z \in F_1$, then $|z - m_i| > \rho r_i$ for $i > 1$ and thus

$$|f_i(z)| < \max_{\nu=1}^{\infty} f_i \frac{1}{r_i^{\nu} \cdot \rho^{\nu}} < \rho^{-1} \| f \|_F .$$

Now

$$\sup_{z \in F_1} |f_2(z) + \ldots + f_{2g}(z)| < \rho^{-1} \| f \|_F .$$

Thus $\sup_{z \in F_1} |f(\infty) + f_1(z)| > \rho^{-1} \| f \|_F$ because

$$\sup_{z \in F_1} |f(z)| > \rho^{-1} \| f \|_F .$$

The last inequality can be seen in the following way:

$$z \in F_1 \text{ if } \gamma_1 z \in F_1'$$

$$\rho^{-1} < \left|\frac{f(z)}{f(\gamma_1 z)}\right| = |c_1| < \rho$$

$$\rho^{-1} |f(\gamma_1 z)| < |f(z)|$$

$$\rho^{-1} \| f \|_F < \| f \|_{F_1} .$$

If $\| f_1 \|_{F_1} \leq |f(\infty)|$, then f has order 0 with respect to the disk B_1

But then $\text{ord}_{B_1'} f(z) = 0$ and $|f(\infty)| > \| f_{g+1} \|_{F_1'}$.

As $\| f_{g+1} \|_{F_1'} = \| f \|_F$ this is not possible.

Therefore

$$\| f_1 \|_{F_1} > |f(\infty)|$$

and this means that $\text{ord}_{B_1} f(z) < 0$. Thus $\text{ord}_{B_1'} f(z) > 0$.

But reasoning as above we see that

$$\sup_{z \in F_1'} |f_i(z)| < \rho^{-1} \| f \|_F$$

if $i \neq g + 1$. Thus also $\text{ord}_{B_1'} f(z) < 0$ which is a contradiction.

(1.5) Consider the analytic mapping

$$\phi = \phi_{g+1} : \Omega^{g+1} \to C .$$

The fibre $\phi^{-1}(1,\ldots, 1)$ is at least one-dimensional and therefore we find $(a_1,\ldots, a_{g+1}) \in \Omega^{g+1}$, not all $a_i \in \Gamma\infty$, such that

$$\phi(a_1,\ldots, a_{g+1}) = 1.$$

Then $f(z) = \prod_{i=1}^{g+1} \Theta(a_i, \infty; z)$ is a Γ-invariant meromorphic function on Ω.

Let $a \in \Omega$ be an arbitrary point. Then

$$h(z) := f(z) - f(a)$$

is a Γ-invariant function with $h(a) = 0$ which has its only pole at $\Gamma\infty$.

Then we find points

$$a_2',\ldots, a_{g+1}' \in \Omega$$

such that

$$h(z) = a_0 \cdot \Theta(a, \infty; z) \cdot \Theta(a_2', \infty; z) \cdot \ldots \cdot \Theta(a_{g+1}', \infty; z),$$

$a_0 \in K^*$.

The means that the divisor

$$\bar{\infty} - \bar{a}$$

is linearly equivalent to

$$\sum_{i=2}^{g+1} (\bar{a}_i - \bar{\infty}).$$

If now $\mathfrak{a} = \sum_{i=2}^{r} \bar{a}_i - \bar{b}_i$ is any divisor on S of degree 0,

We write

$$\mathfrak{a}_1 = \sum_{i=1}^{r} \bar{a}_i - r \cdot \bar{\infty}.$$

$$\mathfrak{a}_2 = r \cdot \bar{\infty} - \sum_{i=1}^{r} \bar{b}_i.$$

such that $\mathfrak{a} = \mathfrak{a}_1 + \mathfrak{a}_2$.

From what we have just proved we can conclude that $\mathfrak{a}_2$ is linearly equivalent to a divisor

$$\sum_{i=1}^{t} (\bar{b}_i' - \bar{\infty}).$$

<u>Proposition:</u> $\phi_r(\Omega^r) = C^*$ for $r \geq g$.

<u>Proof:</u> 1) Let $c \in C^*$ and $f(z)$ be nontrivial automorphic with factor c. Then div $f(z)$ is linearly equivalent to div $(\prod_{i=1}^{r} \Theta(a_i, \infty; z))$ with points $a_i \in \Omega$. We can pick a_1 in such that a way that $f_1(z) = \prod_{i=1}^{r} \Theta(a_1, \infty; z)$ has factor c.

Then $\phi_r(a_1, a_2, \ldots, a_r) = c$.

2) We will show now that $\phi_r(\Omega^r) = \phi_{r-1}(\Omega^{r-1})$ if $r \geq g + 1$.

Let $(a_1, \ldots, a_r) \in \Omega^r$. There is a point $(a_1', \ldots, a_r') \in \Omega^r$, $a_i'\Gamma \neq a_j\Gamma$ for at least one i, such that

$$\phi(a_1, \ldots, a_r) = \phi(a_1', \ldots, a_r').$$

Let $f_1(z) = \prod_{i=1}^{r} \Theta(a_i, \infty; z)$

$$f_2(z) = \prod_{i=1}^{r} \Theta(a_i', \infty; z).$$

Then $f_1(z)$ and $f_2(z)$ are automorphic with the same factor. If one of the a_i or one of the a_i' is contained in $\Gamma\infty$, we are done.

Otherwise there are constants $c_1, c_2 \in K^*$ such that $f(z) = c_1 f_1(z) + c_2 f_2(z)$ has a pole of order $t < r$ at ∞.

But $f(z)$ is not trivial as

$$f(a_i') \neq 0 \text{ for } a_i' \notin \Gamma a_1 \cup \ldots \cup \Gamma a_r.$$

Now $f(z) = \prod_{i=1}^{t} \Theta(b_i, \infty; z)$ as $f(z)$ has its only pole at $\Gamma\infty$.

This shows that

$$\phi_t(b_1, \ldots, b_t) = \phi_r(a_1, \ldots, a_r).$$

§2 Period matrix

(2.1) The quotient

$$Q(\alpha, \beta) = \frac{u_\alpha(z)}{u_\alpha(\beta z)}$$

is the automorphy factor of the form $u_\alpha(z)$ and is thus a non-zero constant $\in K^*$.

As $u_{\alpha_1 a_2}(z) = u_{\alpha_1}(z) \cdot u_{\alpha_2}(z)$ and any automorphy factor is a group homomorphism $\Gamma \to K^*$, $Q(\alpha, \beta)$ is a bimultiplicative form

$$Q : \Gamma \times \Gamma \to K^*.$$

Now $Q(\alpha, \beta) = 1$ if $\alpha \in [\Gamma, \Gamma]$ or if $\beta \in [\Gamma, \Gamma]$ and thus Q can be considered as a bimultiplicative form on $\bar{\Gamma} = \Gamma/[\Gamma, \Gamma]$.

Theorem: Q is symmetric and positive definit, i.e.

$$|Q(\alpha, \alpha)| < 1$$

for any $\alpha \not\equiv \text{id} \bmod [\Gamma, \Gamma]$.

Proof: 1) We first prove symmetry.

By Chap. II, (2.3.6) we have

$$Q(\alpha, \beta) = \frac{u_\alpha(a)}{u_\alpha(\beta a)} = \frac{\Theta(a, \beta a; z)}{\Theta(a, \beta a; \alpha z)} .$$

On the other hand

$$u_\beta(z) = \Theta(a, \beta a; z)$$

$$u_\beta(\alpha z) = \Theta(a, \beta a; \alpha z)$$

which shows that $\dfrac{u_\beta(z)}{u_\beta(\alpha z)} = \dfrac{u_\alpha(a)}{u_\alpha(\beta a)}$.

2) We will show that $|Q(\alpha, \alpha)| \leq 1$ for $\alpha \notin [\Gamma, \Gamma]$ by induction on the length of α. This proof uses a technique interesting also for other questions. It will be given in the next section.

(2.2) length $\alpha = 1$: Then $\alpha = \gamma_i$ or γ_i^{-1}.

Now $|u_i(z)| = \left|\frac{z - m_i}{z - m_i'}\right|$ on F and $\gamma_i z \in \partial B_i'$ iff $z \in \partial B_i$.

Pick $z \in \partial B_i$.

Thus
$$|u_i(z)| = \frac{r_i}{|m_i - m_i'|}$$
$$|u_i(\gamma_i z)| = \frac{|m_i - m_i'|}{r_i'}$$

and $|Q(\gamma_i, \gamma_i)| = \frac{r_i r_i'}{|m_i - m_i'|} < 1$.

length $\alpha > 1$: We may assume without loss of generality that

$\alpha = \gamma_1^{n_1} \ldots \gamma_g^{n_g}$. Then length $\alpha = |n_1| + \ldots + |n_g|$.

Assume also that $n_1 \neq 0$.

We consider now the domain

$$F^* = F \cup \gamma_1(F).$$

$F^* = \mathbb{P} - (B_1 \cup \gamma_1(B_1') \cup B_2 \cup B_2' \cup \ldots \cup B_g \cup B_g' \cup \gamma_1 B_2 \cup \gamma_1 B_2' \cup \ldots \cup \gamma_1 B_g \cup \gamma_1 B_g')$

Case $n_1 = 2k$ even:

In this case let

$$C_1 = B_1, \; C_1' = \gamma_1(B_1')$$
$$C_2 = B_2, \; C_2' = B_2'$$
$$\vdots$$
$$C_g = B_g, \; C_g' = B_g'$$
$$C_{g+1} = \gamma_1(B_2), \; C_{g+1}' = \gamma_1(B_2')$$
$$\vdots$$
$$C_{g+(g-1)} = \gamma_1(B_g), \; C_{2g-1}' = \gamma_1(B_g').$$

Let also
$$\delta_1 = \gamma_1^2$$
$$\delta_2 = \gamma_2$$
$$\vdots$$
$$\delta_g = \gamma_g$$
$$\delta_{g+1} = \gamma_1\gamma_2\gamma_1^{-1}$$
$$\vdots$$
$$\delta_{2g-1} = \gamma_1\gamma_g\gamma_1^{-1}.$$

Then δ_i maps C_i onto $\mathbb{P} - (C_i')^+$ and ∂C_i onto $\partial C_i'$. Thus the subgroup Δ generated by $\delta_1, \ldots, \delta_{2g-1}$ is a Schottky group and F^* is a fundamental domain for Δ with respect to the basis $\{\delta_1, \ldots, \delta_{2g-1}\}$.

Now Δ has index 2 in Γ and $\Gamma = \Delta \cup \Delta\gamma_1$.

$$\alpha = (\gamma_1^2)^k \cdot \gamma_2^{n_2} \quad \ldots \quad \gamma_g^{n_g} = \delta_1^k \delta_2^{n_2} \quad \ldots \quad \delta_g^{n_g} \in \Delta$$

and the length of α with respect to the basis $\{\delta_1, \ldots, \delta_{2g-1}\}$ is $|k| + |n_2| + \ldots + |n_g| <$ length of α with respect to $\{\gamma_1, \ldots, \gamma_g\}$.

$$\text{Also} \quad \gamma_1\alpha\gamma_1^{-1} = (\gamma_1^2)^k \; (\gamma_1\gamma_2\gamma_1^{-1})^{n_2} \quad \ldots \quad (\gamma_1\gamma_g\gamma_1^{-1})^{n_g}$$
$$= \delta_1^k \cdot (\delta_{g+1})^{n_2} \quad \ldots \quad (\delta_{2g-1})^{n_1}$$

has smaller length than length α.

The bimultiplicative form Q with respect to the group Δ is denoted by Q_Δ.

Now

$$Q(\alpha, \alpha) = \prod_{\gamma\in\Gamma} \frac{\left(\frac{z - \gamma a}{z - \gamma\alpha a}\right)}{\left(\frac{\alpha z - \gamma a}{\alpha z - \gamma\alpha a}\right)}$$

$$= \prod_{\gamma\in\Delta} \frac{\left(\frac{z - \gamma a}{z - \gamma\alpha a}\right)}{\left(\frac{\alpha z - \gamma a}{\alpha z - \gamma\alpha a}\right)} \cdot \prod_{\gamma\in\Delta} \frac{\left(\frac{z - \gamma\gamma_1 a}{z - \gamma\gamma_1\alpha a}\right)}{\left(\frac{\alpha z - \gamma\gamma_1 a}{z - \gamma\gamma_1\alpha a}\right)}$$

$$= Q_\Delta(\alpha, \alpha) \cdot Q_\Delta(\gamma_1\alpha\gamma_1^{-1}, \gamma_1\alpha\gamma_1^{-1}).$$

By induction we know that

$$Q_\Delta(\alpha, \alpha)| < 1, \quad Q_\Delta(\gamma_1\alpha\gamma_1^{-1}, \quad \gamma_1\alpha\gamma_1^{-1})| < 1$$

and thus $|Q_\Delta(\alpha, \alpha)| < 1$.

<u>Case $n_1 = 2k_1 + 1$ and $n_2 = 2k_2 + 1$ odd:</u>

If n_2 is even we use the above construction for γ_2 instead of

So we can assume that n_2 is also odd.

Now let

$$\begin{aligned}
C_1 &= B_1, \; C_1' = \gamma_1(B_1')\\
C_2 &= B_2, \; C_2' = \gamma_1(B_2')\\
C_3 &= B_3, \; C_3' = B_3'\\
&\vdots\\
C_g &= B_g, \; C_g' = B_g'\\
\\
C_{g+1} &= B_2', \; C_{g+1}' = \gamma_1(B'_2)\\
C_{g+2} &= \gamma_1(B_3), \; C_{g+2}' = \gamma_1(B_3')\\
&\vdots\\
C_{2g-1} &= \gamma_1(B_g), \; C_{2g-1}' = \gamma_1(B_g')
\end{aligned}$$

and

$$\begin{aligned}
\delta_1 &= \gamma_1^2\\
\delta_2 &= \gamma_1\gamma_2\\
\delta_3 &= \gamma_3\\
&\vdots\\
\delta_g &= \gamma_g\\
\delta_{g+1} &= \gamma_1\gamma_2^{-1}\\
\delta_{g+2} &= \gamma_1\gamma_3\gamma_1^{-1}\\
&\vdots\\
\delta_{2g-1} &= \gamma_1\gamma_g\gamma_1^{-1}.
\end{aligned}$$

Then δ_i maps C_i onto $\mathbb{P} - (C_i')^+$ and ∂C_i onto $\partial C_i'$. The group generated by $\delta_1, \ldots, \delta_{2g-1}$ is a Schottky group and F* is a fundamental domain for Δ with respect to the basis $\{\delta_1, \ldots, \delta_{2g-1}\}$. Also Δ has index 2 in Γ and $\Gamma = \Delta \cup \Delta\gamma_1$.

$$\alpha = \gamma_1^{2k_1} \cdot \gamma_1\gamma_2 \, (\gamma_2\gamma_1^{-1} \cdot \gamma_1\gamma_2)^{k_2} \quad \gamma_3^{n_3} \quad \ldots \quad \gamma_g^{n_g}$$

$$= \delta_1^{k_1} \cdot \delta_2 \cdot (\delta_{g+1} \cdot \delta_2)^{k_2} \cdot \delta_3^{n_3} \quad \ldots \quad \delta_g^{n_g}$$

and the length of α with respect to $\{\delta_1, \ldots, \delta_{2g-1}\}$ is $|k_1| + 1 + 2\,k_2 + |n_3| + \ldots + |n_g|$ which is smaller than $|n_1| + \ldots + |n_g|$.

Also

$$\gamma_1\alpha\gamma_1^{-1} = \gamma_1^{2k_1} \cdot \gamma_1^2\gamma_2\gamma_1^{-1} \cdot \gamma_1\gamma_2^2\gamma_1^{-1})^{k_2} \cdot (\gamma_1\gamma_3\gamma_1^{-1})^{n_3} \cdot \ldots \cdot (\gamma_1\gamma_g\gamma_1^{-1})^{n_g}$$

$$= \delta_1^{2k_1+2} \cdot \delta_{g+1}^{-1} \cdot (\delta_2 \cdot \delta_{g+1}^{-1})^{k_2} \cdot (\delta_{g+2})^{n_3} \quad \ldots \quad (\delta_{2g-1})^{n_g}$$

which shows that the length of $\gamma_1\alpha\gamma_1^{-1}$ with respect to $\{\delta_1, \ldots, \delta_{2g-1}\}$ is

$$|k_1| + 1 + 1 + 2\,|k_2| + |n_3| + \ldots + |n_g|$$

which is smaller than $|n_1| + \ldots + |n_g|$.

Again

$$Q(\alpha, \alpha) = Q_\Delta(\alpha, \alpha) \cdot Q_\Delta(\gamma_1\alpha\gamma_1^{-1}, \gamma_1\alpha\gamma_1^{-1})$$

and $|Q_\Delta(\alpha, \alpha)| < 1$, $|Q_\Delta(\gamma_1\alpha\gamma_1^{-1}, \gamma_1\alpha\gamma_1^{-1})| < 1$ which proves $|Q(\alpha, \alpha)| < 1$.

The proof of the theorem is complete.

(2.3) With respect to the basis $\{\gamma_1,\ldots, \gamma_g\}$ the form Q is given by the matrix

$$(q_{ij})$$

with $q_{ij} = Q(\gamma_i, \gamma_j)$. This matrix is called the period matrix of Γ with respect to $\{\gamma_1,\ldots, \gamma_g\}$.

We identify $\bar{\Gamma} = \Gamma/[\Gamma, \Gamma]$ with $\mathbb{Z}^g$ through

$$\alpha = \gamma_1^{n_1} \ \ldots \ \gamma_g^{n_g} \to (n_1,\ldots, n_g)$$

and set

$$Q(n, m) = Q(\gamma_1^{n_1} \ \ldots \ \gamma_g^{n_g}, \gamma_1^{m_1} \ \ldots \ \gamma_g^{m_g})$$

if $n = (n_1,\ldots, n_g)$, $m = (m_1,\ldots, m_g) \in \mathbb{Z}^g$.

For later purposes we introduce a bimultiplicative, symmetric form $P : \mathbb{Z}^g \times \mathbb{Z}^g \to K^*$ such that

$$P(n, m)^2 = Q(n, m).$$

This can be done by taking square roots $p_{ij} = p_{ji}$ of q_{ij} and then define P through the matrix

$$(p_{ij}).$$

Of course also P is positive definit.

(2.4) Let

$$Q^* : \mathbb{Z}^g \times \mathbb{Z}^g \to \mathbb{R}$$

be defined by

$$Q^*(n, m) := - \log \ |Q(n, m)|.$$

Then Q^* is a $\mathbb{Z}$-bilinear form, symmetric and positive definit which means that

$$Q^*(n, n) > 0$$

for $n \in \mathbb{Z}^g$, $n \neq (0,\ldots, 0)$.

With respect to the standard basis of $\mathbb{Z}^g$ the bilinear form is given by the matrix

$$(v_{ij})$$

if $v_{ij} := -\log |q_{ij}|$.

Q^* determines canonically a $\mathbb{R}$-bilinear form

$$Q^*_{\mathbb{R}} : \mathbb{R}^g \times \mathbb{R}^g \to \mathbb{R}.$$

We now show that this form is again positive definit.

First of all we remark that $Q^*_{\mathbb{R}}$ is positive semi-definit. If $x \in \mathbb{R}^g$, we find a sequence (x_k) converging to x with $x_k \in \mathbb{Q}^g \subseteq \mathbb{R}^g$. As $Q^*_{\mathbb{R}}$ is continuous, we observe that

$$Q^*_{\mathbb{R}}(x, x) = \lim_{k\to\infty} Q^*_{\mathbb{R}}(x_k, x_k) \geq 0.$$

If $Q^*_{\mathbb{R}}$ is not positiv definit, it is degenerate which means that there exists a vector $0 \neq x \in \mathbb{R}^g$ such that $Q_{\mathbb{R}}(x, y) = 0$ for all $y \in \mathbb{R}^g$.

This would have the effect that the columns of the matrix (v_{ij}) are linearly dependent over $\mathbb{R}$. In section (1.4) we have seen that L is discrete in G. So also by definition $-\log |L|$ is discrete in $-\log |C| \subseteq \mathbb{R}^g$, and is thus a lattice in $\mathbb{R}^g$. This lattice is generated by the column vectors of the matrix (v_{ij}) which proves that $Q^*_{\mathbb{R}}$ is indeed positive definit.

Let $x = (\sum_{i=1}^{g} x_i^2)^{\frac{1}{2}}$ be the euclidean length of $x = (x_1, \ldots, x_g) \in \mathbb{R}^g$ and

$$\tilde{M} := \inf_{\|x\| = 1} Q^*_{\mathbb{R}}(x, x) > 0.$$

Then for an arbitrary $x \in \mathbb{R}^g$ we have

$$Q^*_{\mathbb{R}}(x, x) = Q^*_{\mathbb{R}}(\|x\| \cdot \frac{x}{\|x\|}, \|x\| \cdot \frac{x}{\|x\|}).$$

$$= \|x\|^2 \cdot Q^*_{\mathbb{R}}(\frac{x}{\|x\|}, \frac{x}{\|x\|}) \geq \tilde{M} \cdot \|x\|^2.$$

(2.4.1) Therefore we get for $n = (n_1, \ldots, n_g) \in \mathbb{Z}^g$:

$$|Q(n, n)| \leq e^{-\tilde{M}(n_1^2 + \ldots + n_g^2)}$$

and $0 < e^{-\tilde{M}} < 1$.

If $M = \sqrt{e^{-M}} = e^{-\frac{\tilde{M}}{2}}$ then

$$|P(n, n)| \leq M^{n_1^2 + \ldots + n_g^2}.$$

§3 Theta functions

(3.1) We want to show that the formal Laurent series

$$\vartheta(u_1, \ldots, u_g) := \sum_{\substack{n \in \mathbb{Z}^g \\ n=(n_1, \ldots, n_g)}} P(n, n) \cdot u_1^{n_1} \ldots u_g^{n_g}$$

is convergent for all $(u_1, \ldots, u_g) \in (K^*)^g$.

Let C be a constant > 1 and $C^{-1} \leq |u_i| \leq C$.

We have seen in (5.4.1), that there is a constant $M < 1$ such that

$$|P(n, n)| \leq M^{n_1^2 + \ldots + n_g^2}.$$

Thus

$$|P(n, n) \cdot u_1^{n_1} \ldots u_g^{n_g}| \leq C|u_1^{n_1} \ldots u_g^{n_g}| \to 0 \text{ if } n \to \infty.$$

Because: if $k = k(n) = \max_{i=1}^{g} |n_i|$, then

$n_1^2 + \ldots + n_g^2 \geq k^2$ and $|n_1| + \ldots + |n_g| \leq g \cdot k$.

Thus

$$C^{|n_1| + \dots + |n_g|} \cdot M^{n_1^2 + \dots + n_g^2} \leq (C^g)^k \cdot M^{k^2} \leq (C^g \cdot M^k)^k .$$

For almost all n we have

$$C^g \cdot M^{k(n)} < 1.$$

If $n \to \infty$ then $k(n) \to \infty$ and we are done.

(3.2) So $\vartheta(u_1, \dots, u_g)$ is an analytic function on $C = (K^*)^g$ and $\vartheta(1, \dots, 1) \neq 0$ as

$$\vartheta(1, \dots, 1) = \sum_{n \in \mathbb{Z}^g} P(n, n)$$

with $P(0, 0) = 1$ and $|P(n, n)| < 1$ for $n \neq 0$.

If $q_i = (q_{i1}, \dots, q_{ig})$ is the i-th row of the period matrix, then

$$\vartheta(q_i \cdot u) = \vartheta(q_{i1}u_1, \dots, q_{ig}u_g) =$$

$$= \sum_{n \in \mathbb{Z}^g} P(n, n) \cdot q_{i1}^{n_1} \dots q_{ig}^{n_g} \cdot u_1^{n_1} \dots u_g^{n_g}.$$

If $e_i = (0, \dots, 1, \dots, 0)$ is the i-th element of the standard basis of $\mathbb{Z}^g$, then

$$Q(n, e_i) = q_{i1}^{n_1} \dots q_{ig}^{n_g}.$$

Thus

$$P(n + e_i, n + e_i) = P(n, n) \cdot P(n, e_i)^2 \cdot P(e_i, e_i)$$

$$= P(n, n) \cdot Q(n, e_i) \cdot P(e_i, e_i)$$

and $\vartheta(q_i u) = \frac{1}{P(e_i, e_i)} \cdot \sum_{n \in \mathbb{Z}^g} P(n + e_i, n + e_i) \cdot u_1^{n_1} \dots u_g^{n_g}$.

We obtain the functional equation for the theta function $\vartheta(u)$:

$$\vartheta(u) = P(e_i, c_i) \cdot u_i \cdot \vartheta(q_i u).$$

(3.3) We consider $\vartheta(u)$ as an analytic function $f(z)$ on Ω by putting

$$f(z) = \vartheta(u_1(z), \ldots, u_g(z)).$$

Then $f(z) \not\equiv 0$ as $f(\infty) = \vartheta(u_1(\infty), \ldots, u_g(\infty)) = \vartheta(1, \ldots, 1) \neq 0$.

Now $q_{ij} = \dfrac{u_j(z)}{u_j(\gamma_i z)} = \dfrac{u_j(\gamma_i^{-1} z)}{u_j(z)}$ and thus

$$q_i \cdot (u_1(z), \ldots, u_g(z)) = (u_1(\gamma_i^{-1} z), \ldots, u_g(\gamma_i^{-1} z)).$$

Therefore

$$f(\gamma_i^{-1} z) = \vartheta(q_i \cdot u(z)) = \frac{\vartheta(u(z))}{P(e_i, e_i) \cdot u_i(z)}.$$

The functional equation for $f(z)$ is:

$$f(z) = p_{ii} \cdot u_i(z) \cdot f(\gamma_i^{-1} z)$$

or

$$f(\gamma_i z) = \frac{u_i(z)}{p_{ii}} \cdot f(z)$$

as

$$p_{ii} \cdot u_i(\gamma_i z) = p_{ii} \cdot \frac{u_i(z)}{q_{ii}} = \frac{u_i(z)}{p_{ii}}.$$

A point $a \in \Omega$ is a zero of order k for $f(z)$ if and only if αa is a zero of order k for any $\alpha \in \Gamma$. Thus

$$\operatorname{div} f(z) := \sum_{s \in S} \operatorname{ord}_s f(z) \cdot s$$

is a well-defined divisor on S.

<u>Proposition:</u> div $f(z)$ has degree g.

<u>Proof:</u> 1) Assume first that $f(z)$ has no zero on $R = \bigcup_{i=1}^{2g} R_i$.

As

$$\operatorname{ord}_{B_i'} f(z) = - \operatorname{ord}_{B_i} f(\gamma_i z)$$

we get

$$\operatorname{ord}_{B_i} f(z) + \operatorname{ord}_{B'_i} f(z) =$$

$$= \operatorname{ord}_{B_i} f(z) - \operatorname{ord}_{B_i} f(\gamma_i z) =$$

$$= \operatorname{ord}_{B_i} f(z) - \operatorname{ord}_{B_i} u_i(z) - \operatorname{ord}_{B_i} f(z) =$$

$$= - \operatorname{ord}_{B_i} u_i(z) = - 1.$$

Using proposition (2.1.1) we see that the number of zeroes of $f(z)$ counting multiplicities is $- \sum_{i=1}^{2g} \operatorname{ord}_{B_i} f(z) = g$.

2) If $f(z)$ has zeroes on R, we change the radii of B_i into ρr_i and those of B'_i into $\frac{r'_i}{\rho}$ with some ρ close to 1. With respect to the new fundamental domain we can construct R in such a way that $f(z)$ has no zero on R.

(3.4) Let $c = (c_1, \ldots, c_g) \in C$ and

$$\vartheta(c; u) = \vartheta(c_1 u_1, \ldots, c_g u_g).$$

Then

$$\vartheta(c; u) = p_{ii} \cdot c_i \cdot u_i \cdot \vartheta(q_i u).$$

If $f(c; z) = \vartheta(c; u(z))$, then we obtain the functional equation:

$$f(c; \gamma_i z) = c_i \frac{u_i(z)}{p_{ii}} \cdot f(c; z).$$

In the same way as above we see that div $f(c; z)$ is a positive diviso on S of degree g if $f(c; z) \not\equiv 0$.

The quotient

$$h(c; z) = \frac{f(c^{-1}; z)}{f(z)}$$

satisfies the functional equation

$$h(c; \gamma_i z) = c_i^{-1} h(c; z)$$

$$h(c; z)) = c_i h(c; \gamma_i z).$$

Thus if $h(c; z) \not\equiv 0$ it is an automorphic form with factor $c \in C$.

<u>Proposition:</u> Given $c \in C$. Then there is an automorphic form with factor c.

<u>Proof:</u> If $\vartheta(c^{-1}) \neq 0$, then $h(c; z) := \frac{f(c^{-1}; z)}{f(z)}$ is automorphic, non -trivial with factor c, as $h(c; \infty) = \frac{\vartheta(c^{-1})}{\vartheta(1)} \neq 0$.

Also if $c = (c_1, \ldots, c_g)$ with $|c_i| = 1$, then $\vartheta(c) \neq 0$, as $|P(n, n)| \cdot |c_1^{n_1} \ldots c_g^{n_g}| < 1$ for $n \neq 0$.

Thus $h(c; z) \not\equiv 0$ if $c = (c_1, \ldots, c_g)$ and all $|c_i| = 1$.

If c is such that $\vartheta(c^{-1}) = 0$, we find $c' = (c'_1, \ldots, c'_g)$ such that $\vartheta((c')^{-1}) \neq 0$ and $|c'_i| = |c_i|$, because the set of zeroes of $\vartheta(u)$ is an analytic subset.

Let $f_1(z) \not\equiv 0$ be automorphic with factor c' and $f_2(z) \not\equiv 0$ be automorphic with factor $(\frac{c_1}{c'_1}, \ldots, \frac{n_g}{c'_g})$ which is possible to construct as $|\frac{c'_i}{c_i}| = 1$. Then $f_1(z)f_2(z)$ is automorphic with factor c.

(3,5) The subgroup L in C is discrete in C, see (1,4) and thus a lattice. The quotient $\mathfrak{J} := C/L$ is thus an analytic group variety, see [7].

The analytic mapping

$$\phi : \Omega^g \to G$$

induces an analytic mapping

$$\phi : S^g \to \mathfrak{J}$$

because $\phi(\alpha(z_1), z_2, \ldots, z_g) \equiv \phi(z_1, \ldots, z_g) \bmod L$.

Let $\mathcal{P}_g$ be the group of permutations of the set $\{1, 2, \ldots, g\}$.

For $\sigma \in \mathcal{P}_g$, let

$$\sigma(z_1, \ldots, z_g) = (z_{\sigma(1)}, \ldots, z_{\sigma(g)}).$$

Then $\phi(\sigma(z_1,\ldots, z_g)) = \phi(z_1,\ldots, z_g)$ and thus $\phi : S^g \to \mathcal{J}$ induces in a canonical way an analytic mapping

$$\phi : S^{(g)} \to \mathcal{J}$$

where $S^{(g)} = S^g/\mathcal{T}_g$ is the orbit space of the group action of $\mathcal{T}_g$ on S^g.

<u>Proposition:</u> If a point of the fibre $\phi^{-1}(c)$ of the mapping $\phi : S^{(g)} \to \mathcal{J}$ is isolated in the fibre $\phi^{-1}(c)$, then $\phi^{-1}(c)$ consists of a point only.

<u>Proof:</u> Assume that the fibre $\phi^{-1}(c)$ consists of at least two points. Then we find points $(a_1,\ldots, a_g)$ and $(b_1,\ldots, b_g)$ in Ω^g such that the divisors $\bar{a}_1 + \ldots + \bar{a}_g$ and $\bar{b} + \ldots + \bar{b}_g$ are different. Let

$$f_1(z) := \prod_{i=1}^{g} \Theta(a_i, \infty; z)$$

$$f_2(z) := \prod_{i=1}^{g} \Theta(b_i, \infty; z).$$

These two functions are automorphic with factors c_1, c_2 respectively. As $\phi(a_1,\ldots, a_g) \equiv \phi(b_1,\ldots, b_g) \bmod L$ and

$$c_1 = \phi(a_1,\ldots, a_g)$$

$$c_2 = \phi(b_1,\ldots, b_g)$$

we have $c_1 = c_2 \cdot \ell$, $\ell \in L$. If ℓ is the automorphy factor of $u_\alpha(z)$, then

$$\phi(\alpha a_1, a_2,\ldots, a_g) = \phi(b_1,\ldots, b_g)$$

as points in G. So we may assume that

$$c_1 = c_2.$$

Now $f_1(z)$, $f_2(z)$ are linearly independent over K, as $\operatorname{div} f_1(z) \neq \operatorname{div} f_2(z)$.

We next consider the automorphic form

$$f(z) = f_1(z) + ef_2(z)$$

with a constant $e \in K$.

Pick a disk D_i around the point a_i in Ω. Assume that $f_1(z)$ has a zero of order k_i in a_i.

We can choose the disks D_i so small that

$$\frac{f_1(z)}{(z - a_i)^{k_i}}$$

do not have any zero in D_i.

If e satisfies

$$|e| \cdot \|f_2(z)\|_{D_i} < \|f_1(z)\|_{D_i}$$

then f(z) has exactly k_i zeroes in D_i but $\mathrm{ord}_{a_i} f(z) = \min(\mathrm{ord}_{a_i} f_1(z), \mathrm{ord}_{a_i} f_2(z))$ if $\mathrm{ord}_{a_i} f_1(z) \neq \mathrm{ord}_{a_i} f_2(z)$.

As f(z) has only a pole at $\Gamma\infty$ of order $\leq g$, the degree of the divisor of zeroes of f(z) is $\leq g$. Therefore a zero of f(z) must lie in one of the disks D_i.

If we consider the divisor of the zeroes of f(z) as a point in $S^{(g)}$, we see that $\mathcal{K}_g \cdot (\bar{a}_1, \ldots, \bar{a}_g)$ cannot be isolated in the fibre because the disks D_i can be chosen arbitrarily small.

In the course of this proof, we have used the following simple

<u>Lemma:</u> Let $f(z) = \sum_{\nu=0}^{\infty} f_\nu (z - a)^\nu$ be an analytic function in the disk $D = \{z : |z - a| \leq r\}$. Then f(z) has n zeroes in D (counting multiplicities) if and only if

$$|f_n| \cdot r^n > |f_\nu| \cdot r^\nu \quad \text{for } \nu > n$$

$$|f_\nu| \cdot r^\nu \leq |f_n| \cdot r^n \quad \text{for } \nu < n.$$

(3.6) Let W_r be the image $\phi_r(\Omega^r)$ of Ω^r in C. Then $W_r = \phi_r(F^r)$. L and $\phi_r(F^r)$ is bounded in C which means that there is a constant $M > 1$ such that

$$\phi_r(F^r) \subset \{(c_1, \ldots, c_g) \in (K^*)^g : M^{-1} \leq |c_i| \leq M\}.$$

Because: $\phi(a_1,\ldots, a_r) = u(a_1) \ldots u(a_r)$ and

$$\phi(\alpha a_1, a_2,\ldots, a_r) = \frac{u(\alpha a_1)}{u(a_1)} \cdot u(a_1) \ldots u(a_r) \text{ and } \frac{u(\alpha a_1)}{u(a_1)} \in L.$$

Now F^r is an affinoid domain and thus $\phi_r(F^r)$ is a closed subset of C with respect to the topology given by the metric on $(K^*)^g$.

More generally:

Lemma: If $\phi : Y \to X$ is an analytic mapping of affinoid spaces Y and X, then $\phi(Y)$ is closed in X.

Proof: The corresponding K-homomorphism of the affinoid algebras is denoted by

$$\phi^* : A(X) \to A(Y)$$

Let $x_0 \in X$, not in $\phi(Y)$. Let $\mathfrak{m}$ be the maximal ideal of all affinoid functions on X which vanish on x_0. Now

$$\phi^*(\mathfrak{m}) \cdot A(Y) = A(Y).$$

We obtain a relation

$$\sum_{i=1}^{r} h_i \cdot (f_i \circ \phi) \equiv 1$$

with $h_i \in A(Y)$, $f_i \in \mathfrak{m}$.

Let $\|h_i\|_Y \leq M$ and

$$U = \{x \in X : |f_i(x)| < M^{-1}\}.$$

Then $U \cap \phi(Y) = \emptyset$.

Because if $y \in Y$ and $\phi(y) \in U$, then

$$|h_i(y) \cdot f_i(\phi(y))| < M \cdot M^{-1} = 1$$

and thus

$$|1| = |\sum_{i=1}^{r} h_i(y) \cdot (f_i \circ \phi(y))| < \max_{i=1}^{r} |h_i(y) \cdot f_i(\phi(y))|$$

which is a contradiction and so $U \cap \phi(Y) = \emptyset$.

As $\phi(F^r)$ is closed and L discrete we get that

$$W_r = \phi(F^r) \cdot L$$

is also closed in G.

(3.7) The divisor of the automorphic function $\vartheta(u(z))$ has degree g, see (3.3), and therefore we find points $a_1, \ldots a_g \in \Omega$ such that

$$\operatorname{div} \vartheta(u(z)) = \bar{a}_1 + \ldots + \bar{a}_g.$$

Let $\kappa = \phi(a_1, \ldots, a_g) = u(a_1) \cdot \ldots \cdot u(a_g)$.
Take $c \in C$ and look at the quotient

$$\frac{\vartheta(\kappa c^{-1} u(z))}{\vartheta(u(z))}$$

as a function on Ω. We have seen in (3.4) that it is automorphic with factor $c\kappa^{-1}$ if it is not identically zero.

Now if

$$\operatorname{div} \vartheta(\kappa c^{-1} u(z)) = \bar{b}_1 + \ldots + \bar{b}_g$$

then

$$\frac{\vartheta(\kappa c^{-1} u(z))}{\vartheta(u(z))} = \prod_{i=1}^{g} \Theta(b_i, a_i; z) \cdot e(z)$$

where $e(z)$ is automorphic without zeroes.

We can choose b_i in such a way that $e(z)$ is a constant e_o

$$\frac{\vartheta(\kappa c^{-1} u(z))}{\vartheta(u(z))} = e_o \cdot \frac{\prod_{i=1}^{g} \Theta(b_i, \infty; z)}{\prod_{i=1}^{g} \Theta(a_i, \infty; z)} .$$

Thus $\dfrac{\phi(b_1, \ldots, b_g)}{\phi(a_1, \ldots, a_g)} = \kappa^{-1} c.$

But $\phi(a_1, \ldots, a_g) = \kappa$ and thus

$$\phi(b_1, \ldots, b_g) = \kappa \cdot \kappa^{-1} c = c.$$

This result is called the solution of the inversion problem.

(3.8) The theta function $\vartheta(u)$ satisfies the relation

$$\vartheta(u^{-1}) = \vartheta(u)$$

as $P(n, n) = P(-n, -n)$.

Therefore the set N of zeroes of $\vartheta(\kappa^{-1}u)$ is $\{c \in G : \vartheta(\kappa c^{-1}) = 0\}$.

Theorem: (Riemann's vanishing theorem) $N = W_{g-1}$.

Proof: 1) We first prove that $W_{g-1} \subset N$.

Let $c = \phi(a_1, \ldots, a_{g-1}, \infty) \in W_{g-1}$, $c \notin N$.

Now we find points $a_1', \ldots, a_g'$, such that ϕ is locally bianalytic at

$$a' = (a_1', \ldots, a_{g-1}', \infty)$$

and $d = \phi(a') \notin N$ because $\phi^{-1}(N)$ is open in Ω^g.

Let now

$$\operatorname{div}_z \vartheta(\kappa d^{-1} u(z)) = \bar{b}_1 + \ldots + \bar{b}_g.$$

Of course $\bar{b}_i \neq \Gamma\infty$ as $\vartheta(\kappa d^{-1}) \neq 0$ and $u(\infty) = (1, \ldots, 1)$.

We have seen in section (3.7) that

$$\phi(b_1, \ldots, b_g) = d \bmod L.$$

Thus the fibre of $\bar{\phi} : S^{(g)} \to \mathfrak{F}$ above $d \cdot L$ consists of at least two points. Proposition (3.5) shows that this is not possible as ϕ is locally bianalytic at one point of the fibre. This proves $N \subset W_{g-1}$.

2) Define

$$N_r = \{c \in N : c \cdot W_r^{-1} \not\subset N\}.$$

In other words: $c \in N_r$ if the function $\vartheta(\kappa c^{-1} u(z_1) \cdot \ldots \cdot u(z_r))$ is not identically zero in the variables $z_1, \ldots, z_r \in \Omega$.

Of course $N_g = \emptyset$ as $W_g^{-1} = W_g = C$ and $N_o = N$.

Let now $c \in N_{r-1}$, $c \notin N_r$. Then

$$\vartheta(\kappa c^{-1} u(z_1) \cdot \ldots \cdot u(z_{r-1}))$$

is identically zero in $z_1, \ldots, z_{r-1}$, but

$$\vartheta(\kappa c^{-1} (z_1) \ldots u(z_r))$$

is not identically zero in $z_1, \ldots, z_r$.

We take a sequence of points

$$y_n = (y_{n1}, \ldots, y_{nr}) \in \Omega^r$$

converging to $(\infty, \ldots, \infty)$ such that

$$\vartheta(\kappa c^{-1} u(y_{n1}) \ldots u(y_{nr})) \neq 0.$$

Then

$$d_n = \operatorname{div}_z \vartheta(\kappa c^{-1} u(y_{n1}) \ldots u(y_{n,\ r-1}) \cdot u(z)) \equiv$$

$$\equiv \frac{c}{u(y_{n1}) \ldots u(y_{n,\ r-1})} \mod L$$

is contained in W_{g-1} as

$$\vartheta(\kappa c^{-1} u(y_{n,\ r-1})) = 0.$$

Now obviously d_n converges to c and thus $c \in W_{g-1}$ as W_{g-1} is closed in C. This proves $N \subset W_{g-1}$.

(3.9) As we now know that $C^* = C$ we see that $\mathfrak{J} = C/L$ is indeed canonically isomorphic to the group of divisor classes of degree 0 on S. $\mathfrak{J}$ is an abelian variety, see [7].

As the canonical mapping $\phi : S^{(g)} \to \mathfrak{J}$ is locally bianalytic outside a hypersurface and has no points isolated in any fiber we can include through general theorems of p-adic function theory that Φ is birational.

Thus $\mathcal{J}$ coincides with what is algebraically defined to be the Jacobian, see for instance [24].

It is certainly true that also Torelli's theorem holds, see [17].

§4 Analytic differential forms

(4.1) A differential form $\omega = f(z)dz$ on Ω is Γ-invariant, if

$$f(\gamma z)d\gamma z = f(z)dz.$$

If $\gamma(z) = \frac{az+b}{cz+d}$, then

$$\frac{d\gamma z}{dz} = \frac{1}{(cz+d)^2}$$

and

$$f(z) = \frac{1}{(cz+d)^2} f(\gamma z).$$

The differential form is analytic on Ω, if $f(z)$ is analytic on Ω and if $f(z)$ has a zero of order ≥ 2 at ∞.

In section (5.2) of Chap. II, we have introduced the functions

$$w_i(z) = \frac{u_i'(z)}{u_i(z)}.$$

Now $\omega_i(z) = w_i(z)dz = \frac{du_i}{u_i}$ are analytic invariant differentials on Ω which are linearly independent over K.

In (5.2) of Chap. II, we have shown that

$$\operatorname{ord}_{B_k} w_i(z) = \begin{cases} -1 : & k = i \text{ and } k = i+g \\ 0 : & \text{otherwise.} \end{cases}$$

Thus the number of zeroes of $w_i(z)$ in F is 2g.

The divisor of the differential form $\omega_i(z)$ is a divisor on S and

$$\text{ord}\,_\infty \omega_i(z) = \text{ord}\,_\infty w_i(z) - 2$$

as

$$\frac{dz^{-1}}{dz} = -z^{-2} \text{ and thus } \text{ord}\,_\infty dz = -2.$$

If $a \neq \infty$, then $\text{ord}\,_a \omega_i(z) = \text{ord}\,_a w_i(z)$ and thus the degree of div $\omega_i(z)$ is $2g - 2$.

(4.2) Let $\omega(z) = v(z)dz$ be a Γ-invariant differential form on Ω. We want to show that $\omega(z) = \sum_{i=1}^{g} \eta_i \omega_i(z)$, $\eta_i \in K$.

Let $\gamma_i(z) = \dfrac{a_i z + b_i}{c_i z + d_i}$. Then $r_i r_i' = \dfrac{1}{|c_i|^2}$ and $-\dfrac{d_i}{c_i} \in B_i$.

Now $v(z) = \dfrac{1}{(c_i z + d_i)^2} \; v(\gamma_i z)$.

If therefore

$$\sup_{z \in \partial B_i} |v(z)| = \frac{\sigma_i}{r_i}$$

$$\sup_{z \in \partial B_i'} |v(z)| = \frac{\sigma_i'}{r_i'}$$

then $\sigma_i = \sigma_i'$.

Because: $z \in \partial B_i$, $\left|z + \dfrac{d_i}{c_i}\right| = r_i$

$$\gamma_i z \in \partial B_i', \quad \left|\frac{1}{(c_i z + d_i)^2}\right| = \frac{r_i r_i'}{r_i^2} = \frac{r_i'}{r_i}.$$

We may assume that $v(z)$ has no zero on ∂F.

Let now

$$v(z) = \sum_{i=1}^{2g} v_i(z)$$

where $v_i(z)$ is analytic on $\mathbb{P} - B_i$, $1 \leq i \leq 2g$, and $v_i(\infty) = 0$.

Let

$$v_i(z) = \sum_{\nu=1}^{\infty} v_{i\nu} \frac{1}{(z - m_i)^\nu}$$

and $\sigma = \max(\sigma_1, \ldots, \sigma_g)$. We have $\| v_i \|_F \leq \frac{\sigma_i}{r_i}$.

If $k \neq i$ and $z \in F_k$, then $|z - m_i| = |m_k - m_i| \geq \rho \cdot r_k$:

$$|v_{i\nu}| \frac{1}{r_i^\nu} \leq \frac{\sigma_i}{r_i}$$

$$|v_{i\nu}| \cdot \frac{1}{|m_k - m_i|^\nu} = v_{i\nu} \frac{1}{r_i} \cdot \frac{r_i^\nu}{|m_k - m_i|^\nu} \leq \frac{\sigma_i}{r_i} \cdot \frac{r_i}{|m_k - m_i|}$$

as $\frac{r_i}{|m_k - m_i|} < 1$.

Thus

$$|v_{i\nu}| \cdot \frac{1}{|z - m_i|^\nu} \leq \frac{\sigma_i}{\rho \cdot r_k} = \frac{\rho^{-1}\sigma_i}{r_k}.$$

This shows that for all k with $\sigma\rho^{-1} < \sigma_k \leq \sigma$ we have

$$\| \hat{v}_k(z) \|_{F_k} = \| \sum_{i \neq k} v_i(z) \|_{F_k} < \frac{\sigma_k}{r_k}.$$

Now $\| v_k(z) \|_{F_k} = \frac{\sigma_k}{r_k}$ (if $\sigma_k > \sigma\rho^{-1}$)

and $\mathrm{ord}_{B_k} v(z) < 0$, as $\hat{v}_k(z)$ is analytic in B_k^+ and is therefore a power series in $(z - m_k)$.

Also $\mathrm{ord}_{B_k'} v(z) < 0$.

The functional equation

$$v(z) = \frac{1}{(c_i z + d_i)^2} \cdot v(\gamma_i z)$$

shows that

$$\mathrm{ord}_{B_k} v(z) = - \mathrm{ord}_{B_k'} v(z) - 2.$$

Thus we get

$$\mathrm{ord}_{B_k} v(z) = \mathrm{ord}_{B_k'} v(z) = - 1.$$

Now the analytic function $w_k(z)$ has also order -1 with respect to B_k and B'_k and we find $\eta_k \in K$ such that

$$v(z) - \eta_k w_k$$

has order $\neq -1$ with respect to B_k.

If $\operatorname{ord}_{B_k}(v(z) - \eta_k w_k(z)) \geq 0$ then

$$\| v(z) - \eta_k w_k(z) \|_{F_k} \leq \frac{\rho^{-1}\sigma}{r_k}.$$

If $\operatorname{ord}_{B_k}(v - \eta_k w_k) \leq -2$, then

$$\operatorname{ord}_{B'_k}(v - \eta_k w_k) \geq 0 \text{ and}$$

also $\| v(z) - \eta_k w_k(z) \|_{F'_k} \leq \dfrac{\rho^{-1}\sigma}{r'_k}$.

But then also

$$\| v - \eta_k w_k \|_{F_k} \leq \frac{\rho^{-1}\sigma}{r_k}.$$

Let I be the set of all indices k such that $\sigma_k > \rho^{-1}\sigma$ and

$$\tilde{v}(z) = v(z) - \sum_{k \in I} \eta_k \cdot w_k.$$

Now $|\eta_k| \cdot |w_k(z)| = \dfrac{\sigma_k}{r_k}$ for $z \in F_k$ and $|v(z) - \eta_k w_k(z)| < \dfrac{\sigma_k}{r_k}$.

If $\| \tilde{v}(z) \|_{F_i} = \dfrac{\tau_i}{r_i}$, then all $\tau_i \leq \rho^{-1} \cdot \sigma$.

If we are iterating the procedure, we thus can approximate $v(z)$ uniformly of F by linear combinations of $w_1(z), \ldots, w_g(z)$. The details are left to the reader.

Proposition: The space of Γ-invariant analytic differential forms on Ω is a g-dimensional vector space over K, generated by

$$\frac{du_i(z)}{u_i(z)}, \quad 1 \leq i \leq g.$$

For any point a of Ω there is an analytic differential form $\omega(z)$ on Ω with

$$\mathrm{ord}_a \omega(z) = 0.$$

Chapter VII. Automorphisms of Mumford curves

Introduction

An automorphism of a Mumford curve S given by a p-adic Schottky group Γ can be lifted to the Stein domain Ω of ordinary points of Γ and can be continued to an automorphism of $\mathbb{P}(K)$. As any automorphism of the projective line is fractional-linear, one concludes that the automorphism group Aut S is canonically isomorphic to the factor group N/Γ where N is the normalizer of Γ in $PGL_2(K)$. A proof of this result which was first given in [10] will be presented in §1.

In §2 we study the discontinuous group N and show in a most elementary way that $G = N/\Gamma$ acts faithfully on the commutator factor group $\bar{\Gamma}$ of Γ where the operation is induced by inner automorphisms, if the rank g of Γ is greater than 1.

In §3 we will point out that any transformation in G is orthogonal with respect to the definite period form Q defined in Chap. VI, §2. This allows to conclude that G is a finite group if the genus g of S is greater than 1. This result can probably be used to compute the period matrix if the automorphism group of S is large.

In §4 we explain a result of F. Herrlich on the order of the automorphism group of a Mumford curve that much improves the classical Hurwitz estimate. It states that the order of the automorphism group of a Mumford curve is less of equal to $12(g-1)$ if char $K = 0$ and the characteristic p of the residue field $\bar{K}$ of K is different from 2, 3, 5. We will give a full proof of this result for $p = 0$. If $p \neq 0$ the method is the same, but a few additional cases have to be considered, see [21].

§1 Lifting of automorphisms

(1.1) We use the notations of Chap. II, §2.

So F is a good fundamental domain for the Schottky group Γ of rank $g \geq 1$ and $S = S(\Gamma) = \Omega/\Gamma$ is the curve of Γ.

Assume that $\phi : S \to S$ is automorphism of the projective curve S. We want to show that ϕ can be lifted to a bianalytic mapping of Ω into Ω.

Denote by π the projection of Ω onto S and choose a finite covering $\mathfrak{U} = \{U_1, \ldots, U_r\}$ of F through connected affinoid domains U_i such that π is injective on each U_i.

With the notation of Chap. II, §4.1 we can define

$W_i = \{z \in K : r_i \leq |z - m_i| \leq \rho r_i\}$

$W_i' = \{z \in K : r_i' \leq |z - m_i'| \leq \rho \cdot r_i'\}$

$W_o = \{z \in K : |z - m_i| \geq \rho r_i, \ |z - m_i'| \geq \rho r_i' \text{ for all } i\}$.

The collection of these domains gives an affinoid covering $\mathfrak{A}$ that we are looking for.

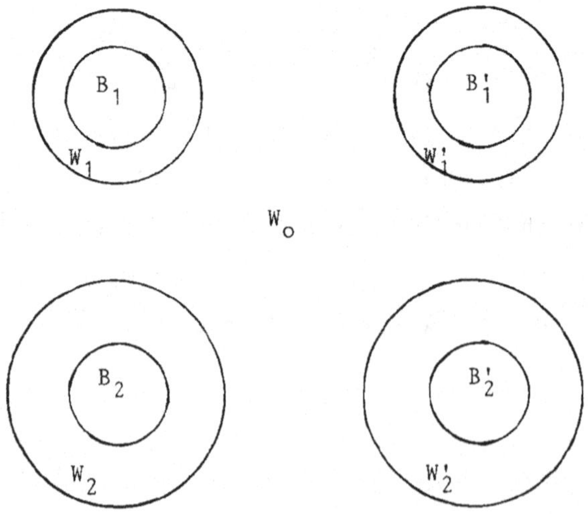

Let τ be the analytic mapping $F \to S$ given through $\phi \circ \pi$. Then the inverse images $\tau^{-1}(\pi(U_i))$ of the affinoid domains $\pi(U_i)$ on S are again affinoid subdomains of F.

Let now $\mathcal{W} = \{V_1, \ldots, V_t\}$ be the covering of F whose members are all the intersections

$$U_i \cap \tau^{-1}(\pi(U_j)).$$

Then we get for any $V_k \in \mathcal{W}$ an index j such that $\tau(V_k) \subseteq \pi(U_j)$.

As now π defines a bianalytic mapping from U_j onto $\pi(U_j)$, we find for any V_k an analytic mapping $\phi_k : V_k \to \Omega$ such that

$$\tau | V_k = \tau \circ \phi_k.$$

We want to show now that we can choose the liftings ϕ_k in such a way that ϕ_k coincides with ϕ_j on the intersection of V_k with V_j.

We can choose the indexing of $\mathcal{W}$ in such a way that

$$X_k := V_1 \cup \ldots \cup V_k \text{ intersects } V_{k+1}.$$

Now X_k is a connected affinoid domain of F and the intersection D_k of X_k with V_{k+1} is also a connected affinoid domain.

Now we prove the lifting problem through induction on k. We may assume that ϕ_i coincides with ϕ_j on $V_i \cap V_j$ for any $i, j \leq k$ and denote by ψ_k the analytic mapping $X_k \to \Omega$ which is induced by $\phi_1, \ldots, \phi_k$.

Now

$$\pi \circ \phi_{k+1} | D_k = \pi \circ \psi_k.$$

As D_k is connected there is a $\gamma \in \Gamma$ such that

$$\psi_k | D_k = \gamma \circ (\phi_{k+1} | D_k).$$

If we replace ϕ_{k+1} by $\gamma \circ \phi_{k+1}$ we see that the above assertion has been proved.

Thus we have obtained an analytic mapping $\tilde{\psi} : F \to \Omega$ such that

$$\pi \circ \psi = \tau .$$

It is a simple exercise to extend $\tilde{\psi}$ to an analytic mapping

$$\psi : \Omega \to \Omega$$

such that ψ is a lifting of ϕ which has the effect that ψ commutes with all $\gamma \in \Gamma$.

From the proposition in (2.3), Chap. II, we see that ψ can be extended to an analytic mapping

$$\psi : \mathbb{P}(K) \to \mathbb{P}(K).$$

As ψ is bianalytic on Ω, it is a fractional-linear transformation.

<u>Theorem:</u> The group Aut S of automorphisms of the curve S is canonically isomorphic to N/Γ, where N is the normalizer of Γ in $PGL_2(K)$.

§2 The normalizer of a Schottky group in $PSL_2(K)$

(2.1) Let N be the normalizer of Γ in $PSL_2(K)$. Thus N consists of all fractional-linear transformations α for which $\alpha\Gamma\alpha^{-1} = \Gamma$. If x is a fixed point of $\gamma \in \Gamma$, $\gamma \neq id$, then $\alpha(x)$ is a fixed point of $\alpha\gamma\alpha^{-1} \in \Gamma$ which shows that α maps the set of all fixed points of α onto itself. Now the set of limit points is the closure of the set of fixed points which shows that α maps the set of limit points of Γ onto itself. And thus α maps the set Ω of ordinary points of Γ onto itself.

Also α induces a map

$$\bar{\alpha} : S \to S$$

because $\alpha(\Gamma a) = \Gamma\alpha a$ as $\alpha\Gamma = \Gamma\alpha$. Obviouly $\overline{\alpha \circ \beta} = \bar{\alpha} \circ \bar{\beta}$ for any $\alpha, \beta \in N$. Next we show that $\bar{\alpha} = id : S \to S$ if and only if $\alpha \in \Gamma$.

If $\alpha \in \Gamma$, then obviously $\alpha(\Gamma a) = \Gamma a$ for all $a \in \Omega$.

On the other hand, if $\bar{\alpha} = \mathrm{id}$ and $a \in \Omega$, then $\alpha(a) = \gamma(a)$ for some $\gamma \in \Gamma$. Thus $(\gamma^{-1}\alpha)(a) = a$ and a is a fixed point of $\gamma^{-1}\alpha$.
If $\gamma^{-1}\alpha \neq \mathrm{id}$, then $\gamma^{-1}\alpha$ has not more than two fixed points. We find a such that $\gamma(D) \cap D = \emptyset$ for all $\gamma \in \Gamma$, $\gamma \neq \mathrm{id}$. If now b is close to a then $b \in \alpha^{-1}(D) \cap D$ and if b is no fixed point of α, we have

$$\alpha(b) \neq \gamma(b)$$

for all $\gamma \in \Gamma$. This shows that $\bar{\alpha}(\Gamma b) \neq \Gamma b$ whenever $\alpha \notin \Gamma$.

(2.2) If $f(z)$ is an non-trivial, automorphic form on Ω with factor $c : \Gamma \to K^*$, then $f(\alpha z)$ is automorphic with factor c^{α}, where $c^{\alpha}(\gamma) = c(\alpha\gamma\alpha^{-1})$.

Proof: $f(\alpha z) = c(\gamma) \cdot f(\gamma\alpha(z))$ for all $\gamma \in \Gamma$.
If $\gamma' = \alpha\gamma\alpha^{-1}$, then

$$\begin{aligned} f(\alpha z) &= c(\alpha\gamma\alpha^{-1}) \cdot f(\alpha\gamma\alpha^{-1} \cdot \alpha z) \\ &= c(\alpha\gamma\alpha^{-1}) \cdot f(\alpha\gamma z) \end{aligned}$$

which proves the assertion.

For the automorphic forms $\Theta(a, b; z)$ we get more precisely:

$$\Theta(a, b; \alpha(z)) = e_o \cdot \Theta(\alpha^{-1}a, \alpha^{-1}b; z)$$

with a constant $e_o \in K^*$.

Proof: $\dfrac{\alpha z - \gamma a}{\alpha z - \gamma b} = \eta_\gamma \cdot \dfrac{z - \alpha^{-1}\gamma a}{z - \alpha^{-1}\gamma b}$

with a constant $\eta_\gamma \in K^*$.

If $\gamma a \neq \alpha\infty$ and $\gamma b \neq \alpha\infty$, then

$$\eta_\gamma = \frac{\alpha\infty - \gamma a}{\alpha\infty - \gamma b}.$$

Let Γ_0 be the set of all γ such that $\gamma a \neq \alpha\infty$ and $\gamma b \neq \alpha\infty$, then

$$\prod_{\gamma\in\Gamma_0} \frac{\alpha z - \gamma a}{\alpha z - \gamma b} = \prod_{\gamma\in\Gamma_0} \frac{\alpha\infty - \gamma a}{\alpha\infty - \gamma b} \cdot \prod_{\gamma\in\Gamma_0} \frac{z - \alpha^{-1}\gamma\alpha\alpha^{-1}a}{z - \alpha^{-1}\gamma\alpha\cdot\alpha^{-1}b}.$$

As $\Gamma - \Gamma_0$ does not contain more than two transformations we get

$$\Theta(a, b; \alpha z) = e_o \cdot \Theta(\alpha^{-1}a, \alpha^{-1}b; z)$$

with some constant $e_o = \prod_{\gamma\notin\Gamma_o} \eta_\gamma \cdot \prod_{\gamma\in\Gamma_o} \frac{\alpha\infty - \gamma a}{\alpha\infty - \gamma b}$.

Of course $e_o = \Theta(a, b; \alpha\infty)$ if $\alpha\infty$ is neither a zero nor a pole of $\Theta(a, b; z)$.

If $f(z)$ is automorphic without zeroes and poles,

then $f(z) = \Theta(a, \beta a; z) = u_\beta(z)$ with $\beta \in \Gamma$.

$$\begin{aligned} \text{Then } f(\alpha z) &= e_o \cdot \Theta(\alpha^{-1}a, \alpha^{-1}\beta\alpha\alpha^{-1}; z) \\ &= e_o \cdot \Theta(a, \alpha^{-1}\beta\alpha a; z) \\ &= e_o \cdot u_{\alpha^{-1}\beta\alpha}(z). \end{aligned}$$

Of course $e_o = f(\alpha\infty)$.

(2.3) Any inner automorphism of N maps Γ onto Γ and the characteristic subgroup $[\Gamma, \Gamma]$ onto $[\Gamma, \Gamma]$. Therefore it induces a group homomorphism $\bar{\Gamma} \to \bar{\Gamma}$.

If $\alpha \in N$, let $\gamma^\alpha := \alpha^{-1}\gamma\alpha$ for any $\gamma \in \Gamma$. Then we have defined a right group action on $\bar{N} := N/\Gamma$ of $\bar{\Gamma} = \Gamma/[\Gamma, \Gamma]$. We will show now that this group action is faithful whenever the rank g of Γ is ≥ 2.

So we pick a $\alpha \in N$, assume that the action of α on $\bar{\Gamma}$ is trivial, which means that $\alpha^{-1}\gamma\alpha\gamma^{-1} \in [\Gamma, \Gamma]$ for all $\gamma \in \Gamma$.

If $f(z)$ is automorphic with factor c, then $f(\alpha^{-1}z)$ is automorphic with factor $c^{\alpha^{-1}}$. As $\alpha^{-1}\gamma\alpha \equiv \gamma \bmod [\Gamma, \Gamma]$, we have

$$c = c^{\alpha^{-1}}.$$

If $f(z) = \Theta(a, b; z)$, then $f(\alpha^{-1}z) = e_o \cdot \Theta(\alpha a, \alpha b; z)$ and therefore

$$h(z) = \frac{\Theta(a, b; z)}{\Theta(\alpha a, \alpha b; z)}$$

is a Γ-invariant meromorphic function on Ω for any $a, b \in \Omega$.

If α has the point a as fixed point, then

$$h(z) = \Theta(\alpha b, b; z).$$

If $\alpha b \notin b\Gamma$, then $h(z)$ is a function on S with a simple pole and a simple zero. Thus

$$h(z) - h(s)$$

is a function with a simple pole at $\bar{b}$ and a simple zero at s. This shows that every divisor of degree 0 is linearly equivalent to the trivial divisor. But the divisor class group is isomorphic to $J = (K^*)^g/L$ which is not trivial for $g > 0$.

If $\bar{\alpha}$ has a fixed point on S, then there is a $\alpha \in \Gamma$ such that $\gamma^{-1}\alpha$ has a fixed point in Ω. This is not possible.

If $\bar{\alpha}$ has no fixed point on S, let $b = \alpha(a)$, then

$$h(z) = \frac{\Theta(a, \alpha a; z)}{\Theta(\alpha a, \alpha^2 a; z)} = \Theta(a, \alpha a; z) \cdot \Theta(\alpha^2 a, \alpha a; z)$$

has a double pole at $\overline{\alpha a}$.

Let $s_1, s_2 \in S$ be two different points and let $h_1(z)$ be a meromorphic function with div $h_1(z) = s_2 + s_2' - 2s_1$ and $h_2(z)$ such that div $h_2(z) = s_1 + s_1' - 2s_2$.

Then div $h_1(z) \cdot h_2(z) = s_1' + s_2' - s_1 - s_2$.

If $s_o \in S$, then div $[h_1h_2(z) - h_1h_2(s_o)] = s_o' + s_o - s_1 - s_2$. This shows that any divisor $s_1 + \dots + s_r - rs_o$ is linearly equivalent to $s - s_o$. Thus the mapping $\phi : \Omega \to J$ is injective and $\dim J = g \geq 2$, this is not possible.

$\bar{N}$ acts faithfully on $\bar{\Gamma} \cong \mathbb{Z}^g$ and thus $\bar{N}$ is canonically isomorphic to a group of linear transformation of $\mathbb{Z}^g$; that is: $\bar{N}$ is a subgroup of $GL_g(\mathbb{Z})$.

§3 Orthogonality

(3.1) We have seen in (2.2) that

$$u_\beta(\alpha z) = e_o \cdot u_{\alpha^{-1}\beta\alpha}(z)$$

if $\beta \in \Gamma$, $\alpha \in N$.

For the bimultiplicative form

$$Q : \Gamma \times \Gamma \to K^*$$

see Chap. VI, §2, we have

$$Q(\beta^\alpha, \gamma^\alpha) = \frac{u_{\alpha^{-1}\beta\alpha}(z)}{u_{\alpha^{-1}\beta\alpha}(\alpha^{-1}\gamma\alpha z)} .$$

The above formula now gives

$$Q(\beta^\alpha, \gamma^\alpha) = \frac{u_\beta(\alpha z)}{u_\beta(\alpha\alpha^{-1}\gamma\alpha z)} = \frac{u_\beta(\alpha z)}{u_\beta(\gamma\alpha z)}$$

$$= \frac{u_\beta(z)}{u_\beta(\gamma z)} = Q(\beta, \gamma).$$

Proposition: Any $\bar{\alpha} \in \bar{N}$ is with respect to the bimultiplicative form Q an orthogonal linear transformation of $\bar{\Gamma}$.

Corollary: $\bar{N}$ is finite.

Proof: Let Q^* be the bilinear form

$$\bar{\Gamma} \times \bar{\Gamma} \to \mathbb{R}$$

given by $Q^*(\beta, \gamma) = -\log Q^*|(\beta, \gamma)|$. Then Q^* induces a bilinear form

$$Q^*_{\mathbb{R}} : \mathbb{R}^g \times \mathbb{R}^g \to \mathbb{R}$$

which is symmetric and positive definit, see Chap. V, (5.4).

Therefore $\bar{N}$ is also a subgroup of the group of all orthogonal $\mathbb{R}$-linear transformations of $Q^*_{\mathbb{R}}$ which is isomorphic to the group of all isometries of the euclidean space $\mathbb{R}^g$. This last group is a compact group and $\bar{N}$ is a discrete subgroup of it. Thus $\bar{N}$ is finite.

§4 On the order of the automorphism group

(4.1) In this paragraph we describe a result on the order of the automorphism groups of Mumford curves that has been recently obtained by F. Herrlich [21].

Theorem: Assume the ground field K to be of characteristic zero and let p be the characteristic of the residue field of K.

Let S be a Mumford curve of genus $g \geq 2$.

Then

$$\text{ord (Aut S)} \leq \begin{cases} 48(g-1) & \text{if } p = 2 \\ 24(g-1) & \text{if } p = 3 \\ 30(g-1) & \text{if } p = 5 \\ 12(g-1) & \text{otherwise.} \end{cases}$$

The estimate of Hurwitz says that ord (Aut S) $\leq 84(g-1)$, see [22], for any curve S of genus $g \geq 2$. So the above result much improves the estimate of Hurwitz for Mumford curves.

We will give a full proof for $p = 0$. If $p \neq 0$ some of the arguments have to be somewhat refined.
The method of proof consists on giving very explicit results on the structure of discontinuous groups and is of great interest aside from the theorem above.

(4.2) Assume that the Mumford curve S is the curve of the p-adic Schottky group Γ of rank $g \geq 2$. We have seen in §1 that Aut S is canonically isomorphic to the factor group N/Γ where N is the normalizer of Γ in $PSL_2(K)$.

We will exploit the fact that N acts without inversion on a tree and use the structure theorem for those groups, see [37], Chap. I, 5.4, Theorem 13.

Let $\mathcal{F}$ be the set of fixed points of all transformations $\alpha \in N$, $\alpha \neq \text{id}$. Then the topological closure $\bar{\mathcal{F}}$ of $\mathcal{F}$ contains the set of limit points for Γ and is a compact set. $\mathcal{F}$ and $\bar{\mathcal{F}}$ are N-invariant subsets because if a is fixed point of β, then $\alpha(a)$ is fixed point of $\alpha\beta\alpha^{-1}$.

Let $\mathcal{T}$ be the tree $T(\bar{\mathcal{F}})$ associated to the compact set $\bar{\mathcal{F}}$, see Chap. I, §2. As $\bar{\mathcal{F}}$ is N-invariant, N acts on $\mathcal{T}$ and the stabilizer of a vertex or an edge of $\mathcal{T}$ is a finite subgroup of N.

We next give a description of $\mathcal{T}$ in terms of classes of affinoid disks.

Let D be an affinoid disk in $K \subseteq \mathbb{P}(K) = K \cup \{\infty\}$, say

$$D = \{z \in K : |z - m| \leq r\}$$

$r \in |K^*|$.

Let D' be another affinoid disk on the projective line.
We call D' equivalent to D if D = D' or if

$$D' = \{z \in \mathbb{P}(K) : |z - a| \geq r\}$$

with $a \in D$.

Each class contains exactly one disk on the affine line K.

Three different points a_0, a_1, a_∞ on $\mathbb{P}(K)$ determine a class of affinoid disks in the following way: let w(z) be the simple rational function on $\mathbb{P}(K)$ which has a simple zero at a_0, a simple pole at a_∞ and which assumes the value 1 at a_1. The class of the disk $D = \{z \in \mathbb{P}(K) : |w(z)| \leq 1\}$ depends only on the equivalence class of the triple (a_0, a_1, a_∞) as defined in Chap. I, (2.1). This shows that $\mathcal{T}$ can be described as follows:

The vertices of $\mathcal{T}$ consist of the classes of these affinoid disks D of K which contain two points a, b $\in \mathcal{F}$ whose distance $|a - b|$ is the radius of D, but which do not contain all of $\mathcal{F}$.

Two affinoid disks D_1, D_2 on K whose classes are in $\mathcal{T}$ are connected by an edge if and only if is a proper maximal subdisk of the other with respect to the disks from $\mathcal{T}$ or if both are maximal disks in $\mathcal{T}$.

(4.2.1) <u>Proposition:</u> N acts without inversions on $\mathcal{T}$.

<u>Proof:</u> Assume that $D_1 \subseteq D_2$ are disks in K whose classes $[D_1]$, $[D_2]$ are in $\mathcal{T}$ and are joined in $\mathcal{T}$ by an edge and that for $\alpha \in N$ we have

$$\alpha \cdot [D_1] = [D_2], \ \alpha[D_2] = [D_1].$$

If $\alpha(D_2) = D_1$, then $\alpha(D_1)$ is a proper subdisk of D_1 which of course does not contain ∞. Thus $[\alpha(D_1)] \neq [D_2]$.

We may assume that

$$D_2 = \{z : |z| \leq 1\}$$
$$D_1 = \{z : |z| < r\}, \ r < 1.$$

Then $\alpha(D_2) = \{z \in \mathbb{P}(K) : |z - m| \geq r\}$ with $m \in D_1$.

Changing the variable z into z - m we see that we may assume that m = 0 and

$$\alpha(D_2) = \{z : |z| \geq r\}.$$

As $0 \notin \alpha(D_1)$ we see that

$$\alpha(D_2) = \{z : |z| \geq 1\}.$$

This shows that any fixed point a of α lies in the strip

$$\{z : r < |z| < 1\}.$$

This shows that the class of the disk

$$D' = \{z : |z| \leq |a|\}$$

is in $\mathcal{T}$ and that D_1 is not a maximal subdisk of D_2 from those in $\mathcal{T}$ which proves that $[D_1]$ and $[D_2]$ are not connected by an edge.

(4.3) Let Δ be a lifting of a maximal subtree Δ' of the quotient graph $\mathcal{T}/N$. Let $\Delta = \{[D_1], \ldots, [D_n]\}$ consist of n vertices and let

$$N_i = \{\alpha \in N : \alpha \cdot [D_i] = [D_i]\}$$

be the stabilizer group of the vertex $[D_i]$. The stabilizer group of an edge η of Δ which connects $[D_i]$ with $[D_j]$ coincides with $N_i \cap N_j$ and is denoted by A_η.

Let $\mathcal{R} = \{\eta_1, \ldots, \eta_r\}$ be a full set of representatives in $\mathcal{T}$ of edges in $\mathcal{T}/N$ which are not in the subtree Δ' and for which one endpoint lies in Δ.

If $[D_i]$ is the endpoint of η_i which lies in Δ then the other endpoint $[D_i']$ of η_i cannot be in Δ because Δ' a maximal subtree of $\mathcal{T}/N$. Let $\gamma_i = \gamma_{\eta_i}$ be a transformation of N such that $\gamma_i[D_i']$ is a vertex of Δ.

Let now $F = N_1 * \ldots * N_n * \langle\gamma_1\rangle * \ldots * \langle\gamma_r\rangle$ be the free product of the groups, $N_1, \ldots, N_n$ and the free group with basis $\gamma_1, \ldots, \gamma_r$. Let $\phi_i : N_i \to F$ be the canonical embedding of N_i into F.

If η is an edge in Δ which joins $[D_i]$ with $[D_j]$, we have canonical embeddings

$$\phi_i : A_\eta = N_i \cap N_j \to N_i \subseteq F$$
$$\phi_j : A_\eta = N_i \cap N_j \to N_j \subseteq F.$$

Let $R_\eta := \{\phi_i(\alpha) \cdot \phi_j(\alpha)^{-1} : \alpha \in A_\eta\}$ and let R be the union of all the sets R_η, η edge in Δ.

For any $\eta \in \mathcal{R}$ which joins $[D_i] \in \Delta$ with $[D_j'] \notin \Delta$, we have

$$\gamma_\eta([D_j']) = [D_j] \in \Delta.$$

The stabilizer subgroup A_η of η is the intersection of N_i with the stabilizer group N_j' of $[D_j']$.

Now $\gamma_\eta A_\eta \gamma_\eta^{-1}$ is contained in the stabilizer group of $\gamma_\eta([D_i']) = [D_j]$.

Let $R_\eta' = \{\phi_i(\alpha) \cdot \phi_j(\gamma_\eta \alpha \gamma_\eta^{-1})^{-1} : \alpha \in A_\eta\}$ and R' be the union of all the R_η', $\eta \in \mathcal{R}$.

The structure theorem for groups acting on trees, see [37], Chap. I, 5.4, Theorem 13 states:

The canonical homomorphism $F \to N$ is surjective and the kernel is the normal subgroup generated by $R \cup R'$.

(4.3.1) <u>Proposition:</u> The genus of the curve S(N) parametrized by N is equal to r.

<u>Proof:</u> The $\mathbb{Z}$-rank of the commutator factor group of N is obviously equal to r. Thus the genus of S(N) is r by Chap. VIII, (4.3).

(4.4) Put v_i = ord N_i. Let the set of edges of Δ be $\{\eta_1, \ldots, \eta_{n-1}\}$ and

$$c_i = \text{ord } A_{\eta_i}.$$

Let $\mathcal{R} = \{\eta_1', \ldots, \eta_r'\}$ and put

$$f_i = \text{ord } A_{\eta_i'}.$$

Let now Γ be a Schottky subgroup of N of finite index j and rank $g \geq 2$.

In [23], Theorem 2, the following formula has been announced for which a proof has been worked out by F. Herrlich.

$$g - 1 = j \cdot \mu$$

with $\mu = \sum_{i=1}^{r} \frac{1}{f_i} + \sum_{i=1}^{n-1} \frac{1}{e_i} - \sum_{i=1}^{n} \frac{1}{v_i}$.

(4.5) The finite subgroups of $PGL_2(K)$ are known. It can be a cyclic group $\mathbb{Z}/m\mathbb{Z}$, a dihedral group $Dh(n)$ of order $2n$, the tetrahedral group Th of order 12, the octahedral group Oh of order 24 or the icosahedral group Ih of order 60.
Any of these groups can occur as stabilizer group of a vertex. The situation is different for stabilizer groups for edges.

(4.5.1) Proposition: Let p denote the characteristic of the residue field $\bar{K}$ of K.

The stabilizer group A of an edge in $\mathcal{T}$ contains a normal subgroup V of order p^s such that the quotient group A/V is cyclic.

Proof: We may assume that the edge η joins the disks $E = \{z : |z| \leq 1\}$ and $D = \{z : |z| \leq r\}$, $r > 1$. If $\alpha \in A_\eta$ and $\alpha(E) \neq E$, then

$$\alpha(E) = \{z \in \mathbb{P}(K) : |z - m| \geq 1\}$$

with $m \in E$.

But then

$$\alpha(D) = \{z : |z - m| \geq r\}$$

which is impossible as $\alpha(E) \subseteq \alpha(D)$.

Thus $\alpha(E) = E$.

Let $\begin{pmatrix} a & b \\ c & d \end{pmatrix} \in SL(2, K)$ such that

$$\alpha(z) = \frac{az + b}{cz + d}.$$

As $\alpha(-\frac{d}{c}) = \infty$ we obtain $|\frac{d}{c}| > 1$ and $|c| < |d| \leq 1$. The class of $\begin{pmatrix} a & b \\ c & d \end{pmatrix}$ in $PSL(2,\bar{K})$ is an upper triangular matrix $\begin{pmatrix} \bar{a} & \bar{b} \\ 0 & \bar{d} \end{pmatrix}$.

Let

$$\rho(\alpha) = \begin{pmatrix} \bar{a} & 0 \\ 0 & \bar{d} \end{pmatrix}.$$

Then ρ is a group homomorphism whose kernel consists or matrices whose trace has class 2 in $\bar{K}$. Thus the order of $\alpha \in A_\eta$ with $\rho(\alpha) = 1$ is a power of p. As $\rho(A_\eta)$ is cyclic the proof is done.

Corollary: The stabilizer group of an edge is either cyclic or a dihedral group $D(p^s)$ for $p \neq 2$. If $p = 2$ it can be also the tetrahedral group.

(4.6) The Schottky group Γ has index j in N and thus the canonical projection $\pi : S(\Gamma) \to S(N)$ of the curves parametrized by Γ (resp. N) has degree j. Using the Riemann-Hurwitz genus formula we get

$$2(g-1) = 2j(r-1) + b$$

where $b = \sum_{i=1}^{t} j(1 - \frac{1}{r_i})$ is the total ramification order of the covering π.

If now $r > 1$, then

$$j \leq g - 1.$$

If $r = 1$, then

$$j(1 - \frac{1}{r_1}) \leq 2(g-1).$$

As $r_1 \geq 2$ we have $(1 - \frac{1}{r_1}) \geq \frac{1}{2}$ and

$$j \leq 4(g-1).$$

(4.7) In order to prove the theorem in (4.1) we have to handle only the case where $r = 0$.

This means that $\mathcal{T}/N \approx \Delta$ is a tree and thus N is a tree product in this case.

In order to prove the theorem we have to show that

$$\mu = \sum_{i=1}^{n-1} \frac{1}{e_i} - \sum_{i=1}^{n} \frac{1}{v_i} \geq \frac{1}{12}.$$

If a is a vertex of Δ, we denote by v_a the order of the stabilizer group N_a of a. If η is an edge of Δ, we denote by e_η the order of the stabilizer group A_η of η.

If B is a subtree of Δ, we define

$$\mu(B) = \sum_{\eta \in \varepsilon(B)} \frac{1}{e_\eta} - \sum_{a \in V(B)} \frac{1}{v_a}$$

where $\varepsilon(B)$ denotes the set of edges of B and $V(B)$ the set of vertices of B. Obviously $\mu = \mu(\Delta)$.

(4.7.1) <u>Proposition:</u> $\mu(B) \leq \mu$.

<u>Proof:</u> We fix a point a in B. The distance dist (a, b) between a and a vertex b of Δ is defined to be the smallest number of edges necessary to join a with b. If now $b \neq a$, we find exactly one point b' of Δ such that b and b' are connected by an edge and dist (a, b') < dist (a, b). Let η_b denote the edge between b' and b. Now b is a point of B if and only if η_b is an edge of B. Thus

$$\mu(\Delta) - \mu(B) = \Sigma \left(\frac{1}{e_{\eta_b}} - \frac{1}{v_b} \right)$$

where the sum is extended over all the vertices of Δ which are not in B. As $e_{\eta_b} \leq v_b$ we clearly have $\mu \geq \mu(B)$.

(4.7.2) <u>Corollary:</u> $\mu \geq \mu(B) + \frac{1}{e_{\eta_b}} - \frac{1}{v_b}$

for any vertex b of Δ which is not in B.

(4.8) A subtree B of Δ is called a segment if B is the subtree generated by two vertices of B. This means there are vertices $a_0, a_1, \ldots, a_m$ in B such that a_{i-1} and a_i are connected by an edge η_i for all i.

a_0 and a_m are called the endpoints of B. They are characterized by the fact that they are endpoints of only one edge of B.

We call such a segment B normal with respect to N if the stabilizer group of any vertex of B is either contained in N_{a_0} or in N_{a_m} and if $A = N_{a_0} \cap N_{a_m}$ is a proper subgroup of N_{a_0} and of N_{a_m}.

(4.8.1) <u>Proposition:</u> $\mu(B) = \frac{1}{e} - \frac{1}{v_{a_0}} - \frac{1}{v_{a_m}}$ where $e = \mathrm{ord}\,(N_{a_0} \cap N_{a_m})$ if B is normal with respect to N.

<u>Proof:</u> Let for the moment $N_i = N_{a_i}$. Let j be the largest index with $N_j \subseteq N_o$. If $\alpha \in N_j$ then α fixes a_0 and a_j, and thus also the path between a_0 and a_j which means

$$N_j \subseteq N_{j-1} \subseteq \cdots \subseteq N_o.$$

Also $N_{j+1} \subseteq N_m$ and more generally

$$N_{j+1} \subseteq N_{j+2} \subseteq \cdots \subseteq N_m.$$

If $i \leq j$ the stabilizer group A_i of the edge η_i between a_{i-1} and a_i coincides with N_i. Thus $e_{\eta_i} =: e_i = v_i := v_{a_i}$ for $i \leq j$. If $i > j +$ the stabilizer group A_i of η_i coincides with N_{i-1}.

Thus

$$e_i = v_{i-1}$$

for $i > j + 1$.

Now all transformations of $N_o \cap N_m$ fix all the edges of B and thus

$$N_o \cap N_m = N_j \cap N_{j+1} = A_{j+1}.$$

Thus ord $(N_o \cap N_m) = e_{j+1}$.

$$\text{Now } \mu(B) = \sum_{i=1}^{m} \frac{1}{e_i} - \sum_{i=o}^{m} \frac{1}{v_i}$$

$$= \frac{1}{e_{j+1}} - \frac{1}{v_o} - \frac{1}{v_m}.$$

(4.8.2) Proposition: If N is infinite, there are normal segments B of Δ.

Proof: If all the groups N_a, $a \in \Delta$, are contained in one group N_{a_o}, then $N = N_{a_o}$ as N is generated by the union of all the N_a, $a \in \Delta$.

Let now a and b be vertices of Δ for which $N_a \not\subseteq N_b$, $N_b \not\subseteq N_a$. Let now b' a point between a and b, closest to a for which

$$N_a \not\subseteq N_{b'}$$

$$N_{b'} \not\subseteq N_a.$$

Then clearly the tree generated by a and b' is a normal segment with respect to N.

(4.9) The case char $\bar{K} = 0$ is simpler to treat because the stabilizer groups of edges are cyclic and because of the following lemma.

(4.9.1) Lemma: If $i \neq j$ and $N_i \cap N_j \neq \{id\}$, then $N_i \cap N_j$ is maximally cyclic in N_i and also in N_j.

Proof: Let $\alpha \in N$, $\alpha^k \in N_i$, $\alpha^k \neq id$. Let $\alpha(z) = \rho \cdot z$, ρ root of unity. Then if N_i is the stabilizer of $[D_i]$, D_i affinoid disk in K, we get

$$\alpha^k(D_i) = D_i.$$

This means that D_i is a disk around 0 because $|\rho - 1| = 1$. But then also $\alpha(D_i) = D_i$ and $\alpha \in N_i$.

We will prove $\mu(\Delta) \geq \frac{1}{12}$ for char $\bar{K} = 0$ by considering different cases:

(4.9.2) If ord $N_b \leq 4$ for all vertices b of Δ, then $\mu(\Delta)$ is a integral multiple of $\frac{1}{12}$ and as $\mu(\Delta) = \frac{g-1}{j} > 0$ we get $\mu(\Delta) \geq \frac{1}{12}$.

(4.9.3) Let B be a normal segment of Δ and let N_1, N_2 be the stabilizer groups of the end points of B. Let $v_1 = \text{ord } N_1$, $v_2 = \text{ord } N_2$, $e = \text{ord } A$. Because of (4.9.2) we may assume that v_1 or v_2 is ≥ 5.

(4.9.4) If N_1 or N_2 is cyclic, then $A = \{id\}$ because of (4.9.1) and

$$\mu(B) \geq 1 - \frac{1}{2} - \frac{1}{5} = \frac{3}{10}.$$

(4.9.5) If N_1 or N_2 is the tetrahedral group Th of order 12, then e = 1, 2 or 3 as there is no element of order $\neq$ 1, 2, 3 in Th.

If e = 2, $\mu(B) \geq \frac{1}{2} - \frac{1}{4} - \frac{1}{12} = \frac{1}{6}$.

If e = 3, $\mu(B) \geq \frac{1}{3} - \frac{1}{6} - \frac{1}{12} = \frac{1}{12}$.

(4.9.6) If N_1 or N_2 is the octahedral group Oh of order 24, then e = 1, 2, 3 or 4. As above we find $\mu(B) \geq \frac{1}{12}$.
If for example e = 4, then v_1 or v_2 is ≥ 8 and $\mu(B) \geq \frac{1}{4} - \frac{1}{8} - \frac{1}{24} = \frac{1}{12}$.

(4.9.7) If N_1 or N_2 is the icosahedral group of order 60, then e = 1, 2, 3 or 5. As above we find $\mu(B) \geq \frac{1}{12}$. If for example e = 5, then v_1 or v_2 is ≥ 10 and $\mu(B) \geq \frac{1}{5} - \frac{1}{10} - \frac{1}{60} = \frac{5}{60} = \frac{1}{12}$.

(4.10) The only case left to consider is the one in which N_1 and N_2 are dihedral groups. Then v_1 or $v_2 \geq 6$.
If e = 2, we get $\mu(B) \geq \frac{1}{2} - \frac{1}{4} - \frac{1}{6} = \frac{1}{12}$.

So let $e \geq 3$.

Let

$$\alpha(z) = \rho \cdot z$$

ρ root of unity, be a generator for A. As A is maximally cyclic in N_1 and N_2 we find that the index of A in N_i is 2 and there are $\sigma_1 \in N_1$, $\sigma_2 \in N_2$ of order 2 which interchange the fixed points of α.

Thus

$$\sigma_1(z) = \frac{b_1}{z}$$

$$\sigma_2(z) = \frac{b_2}{z}.$$

Without loss of generality $b_1 = 1$, $|b| > 1$, $b := b_2$.
Now $\mu(B) = \frac{1}{e} - \frac{1}{2e} - \frac{1}{2e} = 0$.

The segment B is represented by affinoid disks $D_1, \ldots, D_m$ with

$$D_i = \{z : |z| \leq r_i\}$$

and $r_1 < r_2 < \ldots < r_m$.

Now $\sigma_1(D_1) = D_1$ and $\sigma_2\ (D_2) = D_2$ and thus $r_1 = 1$, $r_m = \sqrt{|b|}$.

If every stabilizer group of a vertex of Δ outside B is contained in N_1 or N_2, then $\mu(\Delta) = \mu(B) = 0$ which is impossible as $g \geq 2$.

Therefore there is a normal segment B' between a point [D*] of B and a point [D] outside of B where D is an affinoid disk in K.

Now $0 \notin D$, because $(\sigma_2 \circ \sigma_1)(z) = b \cdot z$ and thus every disk around 0 is N-conjugate to a disk D' around 0 whose radius is ≥ 1 and $\leq \sqrt{|b|}$. But as D' must be in Δ, D would have to be equal to D' which is impossible as [D] is not on B.

From $0 \notin D$ we get that $N_{[D]} \cap N_{[D^*]} = \{id\}$ or $= \{id, \sigma_i\}$.
But then $\mu \geq \mu(B') \geq \frac{1}{12}$.

Chapter VIII. The curve of a discontinuous group and its Jacobian variety

Introduction

A finitely generated discontinuous subgroup N of $PSL_2(K)$ contains a Schottky group Γ as normal subgroup of finite index N. The factor group $G = N/\Gamma$ is canonically a group of automorphisms on the Mumford curve S associated to Γ. The quotient space $T = S/G$ is a projective curve whose rational functions are the G-invariant rational functions of S.

In this chapter we will show how the construction of the divisor class group of degree 0 for a Mumford curve $S = \Omega/\Gamma$ can be extended to curves $T = \Omega/N$ associated to discontinuous groups N.

In §1 we study N-automorphic forms and prove that a form f(z) is a finite product of the basic N-automorphic forms $\tilde{\Theta}(a,b;z)$ if and only if the automorphy factor of f(z) is 1 for any element of finite order of $\bar{N} = N/[N, N]$. In §2 the N-automorphic forms without zeroes and poles are characterized and it is shown for which of them the automorphy factor is a homomorphism on $\tilde{N} = \bar{N}/\bar{N}_o$, $\bar{N}_o$ = subgroup of elements of finite order of $\bar{N}$.

In §3 we introduce the lattice L(N) in the algebraic torus $C(\tilde{N}) := \text{Hom}\ (N, K^*)$ and give the construction of two canonical homomorphisms $V^* : \mathfrak{J}(\Gamma) \to \mathfrak{J}(N)$, $I^* : \mathfrak{J}(N) \to \mathfrak{J}(\Gamma)$ of the analytic tori $\mathfrak{J}(N) = C(\tilde{N})/L(N)$ and $\mathfrak{J}(\Gamma) = C(\bar{\Gamma})/L(\Gamma)$ which are induced by the embedding $I : \Gamma \to N$ and the Verlagerung $V : \bar{N} \to \bar{\Gamma}$.

In §4 we show that $\mathfrak{J}(N)$ is canonically isomorphic to the divisor class group of the curve T and derive that the genus of T is the rank of $\tilde{N}$.

§1 Automorphic forms relative to a discontinuous group

(1.1) Let Γ be a Schottky group of rank $g \geq 2$ and N a subgroup of the group of all normalizers of Γ in $PSL_2(K)$ which contains Γ. Then Γ has finite index in N and N operates on the Stein domain Ω of ordinary points of Γ.

Let S be the Mumford curve associated to Γ. Then G is in a canonical way a group of automorphisms of S, see Chap. VII, §2, and the quotient space

$$T := S/G$$

is a projective curve whose field K(T) of rational functions is the field of G-invariant rational functions on S.
The degree of the canonical covering

$$\pi : S \to T$$

is the order n of the group G which is the index of Γ in N.

A meromorphic function f(z) on Ω is called N-automorphic, if for any $\alpha \in N$ there is a constant $c(\alpha) \in K^*$ such that

$$f(z) = c(\alpha) \cdot f(\alpha z).$$

If f(z) is not identically zero, $c(\alpha)$ is uniquely determined and the mapping $\alpha \to c(\alpha)$ is a group homomorphism $c : N \to K^*$ which is called the automorphy factor of f(z).

Let now $\alpha_1, \ldots, \alpha_n$ be coset representatives of Γ in N and let h(z) be a Γ-automorphic form with factor $c : \Gamma \to K^*$.
Put $f(z) := h(\alpha_1 z) \cdot \ldots \cdot h(\alpha_n z)$.

Proposition: f(z) is N-automorphic with factor $\tilde{c} : N \to K^*$ and $\tilde{c} = c \circ V$ where V is the Verlagerung (tranfer) $\bar{N} \to \bar{\Gamma}$ and $\bar{N}$, $\bar{\Gamma}$ are the respective commutator factor groups of N, Γ.

Proof: Pick $\alpha \in N$. Then there is a permutation σ of $\{1, 2, \ldots, n\}$ such that

$$\alpha_i \alpha = \alpha_{\sigma(i)} \cdot \gamma_i$$

with some $\gamma_i \in \Gamma$.

Now

$$\begin{aligned} f(\alpha z) &= h(\alpha_1 \alpha z) \cdot \ldots \cdot h(\alpha_n \alpha z) \\ &= h(\alpha_{\sigma(1)} \cdot \gamma_1 z) \cdot \ldots \cdot h(\alpha_{\sigma(n)} \cdot \gamma_n z) \\ &= h(\alpha_{\sigma(1)}\gamma_1\alpha_{\sigma(1)}^{-1} \cdot \alpha_{\sigma(1)} z) \cdot \ldots \cdot h(\alpha_{\sigma(n)}\gamma_n\alpha_{\sigma(n)}^{-1}\alpha_{\sigma(n)} z). \end{aligned}$$

As $\alpha_j\gamma_i\alpha_j^{-1} \in \Gamma$ and $h(\gamma z) = \frac{1}{c(\gamma)} \cdot h(z)$ we get

$$f(\alpha z) = \frac{1}{c(\alpha_{\sigma(1)}\gamma_1\alpha_{\sigma(1)}^{-1})} \quad h(\alpha_{\sigma(1)} z) \cdot \ldots \cdot \frac{1}{c(\alpha_{\sigma(n)}\gamma_n\alpha_{\sigma(n)}^{-1})} \cdot h(\alpha_{\sigma(n)} z).$$

As σ is a permutation of $\{1, 2, \ldots, n\}$ we obtain

$$\tilde{c}(\alpha) \cdot f(\alpha z) = f(z)$$

with $\tilde{c}(\alpha) = c(\alpha_{\sigma(1)}\gamma_1\alpha_{\sigma(1)}^{-1}) \cdot \ldots \cdot c(\alpha_{\sigma(n)}\gamma_n\alpha_{\sigma(n)}^{-1})$.

The mapping V which sends α onto the residue class of

$$\alpha_{\sigma(1)}\gamma_1\alpha_{\sigma(1)}^{-1} \cdot \ldots \cdot \alpha_{\sigma(n)}\gamma_n\alpha_{\sigma(n)}^{-1}$$

in the commutator factor group $\bar{\Gamma}$ of Γ is a group homomorphism from N into $\bar{\Gamma}$ and does not depend on the coset representatives chosen to define it. We also consider V as a homomorphism from $\bar{N}$ into $\bar{\Gamma}$. This homomorphism is called Verlagerung or transfer, see [38] for properties of V.

As K^* is commutative the factor c vanishes on the commutator subgroup $[\Gamma, \Gamma]$ of Γ and thus c can be considered as a group homomorphism $c : \bar{\Gamma} \to K^*$. Then clearly

$$\tilde{c}(\alpha) = c(V\alpha).$$

(1.2) In this section we collet a few formulas in connection with the Verlagerung.

We consider $\bar{\Gamma}$ as a left G-module where the operation of G on $\bar{\Gamma}$ is given through $\bar{\gamma} \to {}^{\bar{\alpha}}\bar{\gamma} :=$ class of $\alpha\gamma\alpha^{-1}$ in $\bar{\Gamma}$ for $\gamma \in \Gamma$, $\alpha \in N$ where $\bar{\gamma}$ denotes the residue class of γ in $\bar{\Gamma}$ and $\bar{\alpha}$ denotes the residue class of α in $N/\Gamma = G$.

The norm of an element x of the G-module $\bar{\Gamma}$ is defined to be

$$\Lambda(x) := \sum_{\sigma \in G} \sigma x$$

where the inner operation of $\bar{\Gamma}$ is written additively.

Now the norm Λ is a group homomorphism

$$\bar{\Gamma} \to \bar{\Gamma}$$

and $\Lambda(x) \in \bar{\Gamma}^G :=$ subgroup of G-invariant elements of $\bar{\Gamma}$.

For any $x \in \bar{\Gamma}^G$ we have

$$\Lambda(x) = n \cdot x.$$

The norm Λ is related to the Verlagerung in the following way. The embedding of Γ into N induces a group homomorphism

$$I : \bar{\Gamma} \to \bar{N}.$$

<u>Proposition:</u> $\Lambda = V \circ I$.

<u>Proof:</u> We use the notation of (1.1). If $\alpha \in \Gamma$, then

$$\alpha_i \alpha = \alpha_{\sigma(i)} \cdot \gamma_i$$

with $\gamma_i = \alpha$ and σ is the identity. Then

$$V(\alpha) = \text{residue class of } \alpha_1 \alpha \alpha_1^{-1} \cdot \ldots \cdot \alpha_n \alpha \alpha_n^{-1}$$

in $\bar{\Gamma}$ which is exactly the norm of $\bar{\alpha}$.

<u>Corollary:</u> If we consider V as group homomorphism $\bar{N} \to \bar{\Gamma}$, then the kernel of V consists of the elements of finite order in $\bar{N}$.

The $\mathbb{Z}$-rank of $\bar{N}$ equals the rank of $\bar{\Gamma}^G$.

Proof: As $\bar{\Gamma}$ is free abelian, any element of finite order in $\bar{N}$ is mapped through V onto 0.

On the other hand, if A is the G-module of $\bar{\Gamma}$ generated by all $\sigma x - x$, $\sigma \in G$, $x \in \bar{\Gamma}$, then

$$A \cap \bar{\Gamma}^G = 0$$

$$A \oplus \bar{\Gamma}^G \text{ has rank } g.$$

The first property is true as $\Lambda(\sigma x - x) = 0$ and thus $\Lambda(A) = 0$ while Λ is multiplication by n on $\bar{\Gamma}^G$.

For any $x \in \bar{\Gamma}$ we have

$$\Lambda(x) - nx = \sum_{\sigma \in G} (\sigma x - x) \in \Lambda$$

which shows that $n \cdot \bar{\Gamma} \subset A \oplus \bar{\Gamma}^G$.

Now I maps A onto 0 as $\sigma x - x$ is the class of a commutator $\alpha\gamma\alpha^{-1}\gamma^{-1}$ where α, γ are representatives of σ, x.

The subgroup $I(\bar{\Gamma}^G)$ has the same rank as $\bar{\Gamma}^G$ because $VI(\bar{\Gamma}^G) = \Lambda(\bar{\Gamma}^G) = n \cdot \bar{\Gamma}^G$.

Now $I(\bar{\Gamma}^G)$ has finite index of $\bar{N}$ which proves that the $\mathbb{Z}$-rank of $\bar{N}$ is the rank of $\bar{\Gamma}^G$. Therefore the kernel of V consists only of the elements of finite order.

(1.3) Denote by $\bar{N}_o$ the subgroup of elements of finite order in $\bar{N}$.

Theorem: Let f(z) be a N-automorphic form whose factor of automorphy $\tilde{c}$ is identically 1 on $\bar{N}_o$.
Then there is a Γ-automorphic form h(z) such that

$$f(z) = h(\alpha_1 z) \cdot \ldots \cdot h(\alpha_n z).$$

Proof: The Verlagerung V induces an embedding of $\bar{N}/\bar{N}_o$ into $\bar{\Gamma}^G$. If we consider $\tilde{c}$ as homomorphism from a subgroup of $\bar{\Gamma}$ into K^*, we can extend this homomorphism onto a homomorphism c of the whole $\bar{\Gamma}$ into K^* as K^* is divisible.

If now $\tilde{h}(z)$ is Γ-automorphic with factor c, then the function

$$\tilde{f}(z) = \tilde{h}(\alpha_1 z) \cdot \ldots \cdot \tilde{h}(\alpha_n z)$$

is N-automorphic with factor $c \circ V = \tilde{c}$. Therefore

$$g(z) := \frac{\tilde{f}(z)}{f(z)}$$

is an N-invariant function on Ω.

We now apply the theorem of Tsen which states that algebraic function fields of one variable over algebraically closed fields are quasi-algebraically closed, see [38]. This has the consequence that the Brauer group of K(T) is trivial and that the norm relative to the Galois extension $K(T) \subset K(S)$ is surjective, see [38]. This means there is a rational function $h_1(z)$ on S such that

$$\prod_{\sigma \in G} h_1(\sigma z) = g(z).$$

If we put

$$h(z) = \frac{\tilde{h}(z)}{h_1(z)}$$

then obviously $f(z) = h(\alpha_1 z) \cdot \ldots \cdot h(\alpha_n z)$ and the result has been proved.

(1.4) The function

$$\tilde{\Theta}(a, b; z) := \prod_{\alpha \in N} \frac{z - \alpha(a)}{z - \alpha(b)}$$

is meromorphic on Ω, if $a, b \in \Omega$ and

$$\Theta(a, b; z) = \sum_{i=1}^{n} \Theta(\alpha_i^{-1} a, \alpha_i^{-1} b; z) = \text{const.} \sum_{i=1}^{n} \Theta(a, b; \alpha_i z).$$

Thus $\tilde{\Theta}(a, b; z)$ is N-automorphic and its automorphy factor is 1 on $\bar{N}_o$.

Corollary: Let $f(z)$ be a N-automorphic form whose factor is 1 on $\bar{N}_o$. Then

$$f(z) = \text{const.} \prod_{i=1}^{r} \tilde{\Theta}(a_i, b_i; z)$$

with $a_i, b_i \in \Omega$.

Proof: This is an immediate consequence of the above theorem and of Chap. II, (3.4).

§2 Automorphy factors of forms without zeroes and poles

(2.1) In this section we study the factors of the N-automorphic forms without zeroes and poles on Ω. Any such form is of course Γ-automorphic and it has been proved in Chap. II, (3.3) that

$$f(z) = \text{const. } u_\gamma(z)$$

with $u_\gamma(z) = \Theta(a, \gamma a; z)$.

(2.1.1) Proposition: $u_\gamma(z)$ is N-automorphic if and only if γ mod $[\Gamma, \Gamma]$ is contained in $\bar{\Gamma}^G$.

Proof: For any $\alpha \in N$ we have

$$\begin{aligned} u_\gamma(\alpha z) &= \Theta(a, \gamma a; \alpha z) \\ &= \text{const. } \Theta(\alpha^{-1} a, \alpha^{-1}\gamma a; z), \end{aligned}$$

sec formula (2.2.1) in Chap. VII.

Thus

$$\begin{aligned} u_\gamma(\alpha z) &= \text{const. } \Theta(\alpha^{-1} a, \alpha^{-1}\gamma\alpha\alpha^{-1} a; z) \\ &= \text{const. } \Theta(a, \alpha^{-1}\gamma\alpha a; z) \end{aligned}$$

see formula (2.3.4) in Chap. II.

Thus

$$u_\gamma(\alpha z) = \text{const. } u_{\alpha^{-1}\gamma\alpha}(z).$$

We know from Chap. II, §3, that

$$u_\gamma(z) = u_{\gamma'}(z)$$

if and only if $\gamma \equiv \gamma' \mod [\Gamma, \Gamma]$.

Thus

$$u_\gamma(z) = \text{const.}\ u_\gamma(\alpha z)$$

if and only if

$$\gamma \equiv \alpha^{-1}\gamma\alpha \mod [\Gamma, \Gamma].$$

This means that $u_\gamma(z)$ is N-automorphic if and only if the residue class of γ in $\bar{\Gamma}$ is contained in $\bar{\Gamma}^G$.

(2.2) Let $\tilde{N} = \bar{N}/\bar{N}_o$. Then $\tilde{N}$ is a free abelian group of finite rank. Put

$$C(\bar{N}) = \text{Hom}\ (\bar{N}, K^*)$$
$$C(\tilde{N}) = \text{Hom}\ (\tilde{N}, K^*).$$

Then $C(\tilde{N})$ is canonically a subgroup of $C(\bar{N})$ consisting of those homomorphisms c which are identically 1 on $\bar{N}_o$.

For any $\alpha \in N$ let

$$\tilde{u}_\alpha(z) = \tilde{\Theta}(a, \alpha a; z).$$

This form is independent of a, see the proof of the respective result (2.3.4) of Chap. II, and it has no zeroes or poles. Moreover $\tilde{u}_{\alpha\beta}(z) = \tilde{u}_\alpha(z) \cdot \tilde{u}_\beta(z)$ and thus $\tilde{u}_\alpha(z)$ depends only on the class of α in $\tilde{N}$.

(2.2.1) <u>Proposition:</u> $\tilde{u}_\alpha(z) = u_{V\alpha}(z)$.

<u>Proof:</u> Let $\alpha_1, \ldots, \alpha_n$ be a set of left coset representatives of N modulo Γ.

Now

$$\tilde{u}_\alpha(z) = \prod_{\beta \in N} \frac{z - \beta(a)}{z - \beta\alpha(a)}$$

$$= \prod_{i=1}^{n} \prod_{\gamma \in \Gamma} \frac{z - \gamma\alpha_i a}{z - \gamma\alpha_i \alpha a}.$$

Let σ be the permutation of $\{1, 2, \ldots, n\}$ such that $\alpha_i \alpha \alpha_{\sigma(i)}^{-1} \in \Gamma$, then

$$\tilde{u}_\alpha(z) = \prod_{\gamma \in \Gamma} \prod_{i=1}^{n} \frac{z - \gamma\alpha_{\sigma(i)} a}{z - \gamma\alpha_i \alpha \alpha_{\sigma(i)}^{-1} \alpha_{\sigma(i)} a}$$

$$= \prod_{i=1}^{n} u_{\alpha_i \alpha \alpha_{\sigma(i)}^{-1}}(z)$$

$$= u_\gamma(z)$$

with $\gamma = \sum_{i=1}^{n} \alpha_i \alpha \alpha_{\sigma(i)}^{-1}$ and this is by the definition of the Verlagerung exactly $V(\alpha)$.

(2.3) <u>Proposition:</u> Let $f(z)$ be a N-automorphic form on Ω without zeroes and poles whose automorphy factor is identically 1 on $\bar{N}_o$. Then

$$f(z) = \text{const.}\ \tilde{u}_\alpha(z)$$

with $\alpha \in N$.

<u>Proof:</u> We proved in (1.4) that

$$f(z) = \text{const.} \prod_{i=1}^{r} \tilde{\Theta}(a_i, b_i; z)$$

with $a_i, b_i \in \Omega$.

The point a_i is a zero of $\tilde{\Theta}(a_i, b_i; z)$ unless $b_i \in Na_i$. Therefore there is a permutation σ of $\{1, \ldots, r\}$ such that

$$b_{\sigma(i)} \in Na_i.$$

Now

$$f(z) = \text{const.} \prod_{i=1}^{r} \tilde{\Theta}(a_i, b_{\sigma(i)}; z).$$

If thus $\alpha_i a_i = b_{\sigma(i)}$, we get

$$\tilde{\Theta}(a_i, b_{\sigma(i)}; z) = \tilde{u}_{\alpha_i}(z)$$

and $f(z) = \text{const.} \prod_{i=1}^{r} \tilde{u}_{\alpha_i}(z)$

$= \text{const.}\ \tilde{u}_{\alpha}(z)$

with $\alpha = \alpha_1\alpha_2 \ldots \alpha_r$.

§3 Period lattices

(3.1) The quotient

$$\tilde{Q}(\alpha, \beta) := \frac{\tilde{u}_{\alpha}(z)}{\tilde{u}_{\alpha}(\beta z)}$$

is bimultiplicative in α, $\beta \in N$. As $\tilde{u}_{\alpha}(z) \equiv 1$ if the class of α in $\tilde{N} := \bar{N}/\bar{N}_o$ is trivial and because $Q(\alpha, \beta)$ is as function of β the automorphy factor of $\tilde{u}_{\alpha}(z)$ which is a homomorphism from N into K^* one can consider $\tilde{Q}$ as a bimultiplicative form

$$\tilde{Q} : \tilde{N} \times \tilde{N} \to K^*.$$

<u>Proposition:</u> $\tilde{Q}$ is symmetric and positive definit, i.e.

$$|\tilde{Q}(\nu, \nu)| < 1$$

for any $\nu \in \tilde{N}$, $\nu \neq$ neutral element of $\tilde{N}$.

<u>Proof:</u> The proof that $\tilde{Q}$ is symmetric is the same as the respective proof given for Schottky groups in Chap. VI, (2.1).
One needs the formula

$$\frac{\tilde{u}_{\alpha}(a)}{\tilde{u}_{\alpha}(\beta a)} = \frac{\tilde{\Theta}(a, \beta a; z)}{\tilde{\Theta}(a, \beta a; \alpha z)}.$$

If $\gamma \in \Gamma$ such that γ is congruent modulo $[\Gamma, \Gamma]$ to an element in $\bar{\Gamma}^G$, then

$$VI(\bar{\gamma}) = \bar{\gamma}^n$$

where $\bar{\gamma}$ is the class of γ in $\bar{\Gamma}$, see (1.2).

Now

$$\begin{aligned} \tilde{Q}(I\bar{\gamma}, I\bar{\gamma}) &= \frac{\tilde{u}_\gamma(z)}{\tilde{u}_\gamma(\gamma z)} \\ &= \frac{u_{V\gamma}(z)}{u_{V\gamma}(\gamma z)} \\ &= Q(VI\bar{\gamma}, \bar{\gamma}) \\ &= Q(\bar{\gamma}^n, \bar{\gamma}) = Q(\bar{\gamma}, \bar{\gamma})^n. \end{aligned}$$

As Q is positive definit, we get

$$|Q(\bar{\gamma}, \bar{\gamma})| < 1$$

if $\bar{\gamma}$ is not the zero class. Thus

$$|\tilde{Q}(I\bar{\gamma}, I\bar{\gamma})| < 1$$

if $\bar{\gamma}$ is not the zero class in $\bar{\Gamma}^G$.

Now $I(\bar{\Gamma}^G)$ has finite index in $\tilde{N}$ and thus $\tilde{Q}$ must also be positive definit on $\tilde{N}$.

(3.2) Let

$$C(\tilde{N}) = \text{Hom}\ (\tilde{N}, K^*)$$
$$C(\bar{\Gamma}) = \text{Hom}\ (\bar{\Gamma}, K^*).$$

We consider both group as algebraic tori whose character groups is $\tilde{N}$ and $\bar{\Gamma}$ respectively.

We now have homomorphisms

$$\tilde{\zeta} : \tilde{N} \to C(\tilde{N})$$
$$\zeta : \bar{\Gamma} \to C(\bar{\Gamma})$$

which are defined as follows:

If $\nu \in \tilde{N}$ and $\alpha \in N$ is a representative of ν, then

$$\tilde{\zeta}(\nu) = \text{automorphy factor of } \tilde{u}_\alpha(z).$$

Thus $\tilde{\zeta}(\nu)$ is the homomorphism which maps $\mu \in \tilde{N}$ onto $\tilde{Q}(\nu, \mu)$.

In the same way ζ is defined such that

$$\zeta(\bar{\gamma}, \bar{\gamma}') = Q(\bar{\gamma}, \bar{\gamma}')$$

for $\bar{\gamma}, \bar{\gamma}' \in \bar{\Gamma}$.

We compose the map $I : \bar{\Gamma} \to \bar{N}$ with the residue class map $\bar{N} \to \tilde{N} = \bar{N}/\bar{N}_o$ and obtain a homomorphism

$$I : \bar{\Gamma} \to \tilde{N}$$

which we denote by I again.

The kernel of map $V : \bar{N} \to \bar{\Gamma}$ consists of $\bar{N}_o$ and we get thus a canonical homomorphism

$$V : \tilde{N} \to \bar{\Gamma}$$

which we again denote by V.

Now I and V induce canonical dual homomorphisms

$$I^* : C(\tilde{N}) \to C(\bar{\Gamma})$$

$$V^* : C(\bar{\Gamma}) \to C(\tilde{N})$$

defined as follows:

If $c : \tilde{N} \to K^*$ is a homomorphism, then $I^*(c) = c \circ I$.

If $c : \bar{\Gamma} \to K^*$ is a homomorphism, then $V^*(c) = c \circ V$.

Proposition: The following diagrams are commutative

$$\begin{array}{ccc} \bar{\Gamma} & \overset{\zeta}{\to} & C(\bar{\Gamma}) \\ I\downarrow & & \downarrow V^* \\ \tilde{N} & \overset{\tilde{\zeta}}{\to} & C(\tilde{N}) \end{array} \qquad \begin{array}{ccc} \tilde{N} & \overset{\tilde{\zeta}}{\to} & C(\tilde{N}) \\ V\downarrow & & \downarrow I^* \\ \bar{\Gamma} & \overset{\zeta}{\to} & C(\bar{\Gamma}). \end{array}$$

Proof: 1) We show that the first diagram commutes. Pick $\gamma \in \Gamma$. We have to show that the homomorphism

$$\alpha \to V\alpha \to \frac{u_\gamma(z)}{u_\gamma(V\alpha z)}$$

coincides with

$$\alpha \to \frac{\tilde{u}_\gamma(z)}{\tilde{u}_\gamma(\alpha z)} = \tilde{Q}(\gamma, \alpha)$$

for any $\alpha \in N$.

As $\tilde{Q}(\gamma, \alpha) = \tilde{Q}(\alpha, \gamma)$ the last homomorphism coincides with

$$\alpha \to \frac{\tilde{u}_\alpha(z)}{\tilde{u}_\alpha(\gamma z)} = \frac{u_{V\alpha}(z)}{u_{V\alpha}(\gamma z)} = Q(V\alpha, \gamma).$$

As Q is symmetric we see that this homomorphism coincides with

$$\alpha \to \frac{u_\gamma(z)}{u_\gamma(V\alpha z)}$$

and this is the result we had to prove.

2) We show that the second diagram commutes. Pick $\alpha \in N$. We have to show that the homomorphism

$$\gamma \to \frac{\tilde{u}_\alpha(z)}{\tilde{u}_\alpha(\gamma z)}$$

coincides with

$$\gamma \to \frac{u_{V\alpha}(z)}{u_{V\alpha}(\gamma z)}.$$

But this is immediate as $\tilde{u}_\alpha(z) = u_{V\alpha}(z)$.

(3.3) Let $L(N) := \tilde{\zeta}(\tilde{N})$ and $L(\Gamma) := \zeta(\bar{\Gamma})$.

Then I^* maps $L(N)$ into $L(\Gamma)$ and as the kernel of I^* is finite, this shows that $L(N)$ is a discrete subgroup of the algebraic torus $C(\tilde{N})$, see (1.4) of Chap. VI.

The quotient groups

$$\mathfrak{J}(N) := C(\tilde{N})/L(N)$$
$$\mathfrak{J}(\Gamma) := C(\bar{\Gamma})/L(\Gamma)$$

can be given the structure of analytic spaces, see [7] .
They are called analytic tori. Then I^* and V^* induce analytic mappings

$$V^* : \mathfrak{J}(\Gamma) \to \mathfrak{J}(N)$$
$$I^* : \mathfrak{J}(N) \to \mathfrak{J}(\Gamma)$$

which we denote again by I^*, V^* respctively.

We will make clear in the next paragraph that these mappings are the caninical mappings of the Jacobian varieties and the Picard varieties associated to the covering map $\pi : S \to T$.

§4 Divisor class group

(4.1) Let $\mathcal{D}(S)$ (resp. $\mathcal{D}(T)$) be the group of divisors on S (resp. T) of degree zero. We identify $\mathcal{D}(T)$ with a subgroup of $\mathcal{D}(S)$ in the usual way: if b is a point of T, then b is considered to be the divisor

$$r \cdot \sum_{\pi(a)=b} a$$

where r is the order of the point b with respect to the covering map $\pi : S \to T$. If

$$G_a = \{\sigma \in G : \sigma(a) = a\}$$

then $r = \operatorname{ord} G_a$ for $a \in S$ with $\pi(a) = b$. The action of G on S extends canonically to $\mathcal{D}(S)$. If $\delta = \sum_{i=1}^{t} r_i a_i \in \mathcal{D}(S)$, then $\sigma(\delta) = \sum_{i=1}^{t} r_i \sigma(a_i)$.

So we get a G-module structure on $\mathcal{D}(S)$. Then $\mathcal{D}(T)$ is exactly the norm subgroup of $\mathcal{D}(S)$ with respect to G, i.e.

$$\mathcal{D}(T) = \{\Lambda(\delta) = \sum_{\sigma \in G} \sigma(\delta) \; : \; \delta \in \mathcal{D}(S)\}.$$

If f(x) is a rational function on T, then the divisor $\mathrm{div}_T\, f(x)$ of f(x) of T is thus identified with the divisor $\mathrm{div}_S\, f(x)$ of f(x) on S. This makes sense because there is a coordinate function z for $a \in S$ such that z^r is a coordinate function for $b \in T$ if $b = \pi(a)$ and r the ramification order of b.

Let $\mathcal{H}(S)$ (resp. $\mathcal{H}(T)$) be the group of principal divisors on S (resp. T and $\bar{\mathcal{D}}(S)$ (resp. $\bar{\mathcal{D}}(T)$) the group of divisor classes of degree zero on S (resp. T).

Thus

$$\bar{\mathcal{D}}(S) = \mathcal{D}(S)/\mathcal{H}(S)$$
$$\bar{\mathcal{D}}(T) = \mathcal{D}(T)/\mathcal{H}(T).$$

The embedding $\mathcal{D}(T) \subset \mathcal{D}(S)$ gives rise to a homomorphism

$$i : \bar{\mathcal{D}}(T) \to \bar{\mathcal{D}}(S).$$

The norm homomorphism Λ gives rise to a homomorphism

$$\nu : \bar{\mathcal{D}}(S) \to \bar{\mathcal{D}}(T).$$

(4.2) Denote by $\mathcal{F}^*$ the multiplicative group of all non-zero Γ-automorphic forms modulo the non-zero constants.
For any $f \in \mathcal{F}^*$ one defines div f as in Chap. VI, (1.2), and thus div f is a divisor of degree zero on S, and any divisor $\delta \in \mathcal{D}(S)$ is divisor of a Γ-automorphic form.

Let $u : \bar{\Gamma} \to \mathcal{F}^*$ denote the homomorphism that maps $\bar{\gamma}$ into $u_\gamma(z)$.
Then we have a short exact sequence

$$1 \to \bar{\Gamma} \overset{u}{\to} \mathcal{F}^* \overset{\mathrm{div}}{\to} \mathcal{D}(S) \to 1 \quad (*).$$

If $\alpha \in N$ and $f(z)$ a Γ-automorphic form on Ω, then $f(\alpha z)$ is again Γ-automorphic because

$$f(\alpha\gamma z) = f(\alpha\gamma\alpha^{-1} \cdot \alpha z) = c(\alpha\gamma\alpha^{-1}) \cdot f(\alpha z)$$

if c is the automorphy factor of f.

So N operates on $\mathcal{F}^*$ from the right and as Γ operates trivially on $\mathcal{F}^*$ we get an action of $G = N/\Gamma$ on $\mathcal{F}^*$.

Now u is a G-homomorphism as

$$\begin{aligned} u_\gamma(\alpha z) = \Theta(a, \gamma a; \alpha z) &= \text{const. } \Theta(\alpha^{-1}a, \alpha^{-1}\gamma\alpha\alpha^{-1}a; z) \\ &= \text{const. } \Theta(a, \alpha^{-1}\gamma\alpha a; z) \\ &= \text{const. } u_{\alpha^{-1}\gamma\alpha}(z) \end{aligned}$$

and because the action of G on $\bar{\Gamma}$ is given through inner automorphism action. Also div is a G-homomorphism and thus (*) is a short exact sequence of G-modules.

(4.2.1) <u>Proposition:</u> Let δ be a divisor in $\mathcal{D}(T)$.
Then there is a N-automorphic form whose automorphy factor is in $C(\tilde{N})$ such that

$$\delta = \text{div } f(z).$$

<u>Proof:</u> We denote the norm homomorphism relative to G with Λ.
Then there is a divisor $\delta' \in \mathcal{D}(S)$ such that

$$\delta = \Lambda \cdot \delta'.$$

Let h(z) be Γ-automorphic with div $h(z) = \delta'$. Then put

$$f(z) = \Lambda(h(z)) = \sum_{\sigma \in G} h(\sigma z)$$

modulo constants.

Then div $f(z) = \delta$ as div is a G-homomorphism. Also f(z) is N-automorphic whose factor is in $C(\tilde{N})$ by (1.1).

(4.2.2) Proposition: The divisor of a N-automorphic form whose automorphy factor is in $C(\tilde{N})$ is in $\mathcal{D}(T)$.

Proof: This is immediate by Theorem (1.3) and the proof of (4.2.1).

Theorem: $\bar{\mathcal{D}}(T)$ is canonically isomorphic to $\mathcal{F}(N) = C(\tilde{N})/L(N)$.

Proof: Let $\delta \in \mathcal{D}(T)$ and $f_1(z)$, $f_2(z)$ two N-automorphic forms whose factors are in $C(\tilde{N})$ such that $\delta = \operatorname{div} f_1(z) = \operatorname{div} f_2(z)$.

Then $\frac{f_1(z)}{f_2(z)}$ is N-automorphic without zeroes and poles whose factor is in $C(\tilde{N})$.

By (2.3) we know that

$$\frac{f_1(z)}{f_2(z)} = \tilde{u}_\alpha(z)$$

for some $\alpha \in N$.

Next we consider two divisors δ, $\delta' \in \mathcal{D}(T)$.
Let $f(z)$, $f'(z)$ be N-automorphic forms whose factors are in $C(\tilde{N})$ such that

$$\operatorname{div} f(z) = \delta, \quad \operatorname{div} f'(z) = \delta'.$$

Then δ is linearly equivalent to δ' if and only if the automorphic factor of $f(z)$ coincides with the automorphy factor of $f'(z)$ modulo an automorphy factor of some form $\tilde{u}_\alpha(z)$.
This proves the theorem.

Corollary: The homomorphism $i : \bar{\mathcal{D}}(T) \to \bar{\mathcal{D}}(S)$ is canonically isomorpic to $I^* : \mathcal{F}(N) \to \mathcal{F}(\Gamma)$. The homomorphism $v : \bar{\mathcal{D}}(S) \to \bar{\mathcal{D}}(T)$ is canonically isomorphic to $V^* : \mathcal{F}(\Gamma) \to \mathcal{F}(N)$.

Proof: This is simply seen by comparing the definitions and the canonical isomorphisms $\bar{\mathcal{D}}(T) \simeq \mathcal{F}(N)$ and $\bar{\mathcal{D}}(S) \simeq \mathcal{F}(\Gamma)$.

(4.3) We pick a system of transformations $\alpha_1,\ldots, \alpha_r$ of N whose residue classes $\tilde{\alpha}_1,\ldots, \tilde{\alpha}_r$ in $\tilde{N}$ constitute a basis of the free group $\tilde{N}$. We now identify $C(\tilde{N})$ with the r-fold product $(K^*)^r$ of K^* by letting $c \in C(\tilde{N})$ correspond to the r-tupel $(c(\tilde{\alpha}_1),\ldots, c(\tilde{\alpha}_r)) \in (K^*)^r$.

The mapping

$$\tilde{\Phi} : \Omega \to C(\tilde{N})$$

defined by

$$\tilde{\Phi}(z) = (\tilde{u}_1(z),\ldots, \tilde{u}_r(z))$$

is analytic and $\tilde{\Phi}(\infty) = (1,\ldots, 1)$.

The analog mapping

$$\Phi : \Omega \to C(\bar{\Gamma})$$

relative to the group Γ has been introduced and studied in Chap. II, §5.

It is obvious from (3.2) that

$$\tilde{\Phi} = V^* \circ \Phi.$$

Denote by Ω^r the r-fold product of Ω and by

$$\tilde{\Phi}_r : \Omega^r \to C(\tilde{N})$$

the mapping given through

$$\tilde{\Phi}_r(z_1,\ldots, z_r) = \tilde{\Phi}(z_1) \cdot \tilde{\Phi}(z_2) \cdot \ldots \cdot \tilde{\Phi}(z_r).$$

Again we obtain

$$\tilde{\Phi}_r = V^* \circ \Phi_r$$

if Φ_r denotes the respective mapping relative to the group Γ.

The main property of the mapping $\tilde{\Phi}_r$ is the following:

Let $f(z) = \sum_{i=1}^{r} \tilde{\Theta}(a_i, \infty; z)$. Then the automorphy factor of f(z) is $\tilde{\Phi}_r(a_1,\ldots, a_r)$.

We can now prove as in Chap. II, §5 and (1.5) of Chap. VI that

$$\tilde{\Phi}_r \text{ is surjective}$$

which shows that the genus of T is exactly r.

As L(N) is a discrete subgroup of $C(\tilde{N})$ the quotient group $\mathfrak{J}(N) = C(N)/L(N)$ can be given the structure of an analytic group variety. Now $\tilde{\Phi}_r$ induces an analytic mapping

$$T^{(r)} \to \mathfrak{J}(N)$$

from the symmetric r-fold product of T into $\mathfrak{J}(N)$. From general results one sees as in Chap. VI, (3.9) that $\mathfrak{J}(N)$ is indeed the Jacobian variety of T.

Riemann's vanishing theorem, see Chap. VI, §3, which was proved for Schottky groups also holds for finitely generated discontinuous groups.

§5 Examples:

(5.1) Let ζ be a root of unity of order n in K and let α be the linear transformation $\alpha(z) = \zeta \cdot z$ of $\mathbb{P}$. Let ρ be a real number in $|K^*|$ which is smaller that the absolute value $|\zeta - 1|$.

Let now

$$B_i = \{z \in K : |z - \zeta^{i-1}| \leq \rho\}$$

for $1 \leq i \leq n$.

Let now m be a point of K for which $m\zeta^j$, $0 \leq j \leq n-1$ is not contained in any of the disks B_i and let

$$B_i' = \{z \in K : |z - m \cdot \zeta^{i-1}| \leq \rho\}$$

for $1 \leq i \leq n$.

Then $B_i \cap B_j' = \emptyset$ for all i, j.

Let $\gamma(z)$ be a hyperbolic transformation with fixed points 1 and m which maps ∂B_1 onto $\partial B_1'$. Then

$$\frac{\gamma(z) - m}{\gamma(z) - 1} = t \cdot \frac{z - m}{z - 1}$$

and
$$\gamma(z) = \frac{(m - t) \cdot z + m(t - 1)}{(1 - t) \cdot z + tm - 1}.$$

Let N be the group generated by α and γ. Put $\gamma_1 = \gamma$ and $\gamma_i = \alpha^{i-1}\gamma\alpha^{-i+1}$ for $1 \leq i \leq n$. Then γ_i maps ∂B_i onto $\partial B_i'$ and has the fixed points ζ^{i-1} and $m\zeta^{i-1}$ which shows that the group Γ generated by $\gamma_1, \dots, \gamma_n$ is a Schottky group with good fundamental domain

$$F = \mathbb{P} - \left(\bigcup_{i=1}^{n} \mathring{B}_i \cup \bigcup_{i=1}^{n} \mathring{B}_i'\right)$$

where $\mathring{B}_i$, $\mathring{B}_i'$ are the open disks with radius ρ around the centers ζ^{i-1}, $m\zeta^{i-1}$ respectively.

Clearly Γ is a normal subgroup of N and $G = N/\Gamma$ is cyclic of order n generated by the residue class of α in G.

N is the semi-direct product of a free group of rank n with a cyclic group of order n. Therefore the commutator factor group $\tilde{N}$ is the direct product of a free abelian group $\bar{\Gamma}^G$ with G. Now $\bar{\Gamma}^G$ is the infinite cyclic group generated by $\bar{\gamma}_1 \cdot \bar{\gamma}_2 \cdot \dots \cdot \bar{\gamma}_n$ which shows that $\tilde{N}$ has rank 1 and is generated by the residue class $\tilde{\gamma}_1$ of γ_1 in $\tilde{N}$.

The Mumford curve S of Γ has genus n and the curve T associated to N has genus 1 and is thus an elliptic curve.

The homomorphism $V : \tilde{N} \to \bar{\Gamma}$ is given by $V(\tilde{\gamma}_1) = \bar{\gamma}_1\bar{\gamma}_2 \cdot \dots \cdot \bar{\gamma}_n$. Therefore the Jacobian variety $\mathfrak{J}(N)$ of T is isomorphic to the analytic torus

$$K^*/(q^{\mathbb{Z}})$$

with $q = \tilde{Q}(\tilde{\gamma}_1, \tilde{\gamma}_1) = \dfrac{\tilde{u}_{\gamma_1}(z)}{\tilde{u}_{\gamma_1}(\gamma_1 z)}$

$$= \frac{u_{V\gamma_1}(z)}{u_{V\gamma_1}(\gamma_1 z)} = Q(\bar{\gamma}_1 \cdot \ldots \cdot \bar{\gamma}_n, \bar{\gamma}_1)$$

$$= q_{11} q_{12} \cdot \ldots \cdot q_{1n}$$

if $q_{ij} = Q(\bar{\gamma}_i, \bar{\gamma}_j)$.

The quantities q_{ij} and q depend analytically on the point m and the multiplier t of γ. It seems that choices of m and t can be made such that $\mathfrak{F}(N)$ is isomorphic to any given 1-dimensional analytic torus.

If we identify $C(\bar{\Gamma})$ with $(K^*)^n$ by identifying $c \in C(\bar{\Gamma})$ with $(c(\gamma_1), \ldots, c(\gamma_n))$ then the homomorphism

$$V^* : (K^*)^n \to K^*$$

is given through

$$V^*(c_1, \ldots, c_n) = c_1 c_2 \cdot \ldots \cdot c_n:$$

The homomorphism

$$I^* : K^* \to (K^*)^n$$

is given through $I^*(c) = (c, c, \ldots, c)$.

The elliptic curve T is isomorph to the analytic torus $\mathfrak{F}(N)$ and the isomorphism is induced by the analytic mapping $\tilde{u}_\gamma(z)$.

(5.2) Assume that char $K \neq 2$.

Let t be a parameter in K with $0 < |t| < 1$ and let

$$\gamma_1(z) = t \cdot z$$

$$\frac{\gamma_2(z) - 1}{\gamma_2(z) + 1} = t \cdot \frac{z - 1}{z + 1}.$$

Then $\gamma_2(z)$ is a hyperbolic transformation with fixed points +1 and -1 and the multipliers of γ_1 and γ_2 coincide.

A simple computation gives

$$\gamma_2(z) = \frac{(1 + t)z + (1 - t)}{(1 - t(z + (1 + t)}.$$

The group Γ generated by γ_1, γ_2 is a Schottky group of rank 2 and

$F := \{z \in K : \rho \leq |z| \leq \rho^{-1}, |z - 1| \leq |2|\rho, |z + 1| \geq |2| \rho\}$

is a fundamental domain for Γ where $\rho := \sqrt{|t|}$.

Let $\alpha(z) = \frac{1}{z}$. Then $\alpha^2 = \text{id}$ and

$$\alpha^{-1} \circ \gamma_1 \circ \alpha = \gamma_1^{-1}$$
$$\alpha^{-1} \circ \gamma_2 \circ \alpha = \gamma_2.$$

Let $\beta(z) = -z$. Then $\beta^2 = \text{id}$ and

$$\beta \circ \gamma_1 \circ \beta^{-1} = \gamma_1$$
$$\beta \circ \gamma_2 \circ \beta^{-1} = \gamma_1^{-1}.$$

Let $\sigma(z) = \frac{z - 1}{z + 1}$. Then $\sigma(0) = 1$, $\sigma(1) = 0$, $\sigma(\infty) = -1$, $\sigma(-1) = \infty$ and $\sigma^2 = \text{id}$. Also

$$\sigma \circ \gamma_1 \circ \sigma^{-1} = \gamma_2$$
$$\sigma \circ \gamma_2 \circ \sigma^{-1} = \gamma_1$$

because the fixed points of $\sigma \circ \gamma_1 \circ \sigma^{-1}$ are +1 and -1 and the multiplier of $\sigma \circ \gamma_1 \circ \sigma^{-1}$ is also t.

Moreover

$$\sigma \circ \alpha \circ \sigma^{-1} = \beta$$
$$\sigma \circ \beta \circ \sigma^{-1} = \alpha$$

which shows that $\sigma(\sigma\alpha)\sigma^{-1} = \sigma\beta = \alpha\sigma = (\sigma\alpha)^{-1}$.

As $\sigma\alpha$ is of order 4, we see that the group G generated by α, β, σ is the dihedral group D_4 of order 8.

The normalizer N of Γ in $PGL_2(K)$ is the group generated by G and Γ and is the semi-direct product of G and Γ.

N must be generated by G and Γ because $N/\Gamma \cong G$ is a subgroup of $GL_2(\mathbb{Z})$.

Let now $Q = \begin{pmatrix} q_{11} & q_{12} \\ q_{12} & q_{22} \end{pmatrix}$ be the period matrix with respect to the basis γ_1, γ_2. As the transformations of G on $\bar{\Gamma}$ are orthogonal with respect to Q, we get

$$q_{11} = q_{22}$$

because $\sigma \circ \gamma_1 \circ \sigma^{-1} = \gamma_2$.

Also $q_{12} = Q(\gamma_1, \gamma_2) = Q(\alpha\gamma_1\alpha^{-1}, \alpha\gamma_2\alpha^{-1}) = Q(\gamma_1^{-1}, \gamma_2) = q_{12}^{-1}$.

Thus

$$q_{12}^2 = 1.$$

Now from the Riemann vanishing theorem we can conclude that $q_{12} = 1$ is not possible because then the theta function $\vartheta(u_1, u_2)$ would be a product of two one-parameter theta functions and the zero set of $\vartheta(u_1, u_2)$ would be reducible.

Thus

$$Q = \begin{pmatrix} q(t) & -1 \\ -1 & q(t) \end{pmatrix}$$

where we put $q(t) = q_{11}$. By simple estimates one finds that $q(t)$ depends analytically on t and also that $|q(t)| = |t|$.

Thus we can continue $q(t)$ into $t = 0$ and $q(t)$ is a convergent power series

$$q(t) = \sum_{i=1}^{\infty} a_i \cdot t^i$$

with $|a_1| = 1$ and $\sup_{i=1}^{\infty} |a_i| \leq 1$.

Open question: what are the coefficients a_i?

Let p be a square root of $q(t)$. Then the Riemann theta function is

$$\vartheta(u_1, u_2) = \sum_{(n,m)\in\mathbb{Z}^2} p^{n^2+m^2} (-1)^{nm} \cdot u_1^n \cdot u_2^m.$$

Chap. IX Examples: Groups of Quaternions; Whittaker groups.

Introduction

§1 For any ring (commutative and with identity) we denote by $H(R) = H(\mathbb{Z}) \otimes R$ the Hurwitz-quaternions with coefficients in R. Let p be a prime number $\neq 2$ and let k be a finite extension of $\mathbb{Q}_p$ containing a square root of - 1. Then $H(k) \cong M_2(k)$ the ring of 2×2-matrices with entries in k. In this way $H(\mathbb{Z})[\frac{1}{p}]^*$, the invertible elements of $H(\mathbb{Z})[\frac{1}{p}]$, becomes a subgroup of GL(2, k). Its image Λ in PGL(2, k) is a discontinuous subgroup. The group Λ has interesting congruence subgroups $\Lambda(n)$ $(p\nmid n)$. The genera of the Mumford curves parametrized by Λ and $\Lambda(2)$ are calculated. The geometry of the curves and their stable reductions is made explicit. The cases $p \equiv 1 \bmod (4)$ and $p \equiv 3 \bmod (4)$ have different features.

§2 Let Γ be a subgroup of PGL(2, k) generated by g + 1 elliptic elements $s_o, s_1, \ldots, s_g$ of order 2 (so char. $k \neq 2$). We suppose that the 2g + 2 fixed points are in good position. Then Γ has a subgroup W of index 2 generated by $s_os_1, \ldots, s_os_g$. The group W is called a Whittaker group. It parametrizes a hyperelliptic curve of genus g. Conversely, every hyperelliptic curve, which is a Mumford curve, can be obtained in this way.

§1 Groups of Quaternions

(9.1) The inspiration for this section came from J. P. Serre [37] p. 115. The only case that we will do is that of groups of "ordinary" quaternions. For the convenience of the reader, see [3], we review some results on quaternions. The algebra $H(\mathbb{R})$ of real quaternions is a vectorspace over $\mathbb{R}$ with basis $1, e_1, e_2, e_3$. The bilinear multiplication is given by the formula's:

1 is the unit element

$e_s^2 = -1 \quad s = 1, 2, 3$

$e_ie_j = e_k$ if $i \neq j$ and (i, j, k) is an even permutation of $(1, 2, 3)$

$e_ie_j = -e_k$ if $i \neq j$ and (i, j, k) is an odd permutation of $(1, 2, 3)$.

Let ρ denote the element $\frac{1}{2}(1 + e_1 + e_2 + e_3)$. Then $H(\mathbb{Z})$ the ring of Hurwitz quaternions is given by $\mathbb{Z}\rho + \mathbb{Z}e_1 + \mathbb{Z}e_2 + \mathbb{Z}e_3$. For any ring R (commutative and with identity) we put $H(R) = H(\mathbb{Z}) \otimes_{\mathbb{Z}} R$. $H(R)$ is a free R-module of rank 4 and inherits from $H(\mathbb{Z})$ a ring structure.

The conjugate $\bar{x}$ of a quaternion $x = a_0 + a_1e_1 + a_2e_2 + a_3e_3$ is is $\bar{x} = a_0 - a_1e_1 - a_2e_2 - a_3e_3$ and the norm

$N(x) = x\bar{x} = a_0^2 + a_1^2 + a_2^2 + a_3^2$.

With respect to the norm, $H(\mathbb{Z})$ is a left - and right Euclidean ring. Every element of $H(\mathbb{Z})$ can be written as a product of prime elements and units. An element $x \in H(\mathbb{Z})$ is prime if and only if its norm $N(x) \in \mathbb{Z}$ is a prime number.

$H(\mathbb{Z})^*$, the group of invertible elements of $H(\mathbb{Z})$, is equal to $\{x \in H(\mathbb{Z}) \mid N(x) = 1\}$. A small calculation shows that $H(\mathbb{Z})^*$ consists of 24 elements, namely ρ^α and $\rho^\alpha e_i$ ($i = 1, 2, 3$ and $\alpha = 0, 1, 2, 3, 4, 5$).

The group $H(\mathbb{Z})^*/_{\{\pm 1\}}$ is a group of 12 elements and is isomorphic to A_4.

For every $x \in H(\mathbb{Z})$ one of the elements x, ρx, $\rho^2 x$ lies in $\mathbb{Z} + \mathbb{Z}e_1 + \mathbb{Z}e_2 + \mathbb{Z}e_3$. If $x = a_0 + a_1e_1 + a_2e_2 + a_3e_3$ with all $a_i \in \mathbb{Z}$ and $a_0 + a_1 + a_2 + a_3$ odd then ρx and $\rho^2 x$ do not lie in $\mathbb{Z} + \mathbb{Z}e_1 + \mathbb{Z}e_2 + \mathbb{Z}e_3$.

Let $n \geq 1$ and $k \geq 2$ be integers, then $r_k(n)$ denotes the number of solutions $(a_1, \ldots a_k) \in \mathbb{Z}^k$ with $\Sigma a_i^2 = n$.

A well known formula is $r_4(n) = 8 \sum_{\substack{d|n \\ 4 \nmid d}} d$.

Let n be odd. Then $r_4(n) = 8 \sum_{d|n} d$ and it follows from above that the number of $x \in H(\mathbb{Z})$ with norm equal to n is $3r_4(n) = 24 \sum_{d|n} d$.

(9.2) Let p be a prime number and let k be an extension of $\mathbb{Q}_p$ which contains a square root of -1, denoted by i. If $p = 2$ or $p \equiv 3 \bmod (4)$ then we can take for k a quadratic extension of $\mathbb{Q}_p$. If $p \equiv 1 \bmod (4)$ then $k = \mathbb{Q}_p$ has the required property.

The map $a_0 + a_1e_1 + a_2e_2 + a_3e_3 \mapsto \begin{pmatrix} a_0+a_1 i & a_2+a_3 i \\ -a_2+a_3 i & a_0-a_1 i \end{pmatrix}$ gives an isomorphism $H(k) \tilde{\to} M_2(k)$ = the 2×2-matrices with entries in k.

Let Λ' denote $H(\mathbb{Z}[\frac{1}{p}])^*$ = the group of invertible elements in $H(\mathbb{Z}[\frac{1}{p}])$. Then Λ' is mapped onto a subgroup of $GL(2, k)$. The image of Λ' in $PGL(2, k)$ is denoted by Λ. Clearly $\Lambda \simeq \Lambda' /_{\{\pm p^n | \in \mathbb{Z}\}}$.

Let n be an integer, prime to p. Then $\Lambda'(n)$ denotes the kernel of the group-homomorphism $H(\mathbb{Z}[\frac{1}{p}])^* \to H(\mathbb{Z}/_n)^*$.

Clearly $\Lambda'(n)$ is a normal subgroup of Λ' of finite index.

Let $\Lambda(n)$ denote the image of $\Lambda'(n)$ in $\Lambda \subset PGL(2, k)$.

The groups $\Lambda(n)$, Λ are the groups of quaternions referred to in the title. Their interest for us lies in the following proposition.

(9.3) Proposition: Λ is discrete subgroup of PGL(2, k).

Proof: Let $\gamma_n = \begin{pmatrix} a_n+b_n i & c_n+d_n i \\ -c_n+d_n i & a_n-b_n i \end{pmatrix} \in \Lambda'$ be a convergent sequence in GL(2, k).

The sequences a_n, b_n, c_n, d_n converge to points in $\mathbb{Q}_p$. Hence for some $\alpha \geq 0$, all elements $x_n = p^{\alpha}(a_n + b_n e_1 + c_n e_2 + d_n e_3)$ belong to $H(\mathbb{Z})$.

Their norms form a convergent sequence of powers of p. Hence for $n >> 0$ the norms $N(x_n)$ are constant equal to some p^{β}. This contradicts the fact that $N(x) = p^{\beta}$ has only finitely many solutions in $H(\mathbb{Z})$.

Hence Λ is discrete and according to Ch.I (1.6.4), Λ is a discontinuous group.

(9.4) The case p = 2 turns out to be uninteresting. Namely:

Proposition:

(1) Every element of $H(\mathbb{Z}[\frac{1}{2}])^*$ can uniquely be written as $2^n\varepsilon(1 + e_i)^{\delta}$, where $n \in \mathbb{Z}$; $\varepsilon \in H(\mathbb{Z})^*$; $i = 1, 2, 3$; $\delta = 0$ or 1.

(2) For p = 2 the group Λ is finite.

Proof: (1) implies that Λ has 48 elements. So (2) follows from (1). The uniqueness in (1) is easily verified. Take now $x \in H(\mathbb{Z}[\frac{1}{2}])^*$ then $2^k x = y \in H(\mathbb{Z})$ for some $k \geq 0$. The norm of y is a power of 2. It follows that y is a product of prime elements $a \in H(\mathbb{Z})$ with $N(a) = 2$ and units of $H(\mathbb{Z})$.

A small calculation gives all the possibilities for $a \in H(\mathbb{Z})$, $N(a) = 2$: $\varepsilon(1 + e_i)$ with $\varepsilon \in H(\mathbb{Z})^*$ and $i = 1, 2, 3$.

Any product a_1a_2 of a_1, a_2 with $N(a_i) = 2$ turns out to be 2ε for some $\varepsilon \in H(\mathbb{Z})^*$. From this observation the existence of the expression in (1) follows.

(9.5) From now on we suppose that the prime number p is different from 2. As before $\Lambda'(2)$ denotes the kernel of $\Lambda' = H(\mathbb{Z}[\frac{1}{p}])^* \to H(\mathbb{Z}/_2)^*$ and $\Lambda(2) = \Lambda'(2)/_{\{\pm p^n | n \in \mathbb{Z}\}}$. The groups $\Lambda'(2)$ and $\Lambda(2)$ turn out to be easier to handle that Λ' and Λ.

Put $A = \{z \in H(\mathbb{Z}) | N(z) = p$ and p has image 1 in $H(\mathbb{Z}/_2)\} =$
$= \{z \in H(\mathbb{Z}) \cap \Lambda'(2) | N(z) = p\}$.

One easily verifies the following:

$a_0 + a_1e_1 + a_2e_2 + a_3e_3$ belongs to A if and only if

(i) $a_0, a_1, a_2, a_3 \in \mathbb{Z}$.

(ii) $\Sigma a_i^2 = p$

(iii) $a_0 + a_i \equiv 1 \bmod (2)$ for $i = 1, 2, 3$.

For every $(a_0, a_1, a_2, a_3) \in \mathbb{Z}^4$ with $a_i^2 = p$ there is precisely one element $\varepsilon \in \{1, e_1, e_2, e_3\}$ such that $\varepsilon(a_0 + a_1e_1 + a_2e_2 + a_3e_3)$ satisfies also condition (iii).

Hence A consists of $\frac{1}{4}r_4(p) = 2(1 + p)$ elements.

We divide A into two disjoint sets A_o and A_1.

A_o = the elements of A with $a_0 \neq 0$. Let $y_1, \ldots, y_s$ be the elements of A_o with $a_0 > 0$ and the first $a_i \neq 0$ in any y_i is also positive. Then

$$A_o = \{\pm y_1, \ldots, \pm y_s, \pm \bar{y}_1, \ldots, \pm \bar{y}_s\}.$$

Let A_1 denote the elements of A with $a_0 = 0$. Let $z_1, \ldots, z_t$ be the elements of A_1 such that the first $a_i \neq 0$ in any z_j is positive. Then clearly

$$A_1 = \{z_1, \ldots, z_t, \bar{z}_1, \ldots, \bar{z}_t\}$$

and we note that $\bar{z}_i = - z_i$.

The numbers s and t satisfies: $2s + t = p + 1$ and t is equal to $\frac{1}{2} \# \{(a_1, a_2, a_3) \in \mathbb{Z}^3 | \Sigma a_i^2 = p$ and $a_i \equiv 1 \bmod (2)$ for $i = 1, 2, 3\}$.

One easily calculates the following:

$$t = 0 \qquad \text{if } p \equiv 1 \bmod (4).$$

$$t = \frac{1}{2} r_3(p) \qquad \text{if } p \equiv 3 \bmod (4).$$

A reduced word in $\{y_1, \ldots, y_s, \bar{y}_1, \ldots, \bar{y}_s, z_1, \ldots, z_t\}$ is a product of elements in this set in which no succession of y_i, $\bar{y}_i$ or z_i, z_i occurs.

(9.6) Proposition: Every element in $H(\mathbb{Z}[\frac{1}{p}])^*$ can uniquely be written as $p^n \cdot \varepsilon \cdot$ (reduced word) with $n \in \mathbb{Z}$; $\varepsilon \in H(\mathbb{Z})^*$.

Proof: Let $x \in H(\mathbb{Z}[\frac{1}{p}])^*$. Then for suitable $k \geq 0$ the element $y = p^k x$ belongs to $H(\mathbb{Z})$ and its norm is a power of p. Hence y is a product of prime elements a with $N(a) = p$ and units. The prime elements can be taken in A. So $x = p^n \varepsilon$ (word in $\{y_1, \ldots, y_s, \bar{y}_1, \ldots, \bar{y}_s, z_1, \ldots, z_t\}$). The word can in finitely many steps be transformed into a reduced word. So x has the required expression.

It suffices to show the uniqueness for $x \in H(\mathbb{Z}[\frac{1}{p}])^* \cap H(\mathbb{Z}) = \{y \in H(\mathbb{Z}) \mid N(y) \text{ is a power of } p\}$.

Let $B = \{x \in H(\mathbb{Z}) \mid N(x) = p^k\}$. Then $\# B = 24(1 + p + \ldots + p^k)$ according to (9.1). We also can calculate the number of expressions $p^n \cdot \varepsilon$ (reduced word) belonging to B. The number of reduced words of length u is equal to

$$= \begin{cases} 1 & \text{if } u = 0. \\ 2s + t & \text{if } u = 1. \\ (2s + t)(2s + t - 1)^{u-1} & \text{if } u > 1. \end{cases}$$

For k even we have the expressions $p^n \varepsilon w$ with $n = 0, 1, \ldots, k/2$ and w a reduced word of length $k - 2n$. The total number of expressions is therefore:

$$24(1 + (p + 1)(p + p^2 + \ldots + p^{k-1})) = \# B.$$

For odd k the total number of expressions is

$$24\left((p+1)\left(\sum_{n=o}^{\frac{k-1}{2}} p^{k-1-2n}\right)\right) = \# B.$$

This proves the uniqueness of the expression in (9.6).

(9.7) <u>The groupstructure of Λ, $\Lambda(2)$ and $\Lambda^*(2)$</u>.

(9.7.1) Λ is a semi-direct product of $\Lambda(2)$ and $H(\mathbb{Z}/_2)^* \simeq A_4$.

<u>Proof:</u> By definition we have an exact sequence of goups $1 \to \Lambda(2) \to \Lambda \to H(\mathbb{Z}/_2)^* \to 1$. The group $H(\mathbb{Z}/_2)^*$ is isomorphic to $H(\mathbb{Z})^*/_{\{\pm 1\}}$ and isomorphic to A_4. The map $H(\mathbb{Z})^* \to \Lambda' \to \Lambda$ gives a splitting of the exact sequence and Λ is therefore a semi-direct product.

For $x \in \Lambda'$ we denote by $\hat{x}$ its image in Λ. With this notation:
(9.7.2) $\Lambda(2)$ is generated by $\{\hat{y}_1, \ldots, \hat{y}_s, \hat{z}_1, \ldots, \hat{z}_t\}$. The only relations are $\hat{z}_i^2 = 1$ $(i = 1, \ldots, t)$.

(9.7.3) If $t = 0$ then $\Lambda(2)$ is a free group on $\frac{p+1}{2}$ generators.

(9.7.4) If $t \neq 0$ then we consider the group homomorphism $\phi : \Lambda(2) \to \{\pm 1\}$ given by $\phi(\hat{y}_i) = 1$ and $\phi(\hat{z}_i) = -1$ (all i, j). The kernel of ϕ is denoted by $\Lambda^*(2)$; it is a free group of rank p; free generators for $\Lambda^*(2)$ are

$$\{\hat{y}_1, \ldots, \hat{y}_s, \hat{z}_1\hat{y}_1\hat{z}_1, \ldots, \hat{z}_1\hat{y}_s\hat{z}_1, \hat{z}_1\hat{z}_2, \ldots, \hat{z}_1\hat{z}_t\}.$$

All the statements above follow without much difficulty from (9.6). We have found a subgroup of finite index in Λ which is a Schottky group (Compare Ch. I (3.1)).

The ranks of the various groups coincide with the genus of the Mumford--curve parametrized by the group. (Rank means: the rank of the abelianized group). The rank of $\Lambda(2)$ is s and the rank of $\Lambda^*(2)$ is p.

The rank of Λ is much harder to calculate, as we will see presently.

(9.8) Theorem: The rank of Λ (i.e. the rank of the abelianized group Λ_{ab}) equals

$$\frac{p+1}{24} - \frac{r_3(p)}{48} + \frac{1}{4}\delta_4(p) + \frac{2}{3}\delta_3(p)$$

where

$$\delta_4(p) = \begin{cases} 1 \text{ if } p \equiv 1 \bmod (4) \\ 0 \text{ if } p \equiv 3 \bmod (4) \end{cases} \quad \text{and} \quad \delta_3(p) = \begin{cases} 1 & \text{if } p \equiv 1 \bmod (3). \\ 0 & \text{if } p \equiv 0, 2 \bmod (3). \end{cases}$$

Proof: The rank of Λ_{ab} is calculated using the fact that Λ is a semi-direct product of $\Lambda(2)$ and $H(\mathbb{Z}/_2)^*$. Let $\Lambda(2) \to \Lambda(2)_{ab} = \mathbb{Z}^s \oplus (\mathbb{Z}/_2)^t$ be denoted by $z \to \tilde{z}$. Define the representation Θ_o of $H(\mathbb{Z}/_2)^*$ in $\Lambda(2)_{ab}$ by $\Theta_o(\varepsilon)\tilde{z} = \widetilde{\varepsilon z \varepsilon^{-1}}$. Let Θ be the representation $\Theta_o \otimes 1_{\mathbb{C}}$ of $H(\mathbb{Z}/_2)^*$ on $\Lambda(2)_{ab} \otimes \mathbb{C} = \mathbb{C}^s$.

(9.8.1) Lemma: $\Lambda_{ab} \simeq \Lambda(2)_{ab} \oplus H(\mathbb{Z}/_2)^*_{ab}/B$ where B is the subgroup of $\Lambda(2)_{ab}$ generated by

$$\{\Theta_o(\varepsilon)\tilde{z} - \tilde{z} \mid \varepsilon \in H(\mathbb{Z}/_2)^* \text{ and } \tilde{z} \in \Lambda(2)_{ab}\}.$$

Proof: There is an obvious surjective map $\eta : \Lambda(2)_{ab} \oplus H(\mathbb{Z}/_2)^*_{ab} \to \Lambda_{ab}$. The elements $\Theta_o(\varepsilon)\tilde{z}$ and $\tilde{z}$ have the same image in Λ_{ab}. So $\ker \eta \supseteq B$. The group Λ is generated by $\hat{y}_1, \ldots, \hat{y}_s, \hat{z}_1, \ldots, \hat{z}_t$, and $\varepsilon \in H(\mathbb{Z}/_2)^*$. The relations are $\hat{z}_j^2 = 1$; $\varepsilon\hat{y}_j\varepsilon^{-1} = a_j$; $\varepsilon\hat{z}_j\varepsilon^{-1} = b_j$ where $a_j, b_j \in \Lambda(2)$ and are in $\Lambda(2)_{ab}$ equal to $\Theta_o(\varepsilon)(\tilde{\hat{y}}_j)$ and $\Theta_o(\varepsilon)(\tilde{\hat{z}}_j)$. It follows that $B = \ker \eta$.

(9.8.2) Lemma: rank Λ_{ab} = the number of trivial representations in Θ.

Proof: rank $\Lambda_{ab} = \dim_{\mathbb{C}} \Lambda_{ab} \otimes \mathbb{C} \dim \mathbb{C}^s/_{B \otimes \mathbb{C}}$, where $B \otimes \mathbb{C}$ is the vector-space spanned by

$$\{\Theta(\varepsilon) - z \mid z \in \mathbb{C}^s; \quad \varepsilon \in H(\mathbb{Z}/_2)^*\}.$$

For a representation ψ of $H(\mathbb{F}_2)^*$ in some complex vectorspace V, the linear space W generated by $\{\psi(\varepsilon)z - z | \varepsilon \in H(\mathbb{F}_2)^*;\ z \in V\}$ is invariant under ψ. If V is irreducible then V = W if $\psi \neq 1$ and W = 0 if $\psi = 1$. From this the lemma follows.

Continuation of the proof of (9.8). The basis $a_i = \hat{\tilde{y}}_i (i = 1,\dots, s)$ of $\Lambda(2)_{ab} \otimes \mathbb{C}$ has the property $\Theta(a_i) \in \{a_1,\dots, a_s, -a_1,\dots, -a_s\}$ for i = 1,... s. This follows from the following table for $\Theta_o(\varepsilon)(q_o + q_1e_1 + q_2e_2 + q_3e_3)$.

ε	$\cdot 1$	$\cdot e_1$	$\cdot e_2$	$\cdot e_3$
1	q_o	q_1	q_2	q_3
ρ	q_o	q_3	q_1	q_2
ρ^2	q_o	q_2	q_3	q_1
e_1	q_o	q_1	$-q_2$	$-q_3$
e_2	q_o	$-q_1$	q_2	$-q_3$
e_3	q_o	$-q_1$	$-q_2$	q_3
$e_1\rho$	q_o	q_3	$-q_1$	$-q_2$
$e_2\rho$	q_o	$-q_3$	q_1	$-q_2$
$e_3\rho$	q_o	$-q_3$	$-q_1$	q_2
$e_1\rho^2$	q_o	q_2	$-q_3$	$-q_1$
$e_2\rho^2$	q_o	$-q_2$	q_3	$-q_1$
$e_3\rho^2$	q_o	$-q_2$	$-q_3$	q_1

The number of trivial representations in Θ is equal to $(\chi, 1)$ where χ is the character of Θ.

Further $12\,(\chi, 1) =$

$$\sum_{\varepsilon\in H(\mathbb{F}_2)^*} \# \{a_i \mid \Theta(\varepsilon)a_i = a_i\} - \#\{a_i \mid \Theta(\varepsilon)a_i = -a_i\} =$$

$$= \sum_{a\in\{a_1,\dots,a_s\}} \# \{\varepsilon \mid \Theta(\varepsilon)a = a\} - \# \{\varepsilon \mid \Theta(\varepsilon)a = -a\} =$$

$= \Sigma \# \{\varepsilon \mid \Theta(\varepsilon)a = a\}$ where the last sum is taken over the elements $a \in \{a_1,\dots, a_s\}$ such that $\Theta(\varepsilon)\, a \neq -a$ for all ε.

Let $a \in \{a_1,\dots, a_s\}$ correspond to $q_o + q_1e_1 + q_2e_2 + q_3e_3$. Then $-a$ corresponds to $q_o - q_1e_1 - q_2e_2 - q_3e_3$. The condition $\Theta(\varepsilon)a \neq -a$ for all ε is seen to be equivalent with all $q_i \neq 0$. Inspection of the table gives the following:

For a with all $q_i \neq 0$ one has:

$$\# \{\varepsilon \mid \Theta(\varepsilon)a = a\} = \begin{cases} 3 & \text{if } q_1 = |q_2| = |q_3| \\ 1 & \text{otherwise.} \end{cases}$$

Hence

$$\begin{aligned} 12(\chi, 1) &= \# \{a \in \{a_1,\dots, a_s\} \mid \text{all } q_i \neq 0\} + \\ &\quad + 2\,\#\{a \in \{a_1,\dots, a_s\} \mid q_1 = |q_2| = |q_3|\} = \\ &= \frac{1}{16}\, \# \{(x_0, x_1, x_2, x_3) \in \mathbb{Z}^4 \mid \Sigma x_i^2 = p \text{ and all } x_i \neq 0\} + \\ &\quad + 8\, \# \{(q_o, q_1) \mid q_o, q_1 > 0 \text{ and } q_o^2 + 3q_1^2 = p\} \\ &= \frac{1}{16}\{r_4(p) - 4r_3(p) + 6r_2(p)\} + 8\delta_3(p). \end{aligned}$$

The last equality follows from: $\#\{(q_o, q_1) \mid q_o, q_1 > 0$ and $q_o^2 + 3q_1^2 = p\}$ is equal to 1 if p decomposes in $\mathbb{Z}[\frac{1}{2} + \frac{1}{2}\sqrt{-3}]$ as a product of 2 primes; and is equal to 0 if p is ramified or prime in $\mathbb{Z}[\frac{1}{2} + \frac{1}{2}\sqrt{-3}]$.

Further p decomposes if and only if -3 is a square in $\mathbb{F}_p$. The reciprocity law yields $(\frac{-3}{p}) = (\frac{p}{3}) = \delta_3(p)$.

The formula $\# \{(x_0, x_1, x_2, x_3) \in \mathbb{Z}^4 \mid \Sigma x_i^2 = p$ and all $x_i = 0\} =$ $r_4(p) - 4r_3(p) + 6r_2(p)$ is derived as follows:

Let A be the $(x_0, x_1, x_2, x_3) \in \mathbb{Z}^4$ with $\Sigma\, x_i^2 = p$.

Let B be the $(y_1, y_2, y_3) \in \mathbb{Z}^3$ with $\Sigma\, y_i^2 = p$.

Let C be the $(z_1, z_2) \in \mathbb{Z}^2$ with $\Sigma\, z_i^2 = p$.

Every $c \in C$ gives rise to 6 different elements of A and 3 different elements of B. Every $b \in B$, with all coordinates $\neq 0$, gives rise to 4 different elements of A. The number of elements in A with all coordinates $\neq 0$ is equal to

$$r_4(p) - 6r_2(p) - 4(r_3(p) - 3r_2(p)) = r_4(p) - 4r_3(p) + 6r_2(p).$$

Further we note that $r_2(p) = 8$ if $p \equiv 1 \bmod (4)$ and $r_2(p) = 0$ if $p \equiv 3 \bmod (4)$. So $r_2(p) = 8\ {}_4(p)$. Now the formula (9.8) follows.

(9.9) For the primes $p \leq 100$ we give a table for the numerical data.

primes	rank Λ_{ab}	rank $\Lambda(2)_{ab} = s$	$r_3(p)$
2	0	0	-
3	0	0	8
5	0	3	24
7	1	4	0
11	0	0	24
13	1	7	24
17	0	9	48
19	1	4	24
23	1	12	0
29	0	15	72
31	2	16	0
37	2	19	24
41	0	21	96
43	2	16	24
47	2	24	0
53	2	27	24
59	1	12	72
61	3	31	24
67	3	28	24
71	3	36	0
73	3	37	48
79	4	40	0
83	2	24	72
89	4	45	144
97	4	49	48

(9.10) The geometry of Mumford curves parametrized by $\Lambda(2)$ for primes $p \equiv 1 \bmod (4)$.

Λ and $\Lambda(2)$ are discontinuous subgroup of $PGL(2, \mathbb{Q}_p)$ since $\mathbb{Q}_p$ contains a square root of -1 (denoted by i as usual). The compact set $\mathbb{P}(\mathbb{Q}_p) \subset \mathbb{P}(K)$ is invariant under the action of Λ and so $\mathcal{L}$ = the set of limit points of Λ (or $\Lambda(2)$) is contained in $\mathbb{P}(\mathbb{Q}_p)$.

In fact we will show that $\mathcal{L} = \mathbb{P}(\mathbb{Q}_p)$. A direct calculation of the limit points or of the hyperbolic points of Λ seems to be extremely complicated. Instead we will calculate a fundamental domain for the Schottky group $\Lambda(2)$.

The $s = \frac{p+1}{2}$ generators of $\Lambda(2)$ have the following form

$\gamma_1 = \begin{pmatrix} a+ib & 0 \\ 0 & a-ib \end{pmatrix}$ where $a, b \in \mathbb{N}$ satisfies $a^2 + b^2 = p$ and a is odd.

$\gamma_2 = \begin{pmatrix} a & b \\ -b & a \end{pmatrix}$

$\gamma_3 = \begin{pmatrix} a & ib \\ ib & a \end{pmatrix}$

for $j = 4, \ldots, s$ $\quad \gamma_j = \begin{pmatrix} a_0+a_1 i & a_2+a_3 i \\ -a_2+a_3 i & a_0-a_1 i \end{pmatrix}$ where $a_0, a_1 \in \mathbb{N}$; $a_2, a_3 \in \mathbb{Z}$

satisfy

(i) $\Sigma a_i^2 = p$

(ii) $a_2 \neq 0$ or $a_3 \neq 0$

(iii) $a_0 + a_i$ is odd for $i = 1, 2, 3$.

The fixed points of those hyperbolic transformations are:
0 and ∞ for γ_1; i and $-i$ for γ_2; 1 and -1 for γ_3; for $\gamma_j (j \geq 4)$ the fixed points z_j^+ and z_j^- lie in $\mathbb{Q}_p$ and have absolute value 1. Using that the group $\Lambda(2)$ has no elements of finite order, one can prove that the residues of $0, \pm 1, \pm i, z_j^+, z_j^-$ in $\mathbb{F}_p$ are all different.

For the cyclic groups $\{\gamma_j^n | n \in \mathbb{Z}\}$ a fundamental domain F_j is given by:

$j = 1$ then $F_1 = \{z \in K | \frac{1}{\sqrt{p}} \leq |z| \leq \sqrt{p}\}$

$j = 2$ then $F_2 = \{z \in \mathbb{P}(K) | |z - i| \geq \frac{1}{\sqrt{p}}$ and $|z + i| \geq \frac{1}{\sqrt{p}}\}$

$j = 3$ then $F_3 = \{z \in \mathbb{P}(K) | |z - 1| \geq \frac{1}{\sqrt{p}}$ and $|z + 1| \geq \frac{1}{\sqrt{p}}\}$

$j \geq 4$ then $F_j = \{z \in \mathbb{P}(K) | |z - z_j^+| \geq \frac{1}{\sqrt{p}}$ and $|z - z_j^-| \geq \frac{1}{\sqrt{p}}\}$.

The fundamental domains $F_j (j = 1,..., s)$ are in good position, and it follows that

$$F = \cap F_j = \mathbb{P}(K) - [\{z | |z| > \sqrt{p}\} \cup \{z | |z - a| < \frac{1}{\sqrt{p}}\} \ (a = 0, 1,..., p-1)]$$

is a fundamental domain F, and its translates $\gamma F(\gamma \in \Lambda(2))$, contains no elements of $\mathbb{P}(\mathbb{Q}_p)$. Hence $\mathcal{L} \supset \mathbb{P}(\mathbb{Q}_p)$ and we find $\mathcal{L} = \mathbb{P}(\mathbb{Q}_p)$.

The Mumford curve $X(\Lambda(2) = \mathbb{P}(\mathbb{Q}_p)/_{\Lambda(2)}$ has genus $\frac{p+1}{2}$ and can be obtained from the fundamental domain F by identifying the (p + 1) boundaries pairwise.

The reduction of the affinoid domain F with respect to the affinoid covering:

$\{F_a | a \in \{0, 1,..., p - 1\}\}$ and F_∞, given by

F_∞ : $z \in K$ with $1 \leq |z| \leq p^{1/2}$ and $|z - b| > 1$ for $b = 0. 1,..., p - 1$

F_a : $z \in K$ with $p^{-1/2} \leq |z - a| \leq 1$ and $|z - b| \geq 1$

for $b \in \{0, 1,..., p - 1\}$ and $b \neq a$.

has the form Z:

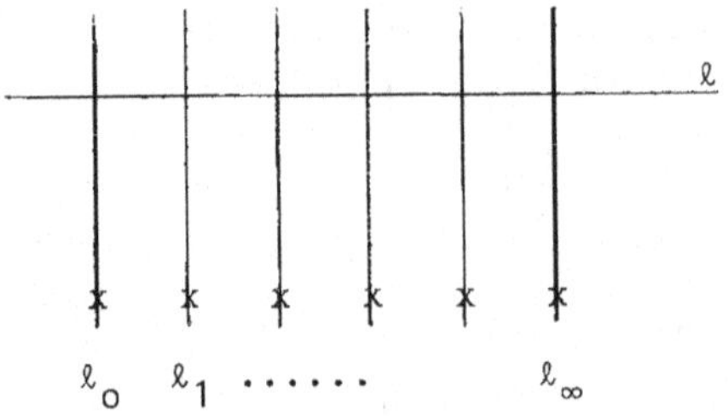

(drawing for p = 5)

In this picture $\ell = \mathbb{P}(\bar{K})$ ($\bar{K}$ is the residue field of K) corresponds to the standard reduction of $\mathbb{P}(K)$. The lines $\ell_0, \ell_1, \ldots, \ell_\infty$ are affine lines intersecting ℓ at the image of 0, 1,..., p - 1, ∞. Further $\ell_0 - \{\ell_1 \cap \ell_0\}$ is the image of the points $z \in F$ satisfies $p^{-1/2} \leq |z| < 1$ etc. The images of the pieces $F_0, F_1, \ldots, F_\infty$ in $X(\Lambda(2))$ form an affinoid covering of the curve $X(\Lambda(2))$. The reduction of $X(\Lambda(2))$ with respect to this covering is obtained from Z by identifying the lines $\ell_0, \ell_1, \ldots, \ell_\infty$ pairwise.

The result is:

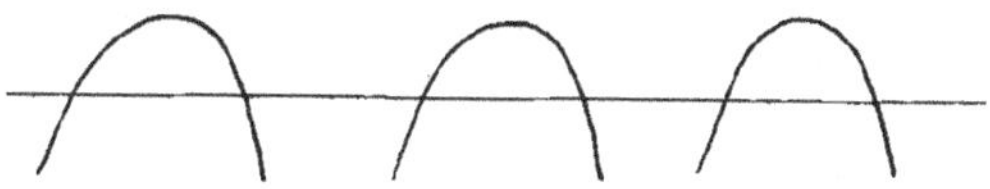

(drawing for p = 5).

This is a pre-stable reduction. The stable reduction is obtained by contracting the $\frac{p+1}{2}$ lines to points. The result is the stable reduction of $X(\Lambda(2))$ consisting of a projective line $\mathbb{P}(\bar{K})$ with $\frac{p+1}{2}$ double points obtained by pairwise identification of the points of $\mathbb{P}(\mathbb{F}_p)$.

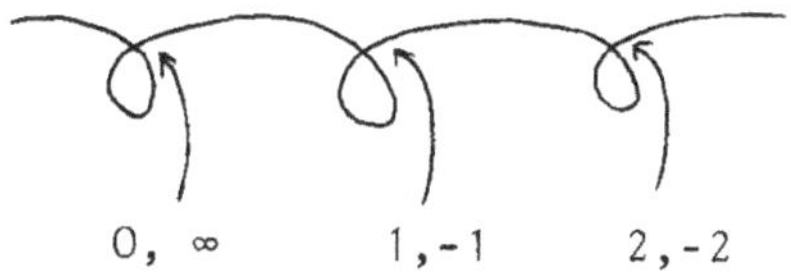

(drawing for p = 5).

Another way to find the stable reduction of $X(\Lambda(2))$ is the following. The canonical reduction $\bar{\Omega}$ of $\Omega = \mathbb{P}(K) - \mathbb{P}(\mathbb{Q}_p)$ is a tree of projective lines. Each line meets p + 1 other lines normally. The group $\Lambda(2)$ acts on $\bar{\Omega}$ and the quotient $\bar{\Omega}/_{\Lambda(2)}$ is the stable reduction of $X(\Lambda(2))$.
The action of $\Lambda(2)$ on $\bar{\Omega}$ can be described as follows: Let ℓ be the line in $\bar{\Omega}$ corresponding to the triple (0, 1, ∞). Then ℓ meets lines $\ell_0, \ell_1, \ldots, \ell_{p-1}, \ell_\infty$ at the points $\bar{0}, \bar{1}, \ldots, \overline{p-1}, \bar{\infty} \in \mathbb{P}(\mathbb{F}_p) \subset \mathbb{P}(\bar{K}) = \ell$.

Some calculation yields: $\{\gamma_1(\ell), \gamma_2(\ell), \dots, \gamma_s(\ell), \gamma_1^{-1}(\ell), \dots, \gamma_s^{-1}(\ell)\} = \{\ell_0, \ell_1, \dots, \ell_{p-1}, \ell_\infty\}$. Hence $\Lambda(2)$ acts transitively on the lines of $\bar{\Omega}$. As a consequence $\ell \to \Omega/_{\Lambda(2)}$ is a surjective map.

Moreover, if $\gamma_j(\ell) = \ell_a$ und $\gamma_j^{-1}(\ell) = \ell_{\eta(a)}$ then the two points $\ell \cap \ell_a = \{\bar{a}\}$ and $\ell \cap \ell_{\eta(a)} = \overline{\eta(a)}$ of are equivalent under the action of $\Lambda(2)$.

Let $\sim$ denote the equivalence relation on $\mathbb{P}(\bar{K}) = \ell$ (or $\mathbb{P}(\mathbb{F}_p)$), generated by $a \sim \eta(a)$. Then $\bar{\Omega}/_{\Lambda(2)} \simeq \mathbb{P}(\bar{K})/_\sim$ as before.

We note that an explicite calculation of η seems to be difficult. From the explicite form of the generators $\gamma_1, \dots, \gamma_s$ of $\Lambda(2)$ one deduces:

$\eta(0) = \infty$; $\eta(i) = -i$; $\eta(1) = -1$ and

$$\eta\left(\frac{a_0 + a_1 i}{-a_2 + a_3 i}\right) = \frac{a_0 - a_1 i}{a_2 - a_3 i} \quad \text{where } a_0 > 0;\ a_1 > 1;\ a_2 \neq 0 \text{ or } a_3 \neq 0;$$
$$a_0 + a_i \text{ odd for } i = 1, 2, 3, \text{ and } \Sigma a_i^2 = p.$$

(9.11) <u>The geometry of the Mumford curves parametrized by Λ for primes $p \equiv 1 \bmod 4$.</u>

The Hurwitz-Riemann formula for the map $X(\Lambda(2)) \to X(\Lambda)$ is $2(\frac{p+1}{2}) - 2 = 12(2g - 2) + v$ where g is the genus of $X(\Lambda)$ and $v = \Sigma(e_z - 1)$. We will try to calculate v explicitely.

First of all,

$$v = \sum_{\varepsilon \in H(\mathbb{F}_2)^*, \varepsilon \neq 1} \# \{z \in X(\Lambda(2)) \mid \varepsilon(z) = z\}.$$

Let $a = 2$, or 3 and $v_a = \sum_{\varepsilon \in H(\mathbb{F}_2)^*,\ \varepsilon \text{ has order } a} \# \{z \in X(\Lambda(2)) \mid \varepsilon(z) = z\}.$

The $v = v_2 + v_3$.

Let F denote the fundamental domain for $\Lambda(2)$ explained in (9.10).

We note that F is invariant under $H(\mathbb{Z})^*/_{\{\pm 1\}} = H(\mathbb{F}_2)^* = \Lambda/_{\Lambda(2)}$.

A fixed point z for some $\varepsilon \neq 1$ corresponds to a point $x \in F$ satisfying $\varepsilon(x) = x$ or $\varepsilon(x) = \gamma_j(x)$ some $j \in \{1,\ldots, s\}$.

If ε has order 3 then one sees that $\varepsilon(x) = x$ must hold.

If ε has order 2 (so $\varepsilon = e_1, e_2$ or e_3) then the fixed points of ε lie in $\mathbb{Q}_p$ (outside F) and so necessarily $\varepsilon(x) = \gamma_j(x)$ for some $j \in \{1,\ldots, s\}$.

Calculation of v_3.

$\rho \sim \begin{pmatrix} i & i \\ -1 & 1 \end{pmatrix}$ and ρ^2 have the fixed points $\frac{1-i}{2} \pm \frac{1}{2}\sqrt{-6i}$.

$(e_1\rho) \sim \begin{pmatrix} i & i \\ 1 & -1 \end{pmatrix}$ and $(e_1\rho)^2$ have the fixed points $\frac{1+i}{2} \pm \frac{1}{2}\sqrt{6i}$.

$(e_2\rho) \sim \begin{pmatrix} -i & i \\ 1 & 1 \end{pmatrix}$ and $(e_2\rho)^2$ have the fixed points $-\frac{1+i}{2} \pm \frac{1}{2}\sqrt{6i}$.

$(e_3\rho) \sim \begin{pmatrix} i & -i \\ 1 & 1 \end{pmatrix}$ and $(e_3\rho)^2$ have the fixed points $\frac{1+i}{2} \pm \frac{1}{2}\sqrt{-6i}$.

Then $\sqrt{i}$ belongs to $\mathbb{Q}_p$ (or $\mathbb{F}_p$) if and only if $p \equiv 1 \bmod (8)$.

Also $\sqrt{2}$ belongs to $\mathbb{Q}_p$ if and only if $p \equiv 1 \bmod (8)$.

So $\sqrt{2i} \in \mathbb{Q}_p$.

Further $\sqrt{3}$ belongs to $\mathbb{Q}_p$ if and only if $(\frac{3}{p}) = (\frac{p}{3}) = 1$.

As before are put $\delta_3(p) = 1$ if $p \equiv 1 \bmod (3)$ and $\delta_3(p) = 0$ if $p \equiv 2 \bmod (3)$. Hence $v_3 = 16\,(1 - \delta_3(p))$.

Calculation of v_2.

We have to solve the equations $e_i(x) = \gamma_j(x)$ ($i = 1, 2, 3$ and $j = 1,\ldots, s$). For $e_1(x) = -x$ and $\gamma_j = \begin{pmatrix} A & B \\ C & D \end{pmatrix}$ the solutions x are $x = -\frac{A+D}{2C} \pm \frac{1}{C}\sqrt{(\frac{A+D}{2})^2 - BC}$. The value of $(\frac{A+D}{2})^2 - BC$ is equal to a^2 for γ_1; $a^2 + b^2 = p$ for γ_2; $a^2 + b^2 = p$ for γ_3; and for $\gamma_j (j \geq 4)$ it is $a_0^2 + a_2^2 + a_3^2 = p - a_1^2$.

Hence the number of roots x lying in F is

$4 + 2 \,\# \{j \geq 4 \mid \text{with } a_1 = 0\} = 4 + 2 \cdot \frac{1}{12}(r_3(p) - 24) = \frac{1}{6} r_3(p)$.

The calculations for e_2 and e_3 are similar. This leads to $v_2 = \frac{1}{2} r_3(p)$.

Substitution of the values of v_2 and v_3 yields:

genus of $X(\Lambda) = \frac{p+1}{24} - \frac{r_3(p)}{48} + \frac{1}{4} + \frac{2}{3}\delta_3(p)$ (just as in (9.8)).

The structure of the stable reduction of $X(\Lambda)$ can again be calculated as $\bar{\Omega}/_{\Lambda}$. If one divides first by the normal subgroup $\Lambda(2)$ then we find: The stable reduction of $X(\Lambda)$ is $\mathbb{P}(\bar{K})/\!\sim\!/H(\mathbb{F}_2)^*$. This is clearly a projective line over $\bar{K}$ with a number of ordinary double points.

The double points of $\mathbb{P}(\bar{K})/_\sim$ are $\{0, \infty\}$, $\{i, -i\}$, $\{1, -1\}$, and the points $\{a, \eta(a)\}$ with $a \in \mathbb{F}_p^*$, $a \neq \pm 1, \pm i$.

Under the action of $H(\mathbb{F}_2)^*$ the first 3 double points disappear. This one easily verifies by calculating the invariants of the complete local ring $K[\![s, t]\!]/(s\,t)$ under the action of the group $\{1, e_1, e_2, e_3\}$. If $p \equiv 1 \bmod (3)$ then the elements of order 3 in $H(\mathbb{F}_2)^*$ fix some $a \in \mathbb{F}_p^*$. If $p \equiv 2 \bmod (3)$ then the fixed points of the elements of order 3 are in $\mathbb{F}_{p^2} - \mathbb{F}_p$. It implies that the number of double points is equal to the number of orbits of $H(\mathbb{F}_2^*)$ in $\mathbb{F}_p^* - \{\pm 1, \pm i\}$. A combinatorial calculation shows that this number is equal to the genus of $X(\Lambda)$ (as it should be!).

(9.12) <u>The case $p \equiv 3 \bmod (4)$.</u>

The group Λ is a discontinuous subgroup of $PGL(2, \mathbb{Q}_p(i))$, where i denotes, as usual, a square root of -1. The set $\mathcal{L}$ of limit points is contained in $\mathbb{P}(\mathbb{Q}_p(i))$. A direct calculation of $\mathcal{L}$ seems rather difficult. In stead we will construct a fundamental domain for $\Lambda^*(2)$.

(9.12.1) The set $B = \{x_0 + ix_1 \mid x_0, x_1 \in \mathbb{Q}_p \text{ and } x_0^2 + x_1^2 = -1\}$ is compact and invariant under Λ. In particular B contains $\mathcal{L}$.

<u>Proof:</u> If $x_0, x_1 \in \mathbb{Q}_p$ satisfy $x_0^2 + x_1^2 = -1$, then $|x_0| > 1$ implies that $|x_0| = |x_1|$ and $(\frac{x_0}{x_1})^2 \equiv -1 \bmod (p)$. This contradicts however $p \equiv 3 \bmod (4)$. Hence B is a compact set.

Let σ denote the $\mathbb{Q}_p$-automorphism of $\mathbb{Q}_p(i)$ given by $\sigma(x_0 + ix_1) = x_0 - ix_1$. Then $B = \{x \in \mathbb{Q}_p(i) \mid \sigma(x) = \frac{-1}{x}\}$. For $\lambda \in \Lambda$ and $x \in \mathbb{Q}_p(i)$ one easily verifies $\sigma\lambda\sigma(x) = e_2\lambda e_2(x)$.

Let now $x \in B$. Then $\sigma\lambda(x) = \sigma\lambda\sigma(\frac{-1}{x}) = e_2\lambda e_2(\frac{-1}{x}) = e_2\lambda(x) = \frac{-1}{\lambda(x)}$. So also $\lambda(x) \in B$.

(9.12.2) We use the notation of (9.7); $\hat{y}_1,\ldots,\hat{y}_s, \hat{z}_1,\ldots,\hat{z}_t$ are generators for $\Lambda(2)$ and $\{\hat{y}_1,\ldots,\hat{y}_s, \hat{z}_1\hat{y}_1\hat{z}_1,\ldots,\hat{z}_1\hat{y}_s\hat{z}_1, \hat{z}_1\hat{z}_2,\ldots,\hat{z}_1\hat{z}_t\}$ is a free base for $\Lambda^*(2)$. Now the new statement:

The p + 1 points $\{y_1(\infty), y_1^{-1}(\infty),\ldots, y_s(\infty), y_s^{-1}(\infty), z_1(\infty),\ldots, z_t(\infty)\}$ have absolute value 1 and and the distance of any two of them is 1. Their residues in $\mathbb{F}_{p^2}$ are the (p + 1) solutions of the equation $a^{p+1} = -1$.

<u>Proof:</u> Put $\lambda = A_o + A_1e_1 + A_2e_2 + A_3e_3 \in \Lambda$ and let a_i denote the residue in $\mathbb{F}_p$ of A_i. The residue $\overline{\lambda(\infty)}$ in $\mathbb{F}_{p^2}$ of $\lambda(\infty)$ has the form

$\overline{\lambda(\infty)} = \frac{a_0 + ia_1}{-a_2 + ia_3}$. The conjugate of $\overline{\lambda(\infty)}$ is $\frac{a_0 - ia_1}{-a_2 - ia_3} = \overline{\lambda(\infty)}^p$, and

$$\overline{\lambda(\infty)}\ \overline{\lambda(\infty)}^p = \frac{a_0 + ia_1}{-a_2 + ia_3} \cdot \frac{a_0 - ia_1}{-a_2 - ia_3} = \frac{a_0^2 + a_1^2}{a_2^2 + a_3^2} = -1.$$

Let $\mu = B_o + B_1e_1 + B_2e_2 + B_3e_3$ and let b_i be the residue of B_i in $\mathbb{F}_p$.

The proof of(9.12.2) will be complete if we can show the following. If $\lambda, \mu \in \{\hat{y}_1, \hat{y}_1^{-1},\ldots, \hat{y}_s, \hat{y}_s^{-1}, \hat{z}_1,\ldots, \hat{z}_t\}$ and $\overline{\lambda(\infty)} = \overline{\mu(\infty)}$ then $\lambda = \mu$.

Using $\frac{a_0 + ia_1}{-a_2 + ia_3} = \frac{b_o + ib_1}{-b_2 + ib_3}$ it follows that

$(A_o - A_1e_1 - A_2e_2 - A_3e_3)(B_o + B_1e_1 + B_2e_2 + B_3e_3) = p\varepsilon$ for some $\varepsilon \in H(\mathbb{Z})$.

Since $\Sigma A_i^2 = \Sigma B_i^2 = p$ it follows that $\varepsilon \in H(\mathbb{Z})^*$. So $\lambda^{-1}\mu = \hat{\varepsilon}$ where $\hat{\varepsilon}$ is the image of ε in $H(\mathbb{Z})^*/_{\{\pm 1\}}$. But since $\lambda, \mu \in \Lambda(2)$ it follows that $\hat{\varepsilon} = 1$ and $\lambda = \mu$.

(9.12.3) The position of the 2p points

$$y_i(\infty),\ y_i^{-1}(\infty),\ z_1y_iz_1(\infty),\ z_1y_i^{-1}z_1(\infty),\ z_1z_j(\infty),\ z_jz_1(\infty)$$

(with $i = 1,\dots, s$ and $j = 2,\dots, t$)

is given by the following reduction of $\mathbb{P}(K)$.

(drawing for $t > 0$).

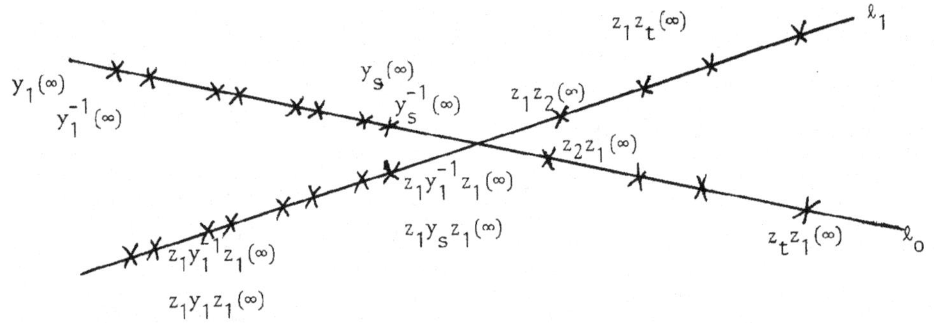

Proof: In the residue field $\mathbb{F}_{p^2}$ we have the following equalities: $y_iz_1(\infty) \equiv y_i(\infty)$; $y_i^{-1}z_1(\infty) \equiv y_i^{-1}(\infty)$ and $z_jz_1(\infty) \equiv z_j(\infty)$. This explains the line ℓ_o. An easy calculation shows that the z_1- images of the points $y_iz_1(\infty)$, $y_i^{-1}z_1(\infty)$, $z_j(\infty)$ (with $i = 1,\dots, s$ and $j = 2,\dots, t$) have all distances equal to p^{-1}. This explains the line ℓ_1.

(9.12.4) Let F denote the complement in $\mathbb{P}(K)$ of the 2p open disks:

$|x - y_i(\infty)| < p^{-1/2}$, $|x - y_i^{-1}(\infty)| < p^{-1/2}$ $\quad i = 1,\dots, s.$

$|x - z_1y_iz_1(\infty)| < p^{-3/2}$, $|x - z_1y_i^{-1}z_1(\infty)| < p^{-3/2}$ $\quad i = 1,\dots, s.$

$|x - z_jz_1(\infty)| < p^{-1/2}$, $|x - z_1z_j(\infty)| < p^{-3/2}$ $\quad j = 2,\dots, t.$

Proposition:

a) F <u>is a fundamental domain for</u> $\Lambda^*(2)$.

b) $\mathcal{L} = \{x_0 + ix_1 | x_0, x_1 \in \mathbb{Q}_p$ <u>and</u> $x_0^2 + x_1^2 = -1\}$.

c) <u>The stable reduction of the curve</u> $X(\Lambda^*(2))$ <u>is</u>

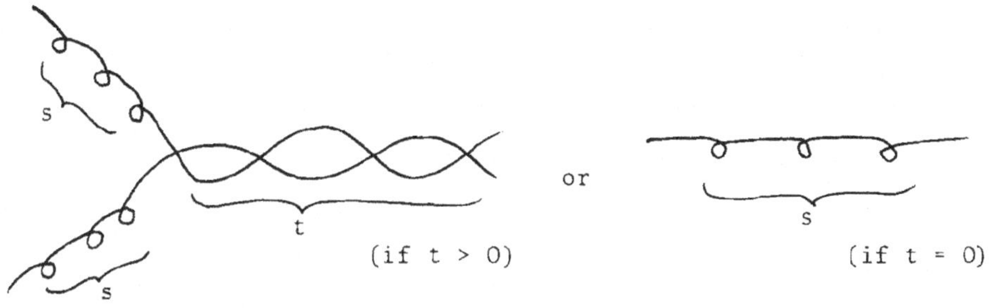

d) <u>The stable reductions of</u> $X(\Lambda(2))$ <u>and</u> $X(\Lambda)$ <u>are rational curves with ordinary double points.</u>

<u>Proof:</u> a) From the geometry of the centres, radii and the norms of the elements $y_i^{\pm 1}$; $z_1 y_i^{\pm 1} z_1$; $z_1 z_j$; $z_j z_1$ the statement follows.

b) The set B has the property $B \cap F = \emptyset$. Hence also for every $\lambda \in \Lambda^*(2)$ one has $B \cap \lambda F = \emptyset$. It follows that $B \subset \mathcal{L}$. Together with (9.12.1) one finds $B = \mathcal{L}$.

c) This follows from the reduction (9.12.3) and Chap. III.

d) Let $t > 0$, then the element z_1 permutes the two components of the stable reduction of $X(\Lambda^*(2))$. So the stable reduction of $X(\Lambda(2))$ is obtained by identifying those 2 components. If one divides the stable reduction of $X(\Lambda(2))$ by the action of $\Lambda/\Lambda(2) = H(\mathbb{F}_2)^*$ one obtains the stable reduction of $X(\Lambda)$. The statement d) follows.

(9.12.5) <u>Remark:</u> As we will see in the next section, $\Lambda^*(2)$ is a Whittaker group if $s = 0$ and parametrizes a hyperelliptic curve.

§2 Whittaker groups

(2.1) Definition of the Whittaker groups.

Given elements $s_o, \ldots, s_g$ in PGL(2, K) of order 2, such that the group Γ generated by them satisfies:

1) Γ is discontinuous

2) Γ is the free product of the groups $\langle s_i \rangle = \{1, s_i\}$.

The kernel W of the grouphomomorphins $\phi : \Gamma \to \{\pm 1\}$ given by $\phi(s_i) = -1$ for all i, is called a Whittaker group. One easily sees that W is a free group on the generators $s_1 s_o, \ldots, s_g s_o$. Since W is a subgroup of a discontinuous group, also W is discontinuous group. So W is a Schottky group of rank g. The groups W and Γ have the same set of ordinary points Γ, since W is of finite index in Γ.

(2.2) Proposition: $\Omega/\Gamma \simeq \mathbb{P}^1$ and Ω/W is a hyperelliptic curve.

Proof: Consider the function $\Theta(a, b; z) = \prod_{\gamma \in \Gamma} \frac{z - \gamma(a)}{z - \gamma(b)}$ where $a, b \in \Omega$ with $a \notin \Gamma b$ and $\infty \notin \Gamma a \cup \Gamma b$.

This product converges uniformly on every affinoid subset of Ω since $\lim |\gamma(a) - \gamma(b)| = 0$. So $\Theta(a, b; z)$ is a meromorphic function Ω. For any $\delta \in \Gamma$ we have $\Theta(a, b; \delta(z)) = c(\delta)\Theta(a, b; z)$ where $c(\delta) \in K^*$. Clearly $c : \Gamma \to k^*$ is a grouphomomorphism and hence $c(\delta) = \pm 1$.

For fixed δ and z the factor $c(\delta)$ depends in a continuous way on a and b. So when a and b are close together $c(\delta) = 1$. It follows that, for a good choice of a and b, the function $c(\delta) = 1$ for all $\delta \in \Gamma$. (Compare with Chap. II for stronger and more general statements).
So Θ is invariant under the group Γ and is a meromorphic function on Ω/Γ having only one pole. Hence Θ induces an isomorphism of Ω/Γ with $\mathbb{P}^1$.

The obvious map $\Omega/W \to \Omega/\Gamma$ has order 2. Hence Ω/W is a hyperelliptic curve.

(2.3) Corollary: If K has characteristic $\neq 2$ then $s_o, \ldots, s_g$ are elliptic elements of order 2. Their fixed points $\{a_o, b_o\}, \ldots, \{a_g, b_g\}$ belong to Ω. The affine equation of Ω/W is:

$$y^2 = \prod_{i=o}^{g} (x - \Theta(a, b; a_i))(x - \Theta(a, b; b_i)).$$

Proof: Let $p \in \Omega$ have as image a ramification point q in Ω/W. The canonical automorphism σ of Ω/W must have q as fixed point. Since Ω is the universal covering of Ω/W the map σ lifts to a $\tau : \Omega \tilde{\to} \Omega$ with $\tau(p) = p$. Clearly $\tau^2 = \text{id}$ and $\tau \in \Gamma$. So τ must be the conjugate of some s_i (i.e. $\tau = ws_iw^{-1}$ with $w \in W$ and $i \in \{0, \ldots, g\}$).
In particular $p \in Wa_o \cup Wb_o \cup \ldots \cup Wb_g$. Since there are exactly $2g + 2$ ramification points in Ω/W it follows that $a_o, b_o, \ldots, a_g, b_g \in \Omega$
Thes rest of the statement in (2.3) is evident.

(2.4) Remark: If K has characteristic 2 then $s_o, \ldots, s_g$ are parabolic elements (of order 2). It is more complicated to calculate in that case the equation definig Ω/W. Work on this is being done by G. van Steen (Univ. of Antwerpen).

(2.5) Remark: For a field K with characteristic $\neq 2$, any elliptic transformation of order 2 is determined by its two fixed points. So in fact the groups Γ and W depend on the $2g + 2$ points $\{a_o, b_o\}, \ldots, \{a_g, b_g\}$ in $\mathbb{P}$.

It seems rather difficult to find a necessary and sufficient condition for the position of those $2g + 2$ points such that the group Γ satisfies

1) Γ is discontinuous
2) Γ is the free product of the groups $\langle s_i \rangle$.

Under the "position" of the $2g + 2$ points we could mean the reduction of $\mathbb{P}$ with respect to this set and the images of the points. For $g = 1$ the situation is rather clear. There are three possible positions:

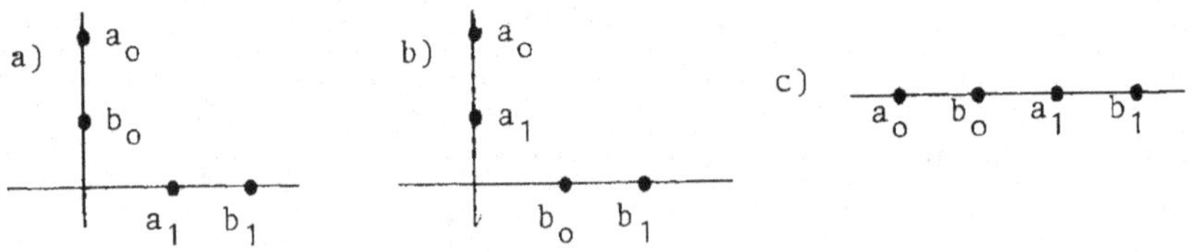

The condition on Γ is equivalent to $s_1 s_o$ is hyperbolic.
We assume further, to simplify, that the characteristic of $\bar{k}$ is $\neq 2$.

Claim: a) is the only good position.

Proof: In case c) we may suppose $a_0 = 0$, $a_1 = \lambda$, $b_o = 1$, $b_1 = \infty$ where $|\lambda| = 1$, $|\lambda - 1| = 1$. Then $s_o(z) = \frac{z}{2z - 1}$ and $s_1(z) = -z + 2\lambda$.

Then $s_1 s_o$ has the matrix $\begin{pmatrix} 1 & -2\lambda \\ +2 & 1-4\lambda \end{pmatrix}$. But this matrix is not hyperbolic.

A similar calculation shows that b) is not a good position and that a) is a good position.

This calculation for $g = 1$ implies a necessary condition for a good position, namely:

(2.5.1) If $\{a_0, b_o\}, \ldots, \{a_g, b_g\}$ are in good position (i.e. char $\bar{k} \neq 2$ and good position means that Γ has the properties 1) and 2)) then the reduction $R : \mathbb{P} \to Z$ with respect to the set $\{a_0, \ldots, b_g\}$ has the property that every line of Z "separates" at most one pair $\{a_i, b_i\}$.

Proof: The word "separates" means that in the corresponding reduction $\mathbb{P} \to \mathbb{P}^1(\bar{K})$ the images of a_i and b_i are different. The statement follows at once from the case $g = 1$ applied to all possibilities $\{a_i, b_i\}$, $\{a_j, b_j\}$.

As we will see later on, a sufficient condition for good position is the following.

(2.5.2) If the reduction $R : \mathbb{P} \to Z$ of $\mathbb{P}$ with respect to the set $\{a_0, \ldots, b_g\}$ has the properties:

1) for each i there is a line L_i in Z containing $R(a_i)$, $R(b_i)$;

2) $L_i \neq L_j$ if $i \neq j$.

3) L_i meets the other components of Z in one point,

then the points are in good position.

(2.5.3) Example: $g = 2$.

The positions for $\{a_0, b_o\}$, $\{a_1, b_1\}$, $\{a_2, b_2\}$ satisfying (2.5.1) are

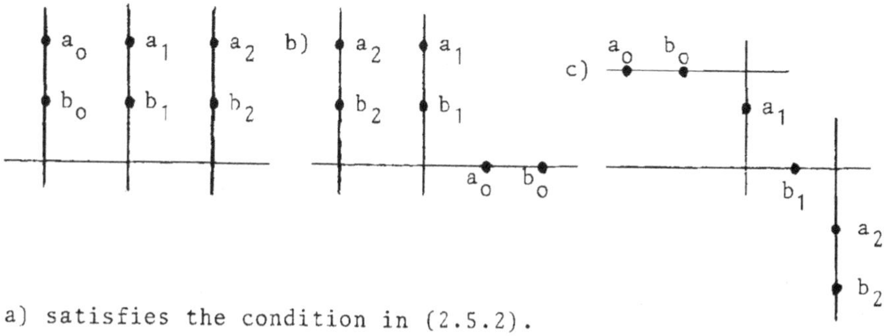

Only a) satisfies the condition in (2.5.2).

In case a) one can calculate that the stable reduction of Ω/W has the form

In case b), one can find examples where the group Γ does not have the right properties. In other cases one finds the stable reduction of Ω/W to be

In case c), one finds examples where the group Γ does not have the right properties. In other cases one finds the stable reduction of Ω/W to be

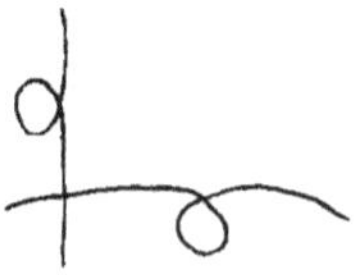

Comparing this with our calculations at the end of Chap V we come to the following Conjecture: The position of the 2g + 2 fixed points of $\{s_o, \ldots, s_g\}$ and the position of the 2g + 2 ramification points of the hyperelliptic curve Ω/W are identical.

(2.6) We want to show the converse of (2.2), that is, we want to show that every totally split hyperelliptic curve X can be parametrized by a Whittaker group (i.e. $X \cong \Omega/W$ for some Whittaker group W).

Let $\Omega \to X$ be the universal covering of X. Then $X \cong \Omega/W$ where W is a Schottky group on g generators.

Let σ denote the canonical automorphism of X with order 2.

Let Γ be the set of all automorphisms of Ω that lift id_X or σ. Clearly Γ is a group and W is a subgroup of index 2. So Γ is discontinuous and we only have to show that Γ is the free product of (g + 1) groups of order 2. We know already that $\Omega/\Gamma \cong X/\{1,\sigma\} \cong \mathbb{P}^1$ and so $\Gamma/[\Gamma,\Gamma]$ (see Chap. II) is a finite group.

Let $x_0 \in X$ be a fixed point of σ and let $\omega_o \in \Omega$ be an original of x_0. There exists a unique lift s_o of σ with $s(\omega_o) = \omega_o$. Clearly $s_o \in \Gamma$ and $s_o^2 = 1$. The element s_o induces an automorphism $\phi : w \mapsto s_o w s_o$ of W and Γ is generated by W and s_o. This ϕ induces an automorphism $\bar{\phi}$ of $W/[W, W] \simeq \mathbb{Z}^g$ of order two. Using that $\Gamma/[\Gamma, \Gamma]$ is finite, one finds that $\bar{\phi}$ must be $-id$. So $\phi(w)w \in [W, W]$ for every $w \in W$.

Undere these circumstances there is a group theoretical proof (see [23]) that W has a free base $\gamma_1, \ldots, \gamma_g$ such that $\phi(\gamma_i) = \gamma_i^{-1}$ $(i = 1, \ldots, g)$. Put $s_1 = \gamma_1 s_o, \ldots, s_g = \gamma_g s_o$. Then $s_o, s_1, \ldots, s_g$ have order 2 and one easily sees that Γ is the free product of the groups $\langle s_i \rangle$. This proves:

Proposition: A totally split, hyperelliptic curve can be parametrized by a Whittaker group.

(2.7) We give another, now very explicit, proof of Prop. (2.6) in the case char $\bar{k} \neq 2$. The group W is the fundamentalgroup of the graph G of a prestable reduction of X. In Chap. V, §3 we have constructed an explicit prestable reduction of X by using the reduction of $\mathbb{P}$ with respect to the set of ramification points V of X. The graph T of $\overline{(\mathbb{P}, V)})$ is a tree. For every vertex $t \in T$ (corresponding to a line in $\overline{(\mathbb{P}, V)})$ we define $n(t)$ to be the number of images of V on the line. Let $T_1, \dots, T_a$ denote the components of $T - \{t\}$ and let $n(S)$, for a subset of vertices of T, denote $\sum_{s \in S} n(s)$. According to Chap. V, §3 the function n has the following properties:

(i) $n(t) \in \{0, 1, 2\}$.

(ii) if $n(t) = 2$ then all $n(T_i)$ are even.

(iii) if $n(t) = 1$ then precisely one $n(T_i)$ is odd.

(iv) if $n(t) = 0$ then at most two $n(T_i)$ is are odd.

The graph G is obtained from T in the following way:

Consider G and T as topological spaces in the usual way.

Then G is the subspaces of $T \times [0, 1]$ given by

a) $T \times \{0\}$ and $T \times \{1\}$ lie in G.

b) if a vertex t has the property $n(t) \neq 0$, or $n(t) = 0$ but some of the $n(T_i)$'s are odd, then $t \times [0, 1] \subset G$.

c) if an edge e (as a topological space $\simeq [0, 1]$) has the property that some component C of $T - \{e\}$ has an odd $n(C)$, then $e \times [0, 1] \subseteq G$.

d) G is the union of the sets described in a), b) and c).

The automorphism σ on X induces an automorphism on $\bar{X}$ and G. The action on G is induced by the map $\phi : T \times [0, 1] \to T \times [0, 1]$ given by $(x, t) \to (x, 1 - t)$.

The action of σ on the fundamentalgroup of G is identical with the automorphism described in (2.6). We take an endpoint t_o of T and we let $(t_o, 0)$ be the base point of the fundamental group of G. A basis of $\pi_1(G, (t_o, 0))$ can be chosen out of the set of closed paths $\{\lambda_t | t \in T\}$. The path λ_t consists of:

1) the unique path in $T \times \{0\}$ from t_o to t;

2) the path from (t, 0) to (t, 1) in $t \times [0, 1]$;

3) the unique path in $T \times \{1\}$ from (t, 1) to $(t_o, 1)$ and

4) the path in $t_o \times [0, 1]$ from $(t_o, 1)$ to $(t_o, 0)$.

Clearly $\sigma(\lambda_t) = \lambda_t^{-1}$ for any $t \in T$. So we have found a basis $\gamma_1, \ldots, \gamma_g$ of the fundamental group $\cong$ W such that $\sigma(\gamma_i) = \gamma_i^{-1}$ $(i = 1, \ldots, g)$. This proves Prop. (2.6).

(2.8) Finally we give a result on combinations of discontinuous groups, which is analogous to the classical case. In particular this result implies (2.5.2).

Let $\Gamma \subset PGL(2, K)$ be a discontinuous group. We will assume that ∞ is an ordinary point of Γ. For our purpose we define a <u>fundamental domain</u> F for Γ to be the complement of open disks $B_1, \ldots, B_n$ in K, such that the corresponding closed disks $B_1^+, \ldots, B_n^+$ are disjoint. Further F must satisfy:

(i) the set $\{\gamma \in \Gamma | \gamma F \cap F \neq \phi\}$ is finite.

(ii) if $\gamma \neq 1$ then $\gamma F \cap F \subseteq \bigcup_{i=1}^{n} (B_i^+ - B_i)$.

(iii) $\bigcup_{\gamma \in \Gamma} \gamma F = \Omega$ the set of ordinary points of Γ.

Let further $\mathring{F}$ denote $\mathbb{P} - (B_1^+ \cup \ldots \cup B_n^+)$.

<u>Proposition: Let</u> $\Gamma_1, \ldots, \Gamma_m$ <u>be discontinuous groups with fundamental domains</u> $F_1, \ldots, F_m$. <u>Suppose that</u> $\mathring{F}_i \supseteq \mathbb{P} - F_j$ <u>for all</u> $i \neq j$. <u>Then the group</u> Γ <u>generated by</u> $\Gamma_1, \ldots, \Gamma_m$ <u>is discontinuous.</u>

<u>Moreover</u> $\Gamma \cong \Gamma_1 * \ldots * \Gamma_m$ <u>(the free product) and</u> $\cap F_i$ <u>is a fundamental domain for</u> Γ.

<u>Proof:</u> Put $F = \cap F_i$ and $\mathring{F} = \cap \mathring{F}_i$. Let $w = \delta_s \delta_{s-1}, \ldots, \delta_1$ be a reduced word in $\Gamma_1 * \ldots * \Gamma_m$, i.e. each $\delta_i \in \cup (\Gamma_j - \{1\})$ and if $\delta_i \in \Gamma_\ell$ then $\delta_{i+1} \notin \Gamma_\ell$. Then $w(\mathring{F}) \subseteq \mathbb{P} - F$ as one easily shows. Hence Γ is equal to $\Gamma_1 * \ldots * \Gamma_m$. Further $w(F) \cap F \neq \emptyset$ implies that $w \in \cup \Gamma_j$. So we have proved the properties (i) and (ii) for F.

Using that $\mathcal{L}_j = \mathbb{P} - \cup \{\gamma F_j | \gamma \in \Gamma_j\}$ = the set of limit points of Γ_j, is compact one finds the following:

For any positive δ there exist finite sets $W_1 \subset \Gamma_1, \ldots, W_m \subset \Gamma_m$ such that the complement of $\bigcup_{\gamma \in W_j} \gamma F_j$ consists of finitely many open disks of radii $\leq \delta$.

Let $\varepsilon > 0$ then there exists an integer $n > 0$ and finite subsets $W_1 \subset \Gamma_1, \ldots, W_m \subset \Gamma_m$ such W = all words in $W_1, \ldots, W_m$ of length $\leq n$ satisfies: $\mathbb{P} - \cup \{\gamma | \gamma \in W\}$ consists of finitely many open disks of radii $\leq \varepsilon$.

So $\mathbb{P} - \cup \gamma F$ is a compact set. This implies with the help of property (i) that Γ is discontinuous. One easily verifies further that $P - \bigcup_{\gamma \in \Gamma} \gamma F$ is infact the set of limit points of Γ.

(2.9) <u>Example:</u> If each $\Gamma_i \cong \mathbb{Z}$ (and so Γ_i is generated by a hyperbolic element) then Γ is a Schottky group on m generators. Let the fundamental domain of Γ_i be given as the complement of two open disks B_i, C_i. Then the situation in (2.8) is identical with the construction in Chap. I, §4.

(2.10) <u>Example:</u> Let the residue field of K have characteristic $\neq 2$. Let s be an elliptic element in PGL(2, K) of order 2 with fixed points a, b in K. Let B be a maximal open disk in K not containing a and b. Then $\mathbb{P} - B$ is a fundamental domain for the group $\langle s \rangle = \{1, s\}$.

Given $s_0, \ldots, s_g$, elliptic elements of order two and for every i a choice of the disk B_i in K such that the corresponding closed disks B_i^+ $(i = 0, \ldots, g)$ are still disjoint. Then we can apply (2.8) and we find a discontinuous group $\Gamma \cong \langle s_0 \rangle * \ldots * \langle s_g \rangle$ with fundamental domain $\mathbb{P} - (B_0 \cup \ldots \cup B_g)$. This is exactly the situation of (2.5.2), where ∞ is properly chosen.

Chapter X. The group $PSL(2, \mathbb{F}_q[t])$.

Introduction

Let k be the Laurent series field $\mathbb{F}_q((\frac{1}{t}))$ and let $K \supset k$ denote some complete, algebraically closed field. The group $PSL(2, \mathbb{F}_q[t]) = \Gamma(1)$ is a discrete subgroup of $PGL(2, k)$. It shares many features with the modular group $PSL(2, \mathbb{Z})$. The set of ordinary points is $K - k$. The quotient $K - k/_{\Gamma(1)}$ is isomorphic to K. For any subgroup Γ of finite index the quotient $K - k/_{\Gamma}$ turns out to be an affine algebraic curve. After adding finitely many parabolic points one obtains a complete non-singular curve $X(\Gamma)$. This curve is a Mumford curve and some of its numerical data are calculated for congruence subgroups of $\Gamma(1)$. It turns out that $\Gamma(1)$ has a rich collection of modular forms. The action of the Hecke-operators and the expansion of the modular forms at ∞ remains still a mystery.

(10.1) The group $PSL(2, \mathbb{F}_q[t])$ is an interesting example of a discontinuous group. It has been studied from various points of view. In J. P. Serre's book [37] its action on trees is studied. Recent work by D. Goss [15] investigates the geometry and the modular function theory for this group. Our treatment is inspired by the preprint of D. Goss.

As usual $\mathbb{F}_q$ denotes the field with $q = p^s$ elements. For convenience we will suppose that $p \neq 2$. Let t be an indeterminante. Then $k = \mathbb{F}_q((\frac{1}{t}))$ denotes the Laurent-series field over $\mathbb{F}_q$ in $\frac{1}{t}$. Let $K \supset k$ denote an algebraically closed and complete field containing k. The valuation on k (and K) is normalized by $|t| = q$.

<u>Lemma:</u> $PSL(2, \mathbb{F}_q[t])$ <u>is a discrete subgroup of</u> $PGL(2, k)$.

Proof: The absolute value of $a \in \mathbb{F}_q[t]$ is equal to $|a| = q^{\text{degree }(a)}$. Let $C > 0$, then the number of elements $\gamma = \begin{pmatrix} a & b \\ c & d \end{pmatrix}$ in $PSL(2, \mathbb{F}_q[t])$ with $\max(|a|, |b|, |c|, |d|) \leq C$ is finite. Therefore the group is discrete.

(10.2) The group $\Gamma(1) = PSL(2, \mathbb{F}_q[t])$ has interesting subgroups; e.g.: for every non-zero ideal I in $\mathbb{F}_q[t]$ the kernel of the group-homomorphism $\Gamma(1) \to PSL(2, \mathbb{F}_q[t]/_I)$ is a normal subgroup of finite index. We denote this kernel by $\Gamma(I)$ or $\Gamma(f)$, where f is a generator for the ideal I. Other subgroups of interest are

$\Gamma_0(I) = \{\gamma \in \Gamma(1) | \gamma \equiv \begin{pmatrix} a & b \\ 0 & d \end{pmatrix} \bmod I\}$ and

$\Gamma_1(I) = \{\gamma \in \Gamma(1) | \gamma \equiv \begin{pmatrix} 1 & b \\ 0 & 1 \end{pmatrix} \bmod I\}$. One has $\Gamma(I) \subset \Gamma_1(I) \subset \Gamma_0(I)$.

Let $f \in \mathbb{F}_q[t]$ be a generator for I and let $f = f_1^{d_1} \ldots f_s^{d_s}$ be its decomposition in to prime factors; $d_i \geq 1$; n_i = degree of f_i; $n = \Sigma n_i d_i$ = the degree of f. Then one can derive the following formulas.

$$[\Gamma(1) : \Gamma(f)] = \tfrac{1}{2}q^{3n} \prod_{i=1}^{s} \left(1 - \frac{1}{q^{2n_i}}\right).$$

$$[\Gamma(1) : \Gamma_1(f)] = \tfrac{1}{2}q^{2n} \prod_{i=1}^{s} \left(1 - \frac{1}{q^{2n_i}}\right).$$

$$[\Gamma(1) : \Gamma_0(f)] = q^{n} \prod_{i=1}^{s} \left(1 + \frac{1}{q^{n_i}}\right).$$

The computation follows easily from $\Gamma(1)/_{\Gamma(f)} \cong \prod_{i=1}^{s} SL(2, \mathbb{F}_q[t]/_{(f_i^{d_i})})$. Compare G. Shimura [40] Ch. I.

(10.3) Any subgroup Γ of $\Gamma(1)$ of finite index contains a subgroup of finite index of $\{\begin{pmatrix} 1 & a \\ 0 & 1 \end{pmatrix} a \in \mathbb{F}_q[t]\}$. This implies that Γ contains parabolic elements and in particular that Γ is not a Schottky group. Using Ch. I, §3, it follows that $\Gamma(1)$ is not finitely generated. Another proof of this fact can be found in J. P. Serre [37] p. 121. In contrast with the earlier examples in this book, $\Gamma(1)$ contains infinite subgroups of parabolic elements. We will study parabolic elements more systematically for discrete subgroups Γ of $PGL(2, k)$.

A point $x \in \mathbb{P}(K)$ is called a <u>parabolic point</u> for Γ if $\Gamma_x = \{\gamma \in \Gamma | \gamma(x) = x\}$ contains infinitely many parabolic elements. According to Ch. I. (1.7), this means that Γ_x contains a normal subgroup of finite index which is conjugated to a group of the form $\{z \to z + a | a \in A\}$ where A is an infinite discrete subgroup of k.

(10.4) <u>Proposition:</u> <u>Let Γ be a subgroup of $\Gamma(1)$ of finite index.</u>

(1) $\mathbb{P}(\mathbb{F}_q(t))$ <u>is the set of all parabolic points of</u> Γ.

(2) $\mathbb{P}(k)$ <u>is the set of all limit points of</u> Γ.

(3) <u>The number of inequivalent parabolic points of</u> Γ <u>is</u> $\leq [\Gamma(1) : \Gamma]$.

<u>Proof:</u> $\Gamma(1)$ and Γ have the same set of parabolic points, since Γ is of finite index in $\Gamma(1)$. Clearly ∞ and its $\Gamma(1)$-conjugates are parabolic points. Hence $\mathbb{P}(\mathbb{F}_q(t)) = \Gamma(1)(\infty)$ consists of parabolic points. For any parabolic element $\gamma \in \Gamma(1)$ the fixed point of γ lies in $\mathbb{P}(\mathbb{F}_q(t))$. This proves (1). The closure of the set of parabolic points consists of limit points. From Ch. I. (1.6.2), statement (2) now follows.

Let $\Gamma\gamma_1, \ldots, \Gamma\gamma_d$ denote the cosets of Γ in $\Gamma(1)$. Then every parabolic point of Γ is equivalent to one of the $\gamma_1(\infty), \ldots, \gamma_d(\infty)$. This proves (3).

The inequality in (3) can be strict, since we have:

$\gamma_i(\infty)$ is equivalent to $\gamma_j(\infty)$ if and only if

$\gamma_i^{-1}\Gamma\gamma_j \cap \{\begin{pmatrix}1 & a\\ 0 & 1\end{pmatrix} a \in \mathbb{F}_q[t]\} \neq \emptyset$.

(10.5) <u>Examples:</u> The actual calculation of the number of inequivalent parabolic points is somewhat cumbersome. We have the formulas for $\Gamma(f)$, $\Gamma_1(f)$, $\Gamma_0(f)$ where $f = f_1^{d_1} \ldots f_s^{d_s}$; n_i = degree (f_i); $n = \Sigma n_i d_i$.

The number of inequivalent parabolic points is:

for $\Gamma(f)$: $$\frac{q^{2n}}{q-1}\prod_{i=1}^{s}\left(1-\frac{1}{q^{2n_i}}\right)$$

for $\Gamma_1(f)$: $$\frac{q^{n}}{q-1}\prod_{i=1}^{s}\left(1-\frac{1}{q^{2n_i}}\right)\left(2+(d_i-1)\left(1-\frac{1}{q^{n_i}}\right)\right)$$

for $\Gamma_0(f)$, where $d_1,\ldots, d_u$ are odd and $d_{u+1},\ldots, d_s$ are even,

$$2^s+\frac{2}{q-1}\left[\left\{\prod_{i=1}^{u}\left(2\,q^{\frac{n_i(d_i-1)}{2}}\right)\prod_{i=u+1}^{s}\left(q^{\frac{n_id_i}{2}}\left(1+\frac{1}{q^{n_i}}\right)\right)\right\}-2^s\right].$$

The last formula is somewhat complicated; special cases are easier to understand.

<u>special case</u> (1). $d_1 = d_2 = \ldots = d_s = 1$. Then the number is 2^s.
A complete set of inequivalent parabolic points for $\Gamma_0(f)$ is given by $\sum_{i\in A}\frac{1}{f_i}$, where A runs through the subsets of $\{1,\ldots, s\}$.

<u>special case</u> (2). $s = 1$ and $d_1 = 2$. Then the number is $2 + 2\,\frac{q^{n_1}-1}{q-1}$.
Representatives for the $\Gamma_0(f)$-equivalence classes of parabolic points are:
0, ∞ and $\frac{\alpha}{f_i}$, where $-\infty <$ degree $(\alpha) < n_1 =$ degree (f_1). We consider here $\frac{\alpha}{f_1}$, $\frac{\alpha'}{f_1}$ to be equal if $\alpha' = \lambda^2\alpha$ with $\lambda\in\mathbb{F}_q^*$.

<u>special case</u> (3). $s = 1$ and $d_1 = 3$; then the number is $2 + 4\,\frac{q^{n_1}-1}{q-1}$.
Representatives are 0, ∞, $\frac{\alpha}{f_1}$, $\frac{\alpha}{f_1^2}$, where α satisfies the same condition as in the special case (2).

We will sketch the proof of the formulas:

For $\Gamma(f)$ it is easily seen that the number of inequivalent parabolic points is $[\Gamma(1) : \Gamma(f)]/_{[\Gamma(1)_\infty:\Gamma(f)_\infty]}$. This implies the formula.

The equivalence classes of parabolic points under $\Gamma(f)$ can be described explicitely. Namely $\mathbb{P}(\mathbb{F}_q(t)) = \{(\alpha,\beta) \mid \alpha,\beta \in \mathbb{F}_q[t]$ and $(\alpha, \beta) = 1\}/_{\mathbb{F}_q^*}$, in which $(\alpha, \beta) \sim (\alpha', \beta')$ if $(\alpha, \beta) = \lambda(\alpha', \beta')$ for some $\lambda \in \mathbb{F}_q^*$.

It follows that $\mathbb{P}(\mathbb{F}_q(t))/_{\Gamma(f)} = \{(\alpha, \beta) \mid \alpha, \beta \in \mathbb{F}_q[t]/_{(f)}$ and $(\alpha, \beta) = 1\}/_{\mathbb{F}_q^*}$.

Let us write $Z(f) = \{(\alpha, \beta) \mid \alpha, \beta \in \mathbb{F}_q[t]/_{(f)}$ and $(\alpha, \beta) = \mathbb{F}_q[t]/_{(f)}\}$. Then $\Gamma_1(f)$ and $\Gamma_0(f)$ (or their quotients $\Gamma_i(f)/_{\Gamma(f)}$) act on $Z(f)$. We have to calculate the number of orbits (taken modulo the action of $\mathbb{F}_q^*$).

One easily sees that $Z(f) = Z(f_1^{d_1}) \times \ldots \times Z(f_s^{d_s})$ and

$$\Gamma_i(f)/_{\Gamma(f)} = \Gamma_i(f_1^{d_1})/_{\Gamma(f_1^{d_1})} \times \ldots \times \Gamma_i(f_s^{d_s})/_{\Gamma(f_s^{d_s})} \quad (i = 0, 1)$$

acts componentwise on $Z(f)$. This gives the reduction to the case where f has only one prime-factor. A careful analysis of the latter case yields the formula's.

(10.6) The set of ordinary points for $\Gamma(1)$ (or Γ of finite index in $\Gamma(1)$) is equal to $K - k$. (D. Goss has given this set the name "the algebraists upper halfplane".) On $K - k$ we have a distance function $d(z) = \min\{|z + \lambda| \mid \lambda \in k\}$ which plays the role of "Im (z)" in the complex case. One has the following result:

<u>Proposition:</u> <u>For any</u> $\begin{pmatrix} a & b \\ c & d \end{pmatrix} \in SL(2, k)$, <u>one has</u> $d\left(\frac{az + b}{cz + d}\right) = \frac{1}{|cz + d|^2} d(z)$.

<u>Proof:</u> Suppose that the formula is correct for $A, A' \in SL(2, k)$ where $A = \begin{pmatrix} a & b \\ c & d \end{pmatrix}$ and $A' = \begin{pmatrix} a' & b' \\ c' & d' \end{pmatrix}$. Then $d(AA'z) = \frac{1}{|cA'z + d|^2} \frac{1}{|c'z + d'|^2} d(z) =$

$$= \frac{d(z)}{|(a'c + d'c)z + b'c + d'd|^2}.$$

Hence the formula is correct for AA'. Now we have only to verify the formula for the generators $\begin{pmatrix}1 & b\\ 0 & 1\end{pmatrix}$, $b \in k$ and $\begin{pmatrix}0 & -1\\ 1 & 0\end{pmatrix}$ of SL(2, k). For $\begin{pmatrix}1 & b\\ 0 & 1\end{pmatrix}$ the formula is obvious. For $\begin{pmatrix}0 & -1\\ 1 & 0\end{pmatrix}$ we have to verify $d(\frac{-1}{z}) = \frac{1}{|z|^2} d(z)$.

Let $\lambda_0 \in k$ be an element such that $d(z) = |z - \lambda_0|$. Then
$|z|^2 d(\frac{-1}{z}) = |z| \min \{|1 + \lambda z| \mid \lambda \in k\}$. Further $|1 + \lambda z| =$
$|(1 + \lambda\lambda_0) + \lambda(z - \lambda_0)| = \max (|1 + \lambda\lambda_0|, |\lambda| d(z))$.
If $|\lambda_0| < |z|$ then $d(z) = |z|$ and we may suppose that $\lambda_0 = 0$.
Then clearly $|z|^2 d(\frac{-1}{z}) = |z| = d(z)$.

If $|z - \lambda_0| < \max (|z|, |\lambda_0|)$ then $|z| = |\lambda_0|$ and the substitution $\lambda = \frac{-1}{\lambda_0}$ gives the minimal value. Hence also $|z|^2 d(\frac{-1}{z}) = d(z)$.

(10.7) A <u>horicycle neighbourhood</u> of ∞ in K - k is by definition a set $U_M = \{z \in K - k \mid d(z) \geq M\}$ where M is some positive real number.

For any point $x \in k$ we take some $\gamma \in PGL(2, k)$ with $\gamma(\infty) = x$.
A horicycle neighbourhood of x in K - k will be a set of the form $\gamma(U_M)$.

Let Γ be a discrete subgroup of PGL(2, k) such that ∞ is a parabolic point of Γ and such that the discrete subgroup $A \subseteq k$, given by $a \in A$ if and only if $(z \mapsto z + a) \in \Gamma$, satisfies $k/_A$ is compact.
This condition is certainly fulfilled for $\Gamma(1)$ and its subgroups of finite index.

<u>Proposition:</u> <u>If</u> Γ <u>is a discrete subgroup of</u> PGL(2, k) <u>satisfying the condition above, then there exists a constant</u> $M > 0$ <u>such that</u>
$\{\gamma \in \Gamma \mid \gamma(U_M) \cap U_M \neq \emptyset\} = \Gamma_\infty$.

<u>Proof:</u> For every double coset $\Gamma_\infty \begin{pmatrix}a & b\\ c & d\end{pmatrix} \Gamma_\infty \in \Gamma_\infty \backslash \Gamma / \Gamma_\infty$ the element $|c|$ does not depend on the choice of $\begin{pmatrix}a & b\\ c & d\end{pmatrix}$. There is a constant $C > 0$ depending only on $|c|$ and Γ such that for suitable $\begin{pmatrix}1 & \alpha\\ 0 & 1\end{pmatrix}$, $\begin{pmatrix}1 & \beta\\ 0 & 1\end{pmatrix} \in \Gamma_\infty$ one has $\begin{pmatrix}a' & b'\\ c' & d'\end{pmatrix} = \begin{pmatrix}1 & \alpha\\ 0 & 1\end{pmatrix}\begin{pmatrix}a & b\\ c & d\end{pmatrix}\begin{pmatrix}1 & \beta\\ 0 & 1\end{pmatrix}$ with $\max (|a'|, |b'|, |c'|, |d'|) \leq C$. This follows easily from the "$k/_A$ is compact" made on Γ.

It follows that for any $D > 0$ the number of double cosets with $|c| \leq D$ is finite and in particular for $D_o > 0$, small enough, the only double coset satisfying $|c| \leq D_o$ is Γ_∞.

Take a number $M > 0$ with $\frac{1}{M^2} \leq D_o$. Let $z \in U_M$, $\gamma(z) \in U_M$, $\gamma \in \Gamma$. Then $d(z) \geq M$ and also $d(\gamma z) = d(\frac{az+b}{cz+d}) = \frac{1}{|cz+d|^2} d(z) \geq M$.

Hence $d(z) \geq M|cz+d|^2 \geq |c|^2 d(z)^2 M$ if $c \neq 0$.

Hence $|c| \leq \frac{1}{M^2}$ if $c \neq 0$. It follows that $c = 0$ and $\gamma \in \Gamma_\infty$

(10.8) We start the investigation of the quotient $K - k/_{\Gamma(1)}$. In (10.8) we calculate it as a set. In (10.9) and (10.10) we show that $K - k/_{\Gamma(1)}$ has a natural structure as analytic space.

Proposition:

(1) If $M > 1$ then $\{\gamma \in \Gamma(1) | \gamma (U_M) \cap U_M \neq \emptyset\} = \Gamma(1)_\infty$.

(2) Every point of $K - k$ is $\Gamma(1)$-equivalent to a point in U_1.

(3) Every point $z \in K - k$ with $d(z) = 1$ is equivalent to a point $z' \in K - k$ with $d(z') = |z'| = 1$.

(4) Points z, z' with $|z| = |z'| = d(z) = d(z') = 1$ are equivalent if and only if $\gamma(z) = z'$ holds for some $\gamma \in PSL(2, \mathbb{F}_q)$.

Proof: (1) follows easily from (10.7).

(2) Let $z \in K - k$ satisfy $d(z) < 1$. For a unique $a_1 \in \mathbb{F}_q[t]$ we have $z_1 = z + a_1$ satisfying $|z_1| < 1$. Let $\lambda \in k$ satisfy $d(z_1) = |z_1 + \lambda|$. Then $\lambda = \alpha t^{-n} + \ldots \in \mathbb{F}_q[\![t^{-1}]\!]$ with $\alpha \neq 0$ and $n > 0$.

Choose $\gamma = \begin{pmatrix} a & b \\ c & d \end{pmatrix} \in \Gamma(1)$ with $c = t^n$ and $d = \alpha$. Then $d(\gamma z_1) = \frac{d(z)}{|cz+d|^2}$ and $|cz + d| < 1$. Hence $d(\gamma z_1) > d(z)$.

If $d(\gamma z_1) \geq 1$ then we are done. If $d(\gamma z_1) < 1$ then for a unique $a_2 \in \mathbb{F}_q[t]$ one has $z_2 = \gamma z_1 + a_2$ has absolute value < 1. We continue this process with z_2 instead of z_1. If the process does not stop then one finds an infinite sequence $z_1, z_2, z_3, \ldots,$ of equivalent points with all $|z_i| < 1$ and $d(z_1) < d(z_2) < \ldots$. This contradicts the discreteness of $\Gamma(1)$.

(3) There exists an $a \in \mathbb{F}_q[t]$ such that $|z + a| = 1$.

(4) Let $z' = \frac{az + b}{cz + d}$ with $\begin{pmatrix} a & b \\ c & d \end{pmatrix} \in \Gamma(1)$. Then $|cz + d| = 1$ and also $|az + b| = 1$ because $|z'| = 1$. It follows that $a, b, c, d \in \mathbb{F}_q$.

(10.9) Construction of the quotient of a horicycle neighbourhood.

Let Γ be a discrete subgroup of $PGL(2, k)$ and let x be a parabolic point of Γ. We suppose that Γ satisfies the condition explained in (10.7) and as a consequence x has a horicycle neighbourhood U such that $\{\gamma \in \Gamma | \gamma U \cap U \neq \emptyset\} = \Gamma_x$. In this section we construct $U/_{\Gamma_x}$ as a k-analytic space and we will show in fact that $U/_{\Gamma_x}$ is isomorphic to $\{z \in K | |z| \geq M\}$ where M is some positive constant.

In order to simplify the notation we suppose that $x = \infty$. The group Γ_∞ has only ∞ as limit point and we will construct $U/_{\Gamma_\infty}$ as subset of $K/_{\Gamma_\infty}$. The elements $\begin{pmatrix} 1 & a \\ 0 & 1 \end{pmatrix}$ in Γ_∞ are identified with $a \in k$.

Let $A \subset k$ be the discrete additive subgroup of k consisting of all $\begin{pmatrix} 1 & a \\ 0 & 1 \end{pmatrix} \in \Gamma_\infty$. Then $\Gamma_\infty/_A$ is a finite cyclic group (Compare Ch. I (1.7)). We will first of all construct $K/_A$. Let $e = e_A : K \to K$ denote the holomorphic function on K defined by $e(z) = z \prod_{\substack{a \in A \\ a \neq 0}} (1 - \frac{z}{a})$.

One easily verifies that e converges on any disk $\{z \in K | |z| \leq R\}$ and that $\|e\|_R = \sup \{|e(z)| \,|\, |z| \leq R\}$ satisfies the formula:

$\|e\|_R = \|z\|_R \prod_{\substack{a \in A \\ a \neq 0}} \|1 - \frac{z}{a}\|_R = R \prod (\frac{R}{\|a\|})$ where the last product is taken over all $a \in A$ with $a \neq 0$ and $|a| < R$. The following lemma shows that $e : K \to K$ is surjective.

(10.9.1) Lemma: Let f be a holomorphic function on $\{z \in K \mid |z| \leq R\}$ which has a zero. Then $\{f(z) \mid z \in K, |z| \leq R\} = \{t \in K \mid |t| \leq \|f\|_R\}$.

Proof: It suffices to take $R = \|f\|_R = 1$. Clearly $\{f(z) \mid z \in K, |z| \leq 1\}$ is contained in $\{t \in K \mid |t| \leq 1\}$. Let $\lambda \in K$ with $|\lambda| \leq 1$ and put $g = f - \lambda$. The residue function $\bar{g} = \bar{f} - \bar{\lambda} \in \bar{K}[z]$ is a non-constant polynominal. So $\bar{g}$ is not invertible. If g is invertible then g^{-1} has norm $\|g^{-1}\| = \|g\|^{-1} = 1$ since the norm is multiplicative. So also $\bar{g}$ would be invertible. This contradiction yields that g has a zero in $\{z \in K \mid |z| \leq 1\}$.

(10.9.2) We continue the study of $e = e_A$. We need the following properties:

$e(z_1 + z_2) = e(z_1) + e(z_2)$ and $\frac{d}{dz} e(z) = 1$.

Proof: By continuity, it suffices to show those properties when A is a finite group. Then $e_A(z) = c \prod_{a \in A} (z - a)$ for some constant $c \neq 0$. Let $f_A(z)$ denote $\prod_{a \in A} (z - a)$.

We claim that the polynomial f_A is a linear combination of $1, z, z^p, z^{p^2}, z^{p^3}, \ldots$. The set A is a vectorspace over $\mathbb{F}_p$ of finite dimension. For dimension 1 one has $f_{\mathbb{F}_p b}(z) = z^p - b^{p-1}z$:

Further if b is $\mathbb{F}_p$-linearly independent of A then $f_{A + \mathbb{F}_p b}(z) = (f_A(z))^p - f_A(b)^{p-1} f_A(z)$. By induction on the dimension the statement on f_A follows and (10.9.2) is proved.

Now $e : K \to K$ is a surjective group homomorphism; its kernel is the set of zero's of e and equal to A. Hence e induces an analytic iso morphism $K/_A \xrightarrow{\sim} K$. Using (10.9.1) and (10.9.2) one sees that $\{z \in K \mid |e(z)| < R\} = A + \{z \in K \mid |z < S\}$ where S is some positive real number depending on R. If R (and hence S) are sufficiently big then $\{z \in K \mid |e(z)| \geq R\}$ is precisely a horicycle neighbourhood U_M of ∞.

It follows that $U_{M/A} \cong \{z \in K \mid |z| \geq R\}$.

Let $\Gamma_{\infty/A}$ have order n, then using the function $e(z)^n$ instead of e, one finds an analytic isomorphism $U_{M/\Gamma_\infty} \xrightarrow{\sim} \{z \in K \mid |z| \geq R^n\}$.

(10.9.3) <u>Example:</u> For (1) the procedure above, with $n = \frac{q-1}{2}$ = the number of squares in $\mathbb{F}_q^*$, one finds an isomorphism

$$U_1/_{\Gamma(1)_\infty} \xrightarrow{\sim} \{z \in K \mid |z| \geq 1\}, \text{ given by } z \to e(z)^{\frac{q-1}{2}},$$

where $e(z) = z \prod_{\substack{a \neq o \\ a \subset \mathbb{F}_q[t]}} (1 - \frac{z}{a})$.

This isomorphism also induces an isomorphism:

$$U_{\sqrt{q}}/_{\Gamma(1)_\infty} \xrightarrow{\sim} \{z \in K \mid |z| \geq q^{\frac{q(q-1)}{4}}\}.$$

(10.1) <u>Construction of the quotient space</u> $K - k/_{\Gamma(1)}$.

Consider the horicycle neighbourhood $U = U_{\sqrt{q}}$ and the affinoid set $V = \{z \in K \mid |z| \leq \sqrt{q}\} - (\{z \in K \mid |z| < \frac{1}{\sqrt{q}}\} + \mathbb{F}_q)$.

The complement of V in $\mathbb{P}(K)$ consists of $(q + 1)$ open disks.

According to (10.8) $U \cup V$ maps surjectively to the quotient set $K - k/_{\Gamma(1)}$; the equivalence relation on U is given by the action of $\Gamma(1)_\infty$; the equivalence relation on V is given by the action of $PSL(2, \mathbb{F}_q)$.

We study now the equivalence relation on V. The group $PSL(2, \mathbb{F}_q)$ acts on all of $\mathbb{P}(K)$ and $\mathbb{P}(K)/_{PSL(2, \mathbb{F}_q)} \xrightarrow{\sim} \mathbb{P}(K)$. This analytic isomorphism is given by the analytic map $f : \mathbb{P}(K) \to \mathbb{P}(K)$ defined by the formula $f(z) = \prod_{\gamma \in PSL(2, \mathbb{F}_q)} \frac{z - \gamma(a)}{z - \gamma(b)}$, where $a, b \in \mathbb{P}(K)$ are inequivalent points for the action of $PSL(2. \mathbb{F}_q)$. Under this map f, the $(q + 1)$ open disks $\{z \in K \mid |z| > \sqrt{q}\}$, $\{z \in K \mid |z - \alpha| < \frac{1}{\sqrt{q}}\}$ $(\alpha \in \mathbb{F}_q)$ are mapped onto one and the same open disk in $\mathbb{P}(K)$.

It follows that $V/_{PSL(2,\mathbb{F}_q)}$ is a closed disk D.

We consider further $W = \{z \in K \mid |z| = \sqrt{q}\}$. The equivalence relation on W is given by the action of the group $\{\begin{pmatrix} a & b \\ 0 & a^{-1}\end{pmatrix} \mid a \in \mathbb{F}_q^*;\ b \in \mathbb{F}_q\} = B$.

Then $W/_B$ maps isomorphically to the boundary ∂D of $D = V/_{PSL(2,\mathbb{F}_q)}$.

Also $W/_B$ maps isomorphically to $\{z \in K \mid |z| = q^{\frac{q(q-1)}{4}}\} \subset U/_{\Gamma(1)_\infty}$.

The gluing of $V/_{PSL(2,\mathbb{F}_q)}$ and $U/_{\Gamma(1)_\infty}$ over $W/_B$ is obviously the analytic space K.

So we have shown:

Theorem: $K - k/_{\Gamma(1)} \tilde{\to} K$ as analytic spaces.

(10.11) Remarks:

(1) The identification in the theorem can also be given by a holomorphic function on K, namely the function $j(z) = \prod_{\gamma\in\Gamma(1)} \frac{z - \gamma(\omega)}{z - \gamma(\infty)}$.

In this formula ω is some element of $K - k$ (which can afterwards be nicely chosen) and a term $\frac{z-\gamma(\omega)}{z-\gamma(\infty)}$ with $\gamma(\infty) = \infty$ means $(1 - \frac{z}{\gamma(\omega)})$.

The convergence of the infinite product has still to be verified.

Take $\pi \in k$, $0 < |\pi| < 1$ and $n \in \mathbb{N}$ and let X_n be the affinoid set $\{z \in K \mid d(z,k) \geq |\pi|^{+n}$ and $|z| \leq |\pi|^{-n}\}$. It suffices to verify the uniform convergence on X_n. This is equivalent to showing

$$\lim_{\gamma\in\Gamma(1)} \left(\sup_{z\in X_n} \left| \frac{z-\gamma(\omega)}{z-\gamma(\infty)} - 1\right|\right) = 0-$$

For $\gamma = \begin{pmatrix} a & b \\ c & d\end{pmatrix} \in \Gamma(1)$ with $c \neq 0$ one has $\left|\frac{z-\gamma(\omega)}{z-\gamma(\infty)} - 1\right| =$

$$\frac{1}{|c\omega + d|\,|cz - a|} \leq \frac{1}{d(\omega,k)\max(|c|,|d|)\cdot d(z,k)\max(|a|,|c|)} \leq$$

$$\leq \frac{|\pi|^{-n}}{d(\omega,k)} \quad \frac{1}{\max(|c|,|d|)\cdot\max(|a|,|c|)}.$$

For $\gamma = \begin{pmatrix} a & b \\ o & d\end{pmatrix}$ one has $\left|\frac{z-\gamma(\omega)}{z-\gamma(\infty)} - 1\right| = \left|\frac{z}{\gamma(\infty)}\right| \leq \frac{|\pi|^{-n}}{|a\omega + b|} \leq$

$\leq \frac{|\pi|^{-n}}{d(\omega,k)} \quad \frac{1}{\max(|a|,|b|)}$. From those inequalities the statement follows.

(2) The construction in (10.9) and (10.10) can be carried out for every subgroup Γ of $\Gamma(1)$ of finite index. One obtains an analytic space $K - k/_\Gamma$ (defined over some extension of k). The analytic space $K - k/_\Gamma$ is obtained by glueing together an affinoid set V with sets of the form $\{z \in K \mid |z| \geq R_i\}$ $(i = 1, \ldots, N)$ where N is the number of inequivalent parabolic points of Γ.

The "completion" $(K - k/_\Gamma)^\wedge$ of $K - k/_\Gamma$ is obtained by glueing together V and $\{z \in \mathbb{P}(K) \mid |z| \geq R_i\}$ $(1 \leq i \leq N)$. The result will be denoted by $X(\Gamma)$, it is a complete analytic space defined over some finite extension of k. It follows as in Chap. III (2.2), that $X(\Gamma)$ is a complete non-singular curve. Further $X(\Gamma)$ is a Mumford curve, since it has a finite covering by affinoid subsets of $\mathbb{P}(K)$. Using the obvious holomorphic surjection $X(\Gamma) \to X(\Gamma(1)) = \mathbb{P}(K)$ we will calculate in some cases the genus of $X(\Gamma)$.

(10.12) <u>The elliptic elements of $\Gamma(1)$.</u>

Let $\gamma \in \Gamma(1)$ be elliptic (or parabolic). Then $\gamma = \begin{pmatrix} a & b \\ c & -a+\alpha \end{pmatrix}$ for some $\alpha \in \mathbb{F}_q$. Since $a(-a+\alpha) - bc = 1$ we have degree (c) $\leq$ degree (a) or degree (b) $\leq$ degree (a). It degree (c) $\leq$ degree (a) then for suitable $\beta \in \mathbb{F}_q[t]$ we find $\begin{pmatrix} 1 & \beta \\ 0 & 1 \end{pmatrix}\begin{pmatrix} a & b \\ c & -a+\alpha \end{pmatrix}\begin{pmatrix} 1 & -\beta \\ 0 & 1 \end{pmatrix} = \begin{pmatrix} a' & b' \\ c' & -a'+\alpha \end{pmatrix}$ with degree a' < degree a. (or a = 0).

If degree (b) $\leq$ degree (a) then $\begin{pmatrix} 1 & 0 \\ \beta & 1 \end{pmatrix}\begin{pmatrix} a & b \\ c & -a+\alpha \end{pmatrix}\begin{pmatrix} 1 & 0 \\ -\beta & 1 \end{pmatrix} = \begin{pmatrix} a' & b' \\ c' & -a'+\alpha \end{pmatrix}$ with degree (a') < degree (a). (or a = 0).

It follows that γ is conjugated to an element $\begin{pmatrix} 0 & b \\ c & \alpha \end{pmatrix}$ and clearly $b, c \in \mathbb{F}_q$. Hence γ is conjugated (in $\Gamma(1)$) with an element of $PSL(2, \mathbb{F}_q)$.

Let A in $SL(2, \mathbb{F}_q)$ have $X^2 - \alpha X + 1$ as characteristic polynomial. If this polynomial is reducible, then A is conjugated to one of the following matrices $\begin{pmatrix} 1 & 0 \\ 0 & 1 \end{pmatrix}$, $\begin{pmatrix} 1 & 1 \\ 0 & 1 \end{pmatrix}$, $\begin{pmatrix} -1 & 0 \\ 0 & -1 \end{pmatrix}$, $\begin{pmatrix} -1 & 1 \\ 0 & -1 \end{pmatrix}$ or $\begin{pmatrix} \mu & 0 \\ 0 & \mu^{-1} \end{pmatrix}$ with $\mu \neq \pm 1$.

If the characteristic polynomial is irreducible then A is conjugated to $\begin{pmatrix} 0 & -1 \\ 1 & \alpha \end{pmatrix}$. An example of the latter case is:

Let ξ denote a generator of $\mathbb{F}^*_{q^2}$ and put $\eta = \xi^{q-1}$. Then $\eta^{q+1} = 1$ and $\eta + \eta^{-1} \in \mathbb{F}_q$.

Hence $\begin{pmatrix} 0 & -1 \\ 1 & \eta+\eta^{-1} \end{pmatrix}$ is an elliptic element of order q + 1. Every $A \in SL(2, \mathbb{F}_q)$ with eigenvalues in $\mathbb{F}_{q^2} - \mathbb{F}_q$ is conjugated to a power of $\begin{pmatrix} 0 & -1 \\ 1 & \eta+\eta^{-1} \end{pmatrix}$.

One finds that every elliptic element of $\Gamma(1)$ with fixed points in K - k is conjugated with $\begin{pmatrix} 0 & -1 \\ 1 & \eta+\eta^{-1} \end{pmatrix}$. Let $\omega_o \in \mathbb{F}_{q^2} - \mathbb{F}_q$. Then ω_o is an elliptic point of K - k of order $\frac{q+1}{2}$ and every elliptic point of $\Gamma(1)$ in K - k is equivalent with ω_o. We will write ε for the unique elliptic point (image of ω_o) in $K - k/_{\Gamma(1)}$.

(10.13) <u>The genus of</u> $X(\Gamma) = (K - k/_{\Gamma})^{\wedge}$.

Let Γ be a subgroup of $\Gamma(1)$ of finite index. Then $\phi : X(\Gamma) \to \mathbb{P}(K) = X(\Gamma(1))$ is separable, has degree $[\Gamma(1) : \Gamma]$ and is only ramified above ∞ and ε(ε is the unique elliptic point of $X(\Gamma(1))$.

The ramification above ∞ is wild and we have to do some work to calculate its contribution in the Hurwitz-Riemann formula.

We have to make an assumption on Γ in order to carry out the calculation. Let $\gamma \in \Gamma(1)$ be a parabolic element. For some $\delta \in \Gamma(1)$ we have $\gamma = \delta \begin{pmatrix} 1 & b \\ 0 & 1 \end{pmatrix} \delta^{-1}$ with $b \in \mathbb{F}_q[t]$. Let $\lambda \in \mathbb{F}_q$ then we define γ^λ to be $\delta \begin{pmatrix} 1 & \lambda b \\ 0 & 1 \end{pmatrix} \delta^{-1}$. One easily verifies that the choice of δ is unimportant. For $\lambda \in \mathbb{F}_p$ the γ^λ is the ordinary power of γ. We make the assumption on Γ : "If $\gamma \in \Gamma$ is parabolic and $\lambda \in \mathbb{F}_q$ then $\gamma^\lambda \in \Gamma$". This assumption is easily verified for $\Gamma(f)$, $\Gamma_1(f)$ and $\Gamma_0(f)$.

(10.13.1) <u>Lemma: Let</u> $\bar{x} \in X(\Gamma)$ <u>lie above</u> ∞, <u>where</u> $\bar{x}$ <u>is the "image" of some parabolic point</u> $x \in \mathbb{P}(\mathbb{F}_q(t))$. <u>Let</u> $[\Gamma(1)_x : \Gamma_x] = vq^w$ <u>with</u> $(v, q) = 1$. <u>Then the contribution</u> $C_{\bar{x}}$ <u>of</u> $\bar{x}$ <u>in the Hurwitz-Riemann</u>

formula equals

$$C_{\bar{x}} = -1 + vq^w + 2v\,\frac{q^w - 1}{q - 1}.$$

Proof: In the computation we may suppose that $x = \infty$. We consider $\psi : K/_{\Gamma_\infty} \cup \{\infty\} - \mathbb{P}(K) \to K/_{\Gamma(1)_\infty} \cup \{\infty\} = \mathbb{P}(K)$. The Hurwitz-Riemann formula for ψ is $-2 = -2vq^w + C_\infty +$ (contribution of the elliptic points).

Let $\Gamma^*_\infty = \{\begin{pmatrix}1 & b\\ 0 & 1\end{pmatrix} \mid \begin{pmatrix}1 & b\\ 0 & 1\end{pmatrix} \in \Gamma_\infty\}$. We can also consider $K/_{\Gamma^*_\infty} \overset{\chi}{\to} K/_{\Gamma(1)_\infty}$. Then $\chi^{+1}(0)$ is isomorphic with $\Lambda = \mathbb{F}_q[t]/_{\Gamma^*_\infty}$, this is a vectorspace over $\mathbb{F}_q$ of dimension w. The action of $\Gamma_\infty/\Gamma^*_\infty$ is multiplication by an element in $\mathbb{F}^*_q$ of order $\frac{q-1}{2v}$.

Further $\psi^{-1}(0)$ is the orbit space of Λ under this action. The point $\{0\}$ is an orbit and has index of ramification v under ψ. The number of the other orbits is $2v\,\frac{q^w - 1}{q - 1}$ and they correspond to points with index of ramification $\frac{q-1}{2}$.

Hence the total contribution of the elliptic points equals:

$2v\,\frac{q^w - 1}{q - 1}\left(\frac{q-1}{2} - 1\right) + (v - 1)$ (since the ramification is tame).

This proves the formula for $C_{\bar{x}}$.

(10.13.2) Proposition: The genus of $X(\Gamma(f))$ is equal to

$$1 + 2\,\frac{(q^n - q - 1)}{(q^2 - 1)q^n}\,[\Gamma(1) : \Gamma(f)] \text{ where } n = \text{degree } (f).$$

Proof: Since $\Gamma(f)$ is normal in $\Gamma(1)$ all the parabolic points of $X(\Gamma(f))$ give the same contribution in the Hurwitz-Riemann formula.

The total contribution above $\infty \in X(\Gamma(1))$ is then:

$$\frac{[\Gamma(1) : \Gamma(f)]}{[\Gamma(1)_\infty : \Gamma(f)_\infty]}\left(\frac{q-1}{2}\,q^n + q^n - 2\right).$$

Put $\mu = [\Gamma(1) : \Gamma(f)]$. Each point of $X(\Gamma(f))$ lying above $\varepsilon \in X(\Gamma(1))$ has index of ramification $\frac{q-1}{2}$. Their number is $\frac{2\mu}{q+1}$ and their total contribution is $\mu - \frac{2\mu}{q+1}$.

Hence $2g - 2 = -2\mu + \dfrac{\mu}{\frac{q-1}{2}q^n}\left(\frac{q-1}{2}q^n + q^n - 2\right) + \mu - \dfrac{2\mu}{q+1}$.

(10.13.3) Proposition: If f is irreducible of degree n, then the genus of $X(\Gamma_o(f))$ is equal to $2\dfrac{q^n - q}{q^2 - 1}$ if n is odd.

Proof: If n is odd then $\Gamma_o(f)$ has no elliptic points and (as before) the contribution of the points above ε is $\mu - \dfrac{2\mu}{q+1}$. Further $X(\Gamma_o(f))$ has two parabolic points, 0 and ∞. Only 0 gives a contribution, namely $-1 + q^n + 2\dfrac{q^n - 1}{q - 1}$. Further $\mu = q^n + 1$. The formula follows.

Remark: It seems rather difficult (but interesting) to find general formula's for the genera of $X(\Gamma_1(f))$ and $X(\Gamma_o(f))$.

(10.14) Modular forms for $PSL(2, \mathbb{F}_q[t]) = \Gamma(1)$.

The most interesting feature of $\Gamma(1)$ is probably the presence of modular forms. In the last part of this chapter we give the basic material.

A modular function for $\Gamma(1)$ is a meromorphic function for $K - k$ satisfying $f\left(\frac{az - b}{cz + d}\right) = (cz + d)^{\ell} f(z)$ for some $\ell \in \mathbb{Z}$ and all $\gamma = \begin{pmatrix} a & b \\ c & d \end{pmatrix}$ in $\Gamma(1)$.

The function f is called a modular form of weight ℓ if moreover:

(i) f is holomorphic on $K - k$ and (ii) f is holomorphic at the parabolic point. The last statement means the following: f is invariant under the group $\{\begin{pmatrix} 1 & b \\ 0 & 1 \end{pmatrix} \mid b \in \mathbb{F}_q[t]\} \cong \mathbb{F}_q[t]$ and as a consequence f induces a holomorphic function F on $U_1/\mathbb{F}_q[t] \tilde{\rightarrow} \{z \in K \mid |z| \geq 1\}$.

The isomorphism is given by the function $e(z) = z \prod_{\substack{\lambda \in \mathbb{F}_q[t] \\ \lambda \neq 0}} (1 - \frac{z}{\lambda})$.

Let u denote $\frac{1}{e}$. Then F (or f) must have the form $\sum_{n \geq 0} c_n u^n$ with $c_n \in K$ and $\lim c_n = 0$.

We put $v_\infty(f)$ = the smallest n with $c_n \neq 0$.

Let $\omega \in K - k$ have image ε in $X(\Gamma(1))$ (for instance, $\omega \in \mathbb{F}_{q^2} - \mathbb{F}_q$). Then ω (and its conjugates) are elliptic points for $\Gamma(1)$ with order $\frac{q+1}{2}$.

We define $v_\varepsilon(f)$ = the order of f at the point ω. (This does not depend on the choice of ω).

<u>Proposition: Let</u> $f \neq 0$ <u>be a modular form of weight</u> ℓ. <u>Then:</u>

$$\sum_{z\in K-k/\Gamma(1)}^{*} v_z(f) + \frac{v_\infty(f)}{(\frac{q-1}{2})} + \frac{v_\varepsilon(f)}{(\frac{q+1}{2})} = \frac{2\ell}{(q^2-1)} .$$

(Σ^* <u>means that</u> ε <u>and</u> ∞ <u>are excluded</u>).

<u>Proof:</u> For convenience we suppose that f has no zeroes on the sets $\{z \in K \mid |z| = \sqrt{q}\}$ and $\{z \in K \mid |z-\alpha| = \frac{1}{\sqrt{q}}\}$ with $\alpha \in \mathbb{F}_q$.

If f does have a zero on one of those sets then we can modify the radii.

Let C be the affinoid set $\{z \in K \mid |z| \leq \sqrt{q};\ |z-\alpha| \geq \frac{1}{\sqrt{q}}$ for all $\alpha \in \mathbb{F}_q\}$ and let C^0 be $\{z \in K \mid |z| < \sqrt{q};\ |z-\alpha| > \frac{1}{\sqrt{q}}$ for all $\alpha \in \mathbb{F}_q\}$.

The non-archimedean analogue of contour-integration is:

$$\sum_{z\in C^0} v_z(f) + \operatorname{ord}_{|z|=\sqrt{q}}(f) + \sum_{\alpha\in\mathbb{F}_q} \operatorname{ord}_{|z-\alpha|=\frac{1}{\sqrt{q}}}(f) = 0.$$

(Compare Chap. II, (3.2). In this formula $v_z(f)$ denotes the order of f at z;

$\operatorname{ord}_{|z|=\sqrt{q}}(f) = n$ if $f = \sum_{i\in\mathbb{Z}} c_i (\frac{z}{\sqrt{t}})^i$ satisfies $|c_n| > |c_i|$ for all $i \neq n$ etc.).

Using $f(\frac{-1}{z}) = (\frac{-1}{z})^{\ell} f(z)$ for $|z| = \frac{1}{\sqrt{q}}$, one finds that

$\operatorname{ord}_{|z| = \frac{1}{\sqrt{q}}}(f) = -\ell + \operatorname{ord}_{|z| = \sqrt{q}}(f)$. And similarly

$\operatorname{ord}_{|z-\alpha| = \frac{1}{\sqrt{q}}}(f) = -\ell + \operatorname{ord}_{|z| = \sqrt{q}}(f)$.

If $z \in C^{o}$ does not lie in $\mathbb{F}_{q^2}$ then f has at z and its $\frac{q(q^2-1)}{2}$ conjugates the same order. If $z \in \mathbb{F}_{q^2} - \mathbb{F}_q$ then z has $q^2 - q$ conjugates in C^{o}. It follows that

$$\sum_{z \in C^{o}} v_z(f) = (q^2 - q)v_{\varepsilon}(f) + \frac{q(q^2-1)}{2} \sum^{*}_{(z \in C^{o}/\Gamma(1) \text{ and } z \neq \varepsilon)} v_z(f).$$

Now we consider f as a function on $D = U_{\sqrt{q}} \Big/ \{ \begin{pmatrix} 1 & b \\ 0 & 1 \end{pmatrix} \mid b \in \mathbb{F}_q[t] \}$.

This is again the function F.

The boudary $\{z \in K \mid |z| = \sqrt{q}\}$ maps onto the boundary $|u| = q^{q/2}$ of $D = \{u \in K \mid |u| \geq q^{q/2}\}$. The map is given by e, which has order q.

Hence $\operatorname{ord}_{|z| = \sqrt{q}}(f) = q \operatorname{ord}_{|u| = q^{q/2}}(F)$.

Further $\operatorname{ord}_{|u| = q^{q/2}}(F) = v_{\infty}(f) + \frac{q-1}{2} \sum_{a \in U_{\sqrt{q}}/\Gamma(1)} v_a(f)$.

Combining these results one finds the formula of the proposition.

(10.15) It is clear from (10.14) that the weight ℓ of a modular form must be ≥ 0. Using the matrix $\begin{pmatrix} -1 & 0 \\ 0 & -1 \end{pmatrix} \in \Gamma(1)$ one finds that ℓ must be even. Let M_s denote the vector space of modular forms of weight 2s. For any $f \in M_s$ we have the formula:

$$\sum^{*}_{z \in K-k/\Gamma(1)} v_z(f) + \frac{v_{\infty}(f)}{(\frac{q-1}{2})} + \frac{v_{\varepsilon}(f)}{(\frac{q+1}{2})} = \frac{s}{(\frac{q-1}{2})(\frac{q+1}{2})}.$$

It follows at once that $\dim M_s < \infty$. In the sequel we will calculate this dimension and give an explicit base for M_s.

We consider the following Eisenstein series

$$G_{n,m}(z) = \Sigma^* \frac{1}{(cz+d)^{n(q-1)+2m}} u\left(\frac{az+b}{cz+d}\right)^m$$

in which: $n \geq 1$ and $m \geq 0$

The summation is taken over all $c, d \in \mathbb{F}_q[t]$ with g. c. d. $(c, d) = 1$.

$u(z) = \frac{1}{e(z)}$ and $e(z) = z \prod_{\substack{f \neq o \\ f \in \mathbb{F}_q[t]}} (1 - \frac{z}{f})$ as before.

$u(z) = \sum_{f \in \mathbb{F}_q[t]} \frac{1}{z+f}$ since $\frac{1}{dz} e(z) = 1$

for every pair (c, d) one pair (a, b) is chosen such that $\begin{pmatrix} a & b \\ c & d \end{pmatrix} \in SL(2, \mathbb{F}_q[t])$. The value of $u(\frac{az+b}{cz+d})$ is independent of the choice of (a, b).

First we will show the convergence of the expression. That is, we will show that the sum is uniformly convergent on any affinoid in $K - k$ of the form $|z| \leq M$ and $d(z) \geq \delta > 0$.

(10.15.1) Lemma: $|e(z)| \geq d(z)$ for every $z \in K - k$.

Proof: We may shift z over an element of $\mathbb{F}_q[t]$. After that operation $d(z) = |z - \lambda_o|$ for some $\lambda_o \in k$ with $|\lambda_o| < 1$.

Then $|e(z)| = |z| \prod_{\substack{f \neq o \\ f \in \mathbb{F}_q[t]}} |1 - \frac{z}{f}|$. Further $|1 - \frac{z}{f}| = |\frac{z-f}{f}| =$

$$= \frac{|z - \lambda_o + \lambda_o - f|}{|f|} = \frac{\max(|z - \lambda_o|, |\lambda_o - f|)}{|f|} \geq 1.$$

Hence $|e(z)| \geq |z| \geq d(z)$.

(One can show more, namely: if $d(z) \geq 1$ then $|e(z)| \sim q^{d(z)}$).

Using the lemma we find $|u(\frac{az+b}{cz+d})| \leq \frac{1}{d(\frac{az+b}{cz+d})} = \frac{|cz+d|^2}{d(z)}$.

For any term of $G_{n,m}$ one has the estimate:

$$\left|\frac{1}{(cz+d)^{n(q-1)+2m}} \quad u\left(\frac{az+b}{cz+d}\right)^m\right| \leq \frac{1}{d(z)^m} \quad \frac{1}{|cz+d|^{n(q-1)}}.$$

From this the uniform convergence on "$|z| \leq M$ and $d(z) \geq \delta > 0$" follows.

For z with $d(z) \geq 1$ one finds $|G_{n,m}(z)| \leq \max\limits_{(c,d)} \dfrac{1}{|cz+d|^{n(q-1)}}$, and

$$|cz+d| = \begin{cases} 1 & \text{if } c = 0 \\ |c|\,|z + \frac{d}{c}| \geq |c| \geq 1 & \text{if } c \neq 0. \end{cases}$$

So the holomorphic function $G_{n,m}$ is bounded on $\{z \in K \mid d(z) \geq 1\}$.

For $\gamma = \begin{pmatrix} a' & b' \\ c' & d' \end{pmatrix} \in SL(2, \mathbb{F}_q[t])$ one easily verifies .

$$G_{n,m}\left(\frac{a'z+b'}{c'z+d'}\right) = (c'z+d')^{n(q-1)+2m} \quad G_{n,m}(z)$$

It follows that $G_{n,m}$ is a modular form or weight $n(q-1) + 2m$.

(10.15.2) <u>Proposition: If</u> $n \geq m$ <u>then</u> $G_{n,m}$ <u>is a non-zero modular form of weight</u> $n(q-1) + 2m$.

<u>Proof:</u> If suffices to show that $G_{n,m}(\sqrt{t}) \neq 0$. The sum Σ^* in the expression of $G_{n,m}(\sqrt{t})$ can be split into several parts.

$$\text{Part A} = \sum_{c=o,\, d\in\mathbb{F}_q^*} \frac{1}{d^{n(q-1)}} \left(\frac{u\left(\frac{d^{-1}\sqrt{t}}{d}\right)}{d^2}\right)^m = -\,u(\sqrt{t})^m.$$

We have used here $u(\lambda z) = \lambda^{-1} u(z)$ for any $\lambda \in \mathbb{F}_q^*$.

Further $|A| = \dfrac{1}{|\sqrt{t}|^{qm}}$.

$$\text{Part B} = \sum_{c\in\mathbb{F}_q^*,\, d\in\mathbb{F}_q} \frac{1}{(c\sqrt{t}+d)^{n(q-1)}} \left(\frac{u\left(\frac{-c^{-1}}{c\sqrt{t}+d}\right)}{(c\sqrt{t}+d)^2}\right)^m \qquad \text{equals}$$

$$(-1)^{m+1} \sum_{d\in\mathbb{F}_q} \frac{u\left(\frac{1}{\sqrt{t}+d}\right)^m}{(\sqrt{t}+d)^{n(q-1)+2m}}.$$

The formula for e yields $e(\frac{1}{\sqrt{t} + d}) = \frac{1}{\sqrt{t} + d}(1 + \delta)$ with $|\delta| < 1$ and hence $u(\frac{1}{\sqrt{t} + d}) = (\sqrt{t} + d)(1 + \delta)$ with $|\delta| < 1$.

So B is equal to

$$(-1)^{m+1} \sum_{d \in \mathbb{F}_q} \frac{1}{(\sqrt{t} + d)^{n(q-1)+m}} + \delta \text{ with } |\delta| < \frac{1}{|\sqrt{t}|^{n(q-1)+m}}.$$

And since $|\sqrt{t}| > |d|$ for all $d \in \mathbb{F}_q$ we also have

$$B = (-1)^{m+1} \sum_{d \in \mathbb{F}_q} \frac{1}{(\sqrt{t})^{n(q-1)+m}} + \delta \text{ with } |\delta| < \frac{1}{|\sqrt{t}|^{n(q-1)+m}}.$$

The sum in this expression is zero; it follows that

$$|B| < \frac{1}{|\sqrt{t}|^{n(q-1)+m}} \leq \frac{1}{|\sqrt{t}|^{qm}} = |A|.$$

$$\text{Part } C = \sum_{(c,d)}{}' \frac{1}{(c\sqrt{t} + d)^{n(q-1)}} \left(\frac{u(\frac{a\sqrt{t} + b}{c\sqrt{t} + d})}{(c\sqrt{t} + d)^2} \right)^m$$

where only pairs (c, d) are considered with gcd(c, d) = 1 and degree (c) + degree (d) $\geq$ 1.

The choice of (a, b) is made such that $ad - bc = 1$, and degree (a) < degree (c) and degree (b) < degree (d).

It follows that $\left|\frac{a\sqrt{t} + b}{c\sqrt{t} + d}\right| = \frac{\max(|a|q^{1/2}, |b|)}{\max(|c|q^{1/2}, |d|)} \leq \frac{1}{q} < 1.$

Then $\left|u(\frac{a\sqrt{t} + b}{c\sqrt{t} + d})\right| = \frac{1}{\left|e(\frac{a\sqrt{t} + b}{c\sqrt{t} + d})\right|} = \left|\frac{c\sqrt{t} + d}{a\sqrt{t} + b}\right|$.

Each term in the expression of C has absolute value

$$\frac{1}{|c\sqrt{t} + d|^{n(q-1)+m}|a\sqrt{t} + d|^m} < \frac{1}{|\sqrt{t}|^{qm}} = |A|.$$

Finally $G_{n,m}(\sqrt{t}) = A + B + C$ and $|B| < |A|$, $|C| < |A|$.

Hence $G_{n,m}(\sqrt{t}) \neq 0$.

(10.16) Theorem: Let M_s denote the space of modular forms of weight 2s.

Then

(1) $\dim M_s = \left[\dfrac{s}{(\frac{q^2-1}{4})} \right] + \delta(s)$, where

$\delta(s) = 1$ or 0 according the whether $s' = s - \left[\dfrac{s}{(\frac{q^2-1}{4})} \right] (\frac{q^2-1}{4})$

lies in the semigroup $\{\alpha \frac{q-1}{2} + \beta \frac{q+1}{2} \mid \alpha \geq 0, \beta \geq 0\}$ or not.

(2) M_s has basis $\{G_{1,o}^{\alpha}\, G_{1,1}^{\beta} \mid \alpha \frac{q-1}{2} + \beta \frac{q+1}{2} = s\}$.

(3) The graded algebra $\bigoplus_{s \geq o} M_s = K[G_{1,o}, G_{1,1}] \cong K[X, Y]$ is a polynomial ring in two variables with weights $\frac{q-1}{2}$ and $\frac{q+1}{2}$.

Proof: The formula of Prop. (10.14) yields:

$v_{\varepsilon}(G_{1,o}) = 1$, $v_{\infty}(G_{1,o}) = 0$, $v_a(G_{1,o}) = 0$ for all other $a \in K - k/_{\Gamma(1)}$.

$v_{\varepsilon}(G_{1,1}) = 0$, $v_{\infty}(_{1,1}) = 1$, $v_a(G_{1,1}) = 0$ for all other a.

Let $\Delta = (G_{1,o})^{\frac{q+1}{2}} - c(G_{1,1})^{\frac{q-1}{2}}$ where $c \in K$ is chosen such that

$\Delta(\sqrt{t}) = 0$. Then Δ has weight $\frac{q^2-1}{2}$. It follows from (10.14) that

$v_{\sqrt{t}}(\Delta) = 1$ and $v_a(\Delta) = 0$ for all other a.

Let $M_s^o = \{f \in M_s \mid f(\sqrt{t}) = 0\}$. If $\dim M_s \geq 1$ then $\dim M_s^o = \dim M_s - 1$.

Further $\Delta M_s = M^o_{s+\frac{q^2-1}{4}}$. So ist suffices to veriy (1) for $0 \leq s < \frac{q^2-1}{4}$.

If s does not have the form $\alpha \frac{q+1}{2} + \beta \frac{q-1}{2}$ for $\alpha, \beta \geq 0$ then (10.14) implies $M_s = 0$.

If $s = \alpha \frac{q+1}{2} + \beta \frac{q-1}{2}$ then since $s < \frac{q^2-1}{4}$ the α, $\beta \geq 0$ are unique. Hence $\dim M_s \leq 1$. In fact $\dim M_s = 1$ since $G_{10}^{\beta} G_{11}^{\alpha} \neq 0$ and lies in M_s. (2) and (3) are easy consequences of (1).

(10.1)) <u>Modular forms for</u> $PGL(2, \mathbb{F}_q[t])$.

A holomorphic function f is a modular form of weight 2 s for $PGL(2, \mathbb{F}_q[t])$ if

(i) $f(\frac{az-b}{cz+d}) = \{\frac{(cz+d)^2}{ad-bc}\}^s f(z)$ for all $\begin{pmatrix} a & b \\ c & d \end{pmatrix} \in GL(2, \mathbb{F}_q[t])$.

(ii) f is "holomorphic" at the parabolic point ∞.

Let $\lambda_o \in \mathbb{F}_q^* - (\mathbb{F}_q^*)^2$. Then f is a modular form of weight 2 s for $PGL(2, \mathbb{F}_q[t])$ if and only if $f \in M_s$ and $f(\lambda_o z) = \lambda_o^{-s} f(z)$.

Define $\phi : M_s \to M_s$ by $\phi(f) = \lambda_o^s f(\lambda_o z)$. Then $f \in M_s$ is modular for $PGL(2, \mathbb{F}_q[t])$ if and only if $\phi(f) = f$.

An easy calculation yields $\phi(G_{n,m}) = (-1)^n G_{n,m}$. It follows that $\phi^2 = id$. Let $M_s = M_s^+ \oplus M_s^-$ with $M_s^+ = \{f \in M_s | \phi(f) = f\}$ and $M_s^- = \{f \in M_s | \phi(f) = -f\}$. Since $\phi(G_{10}) = -G_{10}$ and $\phi(G_{11}) = -G_{11}$ we find that M_s^+ has as basis the set $\{G_{10}^{\alpha} G_{11}^{\beta} | \alpha + \beta \equiv 0(2)$ and $\alpha \frac{q-1}{2} + \beta \frac{q+1}{2} = s\}$.

An easy calculation gives further:

$$\bigoplus_{s \geq o} M_s^+ = K[G_{1,o}^2\ G_{10}G_{11},\ G_{11}^2] \cong K[x, y, z] / _{(xz - y^2)} .$$

(1018) <u>Concluding Remarks:</u>

(1) The function j in (10.11) part (1), with the choice $\omega \in \mathbb{F}_{q^2} - \mathbb{F}_q$, is up to a constant equal to $G_{1,o}^{\frac{q+1}{2}} \Big/ G_{1,1}^{\frac{q-1}{2}}$.

(2) The expansion of modular forms at ∞, as series $\sum_{n>0} c_n u^n$, seems to be difficult to make explicit. In D. Goss [13], some steps towards an explicit calculation are taken. In particular it would be interesting to know the expansion of j and Δ (Δ is the modular form introduced in the proof of (10.16)).

(3) One can easily define Hecke-operators, acting on $\bigoplus_{s\geq 0} M_s$.

A possible definition would be:

For every monic polynomial $h \in \mathbb{F}_q[t]$ of degree ≥ 1 and $f \in M_s$ the Hecke-operator T(h) is given by.
$T(h)(f) = \Sigma f(\frac{az+b}{d})$ where the sum is taken over all monic polynomials a, d with ad = h and all $b \in \mathbb{F}_q[t]$ with degree (b) < degree (d). The Hecke operators satisfy the rule $T(h_1h_2) = T(h_1)T(h_2)$ for <u>all</u> h_1, h_2. This is unlike the classical case. We ignore the explicit action of T(h) on $\oplus M_s$.

(4) The modular forms can be derived as functions on lattices of rank 2 (i.e. a discrete $\mathbb{F}_q[t]$ - submodule of K of rank 2). Unlike the classical case those lattices do not correspond to elliptic curves. Here seems to be another mystery.

References

[1] Abhyankar, S.: Local Analytic Geometry, Academic Press, New York and London, 1964.

[2] Berger, R., Kiehl, R., Kunz, E., Nastold, H.J.: Differentialrechnung in der analytischen Geometrie, Lecture Notes in Math. 38, Springer-Verlag 1967.

[3] Blanchard, A.: Les corps non-commutatifs, Paris, Presses Universitaires France, 1972, coll. Sup..

[4] Bosch, S.: Eine bemerkenswerte Eigenschaft der formellen Fasern affionider Räume, Math. Ann. 229, 25-45, 1977.

[5] Deligne, P. Mumford, D.: The irreducibility of the space of curves of a given genus, Publ. I. H. E. S. No. 36, 1969.

[6] Fulton, W.: Hurwitz schemes and irreducibility of moduli of algebraic curves, Annals of Math. Ser. II 90, 542-575, 1969.

[7] Gerritzen. L.: On Non-Archimedean Representations of Abelian Varieties, Math. Ann. 169, 323-346 (1972).

[8] Gerritzen, L.: Zur nichtarchimedischen Uniformisierung von Kurven, Math. Ann. 210, 321-337 (1974).

[9] Gerritzen, L.: Unbeschränkte Steinsche Gebiete von $\mathbb{P}_1$ und nichtarchimedische automorphe Formen, J. reine angew. Math. 297, 21-34 (1978).

[10] Gerritzen, L.: On automorphism groups of p-adic Schottky curves, J. d' Analyse ultramétrique (Y. Amice, D. Barsky, P. Robba) 1976/77.

[11] Gerritzen, L.: On the Jacobian variety of a p-adic Schottky curve, Proceedings of the Conference on p-adic Analysis, Univ. Nijmegen 1978.

[12] Gerritzen, L. Grauert, H.: Die Azyklizität der affinoiden Überdeckungen, Global Analysis Papers in Honor of K. Kodaira, Univ. of Tokyo Press 1969.

[13] Goss, D.: Modular forms for $\mathbb{F}_r[t]$, to appear.

[14] Goss, D.: π-adic Eisenstein Series for Function Fields, Compositio math. (to appear).

[15] Goss, D.: The Algebraist's Upper Half Plane, to appear.

[16] Grauert, H. Remmert, R.: Nichtarchimedische Funktionentheorie, Arbeitsgemeinschaft f. Forschung des Landes Nordrhein-Westfalen, Wiss. Abh. Bd. 33, 393-476, Opladen, Westdeutscher Verlag (1966)

[17] Griffiths, Ph. Harris, J.: Principles of Algebraic Geometry, John Wiley & Sons, New York Chichester Brisbane Toronto, 1978.

[18] Grothendieck, A.: Groupes de Monodromie en Géométrie Algébrique, SGA 7; Exposé IX, Lecture Notes in in Math. 288, Springer-Verlag 1972.

[19] Gruson, L.: Fibrés vectoriels sur un polydisque ultramétrique, Ann. Scient. Ec. Norm. Sup. 4e serie, t.1, p. 45-89, 1968.

[20] Herrlich, F.: The automorphisms of p-adic Schottky curves of genus 2, Proceedings of the Conference on p-adic Analysis. Report 7806 Math. Inst., Kath. Univ. Nijmegen 1978.

[21] Herrlich, F.: Die Ordnung der Automorphismengruppe einer p-adischen Schottkykurve, to appear in Math. Ann..

[22] Hurwitz, A.: Über algebraische Gebilde mit eineindeutigen Transformationen in sich, Math. Ann. 41, 403-442 (1893).

[23] Karrass, A., Pietrowski, A., Solitar, D.: Finite and infinite cyclic extensions of free groups. J. Austr. Math. Soc. 16, 458-466 (1973).

[24] Lang, S.: Abelian Varieties, Interscience Publ. New York, 1959.

[25] Lehner, J.: Discontinuous groups and automorphic functions. Amer. Math. Soc. Providence, R. I. 1964.

[26] Manin, Yu.: p-adic Automorphic Functions, Itogi Nauki i Tekhniki, Sovremennye Problemy Matematiki, Vol. 3, pp. 5-92, 1974.

[27] Manin, Yu, Drinfeld, V.G.: Periods of p-adic Schottky groups, J. reine angew. Math. 262/263, 239-247 (1973).

[28] Mumford, D.: An analytic construction of degenerating curves over complete local fields, Compositio Math. 24, 129-174 (1972).

[29] Myers, J. F.: p-adic Schottky groups, Thesis, Harvard Univ. 1973.

[30] Nagata, M.: Local rings, Interscience Publ. New York London, 1962.

[31] van der Put, M.: Rigid Analytic Spaces, Journées d'analyse ultramétrique, 1975, Marseille-Luminy.

[32] van der Put, M.: Schottky groups and Schottky curves, Algebraic Geometry 1978, Lecture Notes in Math. 732, 518-526.

[33] van der Put, M.: Discontinuous groups, Proceedings of the of the Conference on p-adic Analysis, Math. Inst., Kath. Univ. Nijmegen, 1978.

[34] Roquette, P.: Analytic theory of elliptic functions over local fields, Hamburger Math. Einzelschriften, Neue Folge, Heft 1, Vandenhoeck & Ruprecht in Göttingen 1970.

[35] Schottky, F.: Über eine spezielle Funktion, welche bei einer bestimmten linearen Änderung ihres Arguments unverändert bleibt, J. reine angew. Math. 101, 227-272 (1887).

[36] Selberg, A.: On discontinuous groups in higher-dimensional symmetric spaces - contributions to function theory, (International Coll. Functions Theory, Bombay 1960) p. 147-164, Tata Institute of Fundamental Research, Bombay 1960.

[37] Serre, J. P.: Arbres, Amalgames, SL_2, Astérisque, n° 46, 1977, Paris.

[38] Serre, J. P.: Corps Locaux, Hermann, Paris 1968, 2. Ed..

[39] Serre, J. P.: Cours d'arithmétique, Paris, Presses Universitaires de France, 1970, Coll. Sup..

[40] Shimura, G. L.: Introduction to the arithmetic theory of automorphic functions, Iwanami Shoten and Princeton University Press, 1971.

[41] Siegel, C. L.: Topics in Complex function Theory. Wiley & Sons, Inc, New York, London, Sydney, Toronto, 1971.

[42] Tate, J.: Rigid Analytic Spaces, Invent. math. 12, 257-289 (1971).

[43] Bosch, S.: Zur Kohomologietheorie rigid analytische Räume. Manuscripta. Math. 20, 1-12, (1977).

Subject index

Symbols:

Vol. 670: Fonctions de Plusieurs Variables Complexes III, Proceedings, 1977. Edité par F. Norguet. XII, 394 pages. 1978.

Vol. 671: R. T. Smythe and J. C. Wierman, First-Passage Perculation on the Square Lattice. VIII, 196 pages. 1978.

Vol. 672: R. L. Taylor, Stochastic Convergence of Weighted Sums of Random Elements in Linear Spaces. VII, 216 pages. 1978.

Vol. 673: Algebraic Topology, Proceedings 1977. Edited by P. Hoffman, R. Piccinini and D. Sjerve. VI, 278 pages. 1978.

Vol. 674: Z. Fiedorowicz and S. Priddy, Homology of Classical Groups Over Finite Fields and Their Associated Infinite Loop Spaces. VI, 434 pages. 1978.

Vol. 675: J. Galambos and S. Kotz, Characterizations of Probability Distributions. VIII, 169 pages. 1978.

Vol. 676: Differential Geometrical Methods in Mathematical Physics II, Proceedings, 1977. Edited by K. Bleuler, H. R. Petry and A. Reetz. VI, 626 pages. 1978.

Vol. 677: Séminaire Bourbaki, vol. 1976/77, Exposés 489–506. IV, 264 pages. 1978.

Vol. 678: D. Dacunha-Castelle, H. Heyer et B. Roynette. Ecole d'Eté de Probabilités de Saint-Flour. VII-1977. Edité par P. L. Hennequin. IX, 379 pages. 1978.

Vol. 679: Numerical Treatment of Differential Equations in Applications, Proceedings, 1977. Edited by R. Ansorge and W. Törnig. IX, 163 pages. 1978.

Vol. 680: Mathematical Control Theory, Proceedings, 1977. Edited by W. A. Coppel. IX, 257 pages. 1978.

Vol. 681: Séminaire de Théorie du Potentiel Paris, No. 3, Directeurs: M. Brelot, G. Choquet et J. Deny. Rédacteurs: F. Hirsch et G. Mokobodzki. VII, 294 pages. 1978.

Vol. 682: G. D. James, The Representation Theory of the Symmetric Groups. V, 156 pages. 1978.

Vol. 683: Variétés Analytiques Compactes, Proceedings, 1977. Edité par Y. Hervier et A. Hirschowitz. V, 248 pages. 1978.

Vol. 684: E. E. Rosinger, Distributions and Nonlinear Partial Differential Equations. XI, 146 pages. 1978.

Vol. 685: Knot Theory, Proceedings, 1977. Edited by J. C. Hausmann. VII, 311 pages. 1978.

Vol. 686: Combinatorial Mathematics, Proceedings, 1977. Edited by D. A. Holton and J. Seberry. IX, 353 pages. 1978.

Vol. 687: Algebraic Geometry, Proceedings, 1977. Edited by L. D. Olson. V, 244 pages. 1978.

Vol. 688: J. Dydak and J. Segal, Shape Theory. VI, 150 pages. 1978.

Vol. 689: Cabal Seminar 76–77, Proceedings, 1976–77. Edited by A.S. Kechris and Y. N. Moschovakis. V, 282 pages. 1978.

Vol. 690: W. J. J. Rey, Robust Statistical Methods. VI, 128 pages. 1978.

Vol. 691: G. Viennot, Algèbres de Lie Libres et Monoïdes Libres. III, 124 pages. 1978.

Vol. 692: T. Husain and S. M. Khaleelulla, Barrelledness in Topological and Ordered Vector Spaces. IX, 258 pages. 1978.

Vol. 693: Hilbert Space Operators, Proceedings, 1977. Edited by J. M. Bachar Jr. and D. W. Hadwin. VIII, 184 pages. 1978.

Vol. 694: Séminaire Pierre Lelong – Henri Skoda (Analyse) Année 1976/77. VII, 334 pages. 1978.

Vol. 695: Measure Theory Applications to Stochastic Analysis, Proceedings, 1977. Edited by G. Kallianpur and D. Kölzow. XII, 261 pages. 1978.

Vol. 696: P. J. Feinsilver, Special Functions, Probability Semigroups, and Hamiltonian Flows. VI, 112 pages. 1978.

Vol. 697: Topics in Algebra, Proceedings, 1978. Edited by M. F. Newman. XI, 229 pages. 1978.

Vol. 698: E. Grosswald, Bessel Polynomials. XIV, 182 pages. 1978.

Vol. 699: R. E. Greene and H.-H. Wu, Function Theory on Manifolds Which Possess a Pole. III, 215 pages. 1979.

Vol. 700: Module Theory, Proceedings, 1977. Edited by C. Faith and S. Wiegand. X, 239 pages. 1979.

Vol. 701: Functional Analysis Methods in Numerical Analysis, Proceedings, 1977. Edited by M. Zuhair Nashed. VII, 333 pages. 1979.

Vol. 702: Yuri N. Bibikov, Local Theory of Nonlinear Analytic Ordinary Differential Equations. IX, 147 pages. 1979.

Vol. 703: Equadiff IV, Proceedings, 1977. Edited by J. Fábera. XIX, 441 pages. 1979.

Vol. 704: Computing Methods in Applied Sciences and Engineering, 1977, I. Proceedings, 1977. Edited by R. Glowinski and J. L. Lions. VI, 391 pages. 1979.

Vol. 705: O. Forster und K. Knorr, Konstruktion verseller Familien kompakter komplexer Räume. VII, 141 Seiten. 1979.

Vol. 706: Probability Measures on Groups, Proceedings, 1978. Edited by H. Heyer. XIII, 348 pages. 1979.

Vol. 707: R. Zielke, Discontinuous Čebyšev Systems. VI, 111 pages. 1979.

Vol. 708: J. P. Jouanolou, Equations de Pfaff algébriques. V, 255 pages. 1979.

Vol. 709: Probability in Banach Spaces II. Proceedings, 1978. Edited by A. Beck. V, 205 pages. 1979.

Vol. 710: Séminaire Bourbaki vol. 1977/78, Exposés 507–524. IV, 328 pages. 1979.

Vol. 711: Asymptotic Analysis. Edited by F. Verhulst. V, 240 pages. 1979.

Vol. 712: Equations Différentielles et Systèmes de Pfaff dans le Champ Complexe. Edité par R. Gérard et J.-P. Ramis. V, 364 pages. 1979.

Vol. 713: Séminaire de Théorie du Potentiel, Paris No. 4. Edité par F. Hirsch et G. Mokobodzki. VII, 281 pages. 1979.

Vol. 714: J. Jacod, Calcul Stochastique et Problèmes de Martingales. X, 539 pages. 1979.

Vol. 715: Inder Bir S. Passi, Group Rings and Their Augmentation Ideals. VI, 137 pages. 1979.

Vol. 716: M. A. Scheunert, The Theory of Lie Superalgebras. X, 271 pages. 1979.

Vol. 717: Grosser, Bidualräume und Vervollständigungen von Banachmoduln. III, 209 pages. 1979.

Vol. 718: J. Ferrante and C. W. Rackoff, The Computational Complexity of Logical Theories. X, 243 pages. 1979.

Vol. 719: Categorial Topology, Proceedings, 1978. Edited by H. Herrlich and G. Preuß. XII, 420 pages. 1979.

Vol. 720: E. Dubinsky, The Structure of Nuclear Fréchet Spaces. V, 187 pages. 1979.

Vol. 721: Séminaire de Probabilités XIII. Proceedings, Strasbourg, 1977/78. Edité par C. Dellacherie, P. A. Meyer et M. Weil. VII, 647 pages. 1979.

Vol. 722: Topology of Low-Dimensional Manifolds. Proceedings, 1977. Edited by R. Fenn. VI, 154 pages. 1979.

Vol. 723: W. Brandal, Commutative Rings whose Finitely Generated Modules Decompose. II, 116 pages. 1979.

Vol. 724: D. Griffeath, Additive and Cancellative Interacting Particle Systems. V, 108 pages. 1979.

Vol. 725: Algèbres d'Opérateurs. Proceedings, 1978. Edité par P. de la Harpe. VII, 309 pages. 1979.

Vol. 726: Y.-C. Wong, Schwartz Spaces, Nuclear Spaces and Tensor Products. VI, 418 pages. 1979.

Vol. 727: Y. Saito, Spectral Representations for Schrödinger Operators With Long-Range Potentials. V, 149 pages. 1979.

Vol. 728: Non-Commutative Harmonic Analysis. Proceedings, 1978. Edited by J. Carmona and M. Vergne. V, 244 pages. 1979.

Vol. 729: Ergodic Theory. Proceedings, 1978. Edited by M. Denker and K. Jacobs. XII, 209 pages. 1979.

Vol. 730: Functional Differential Equations and Approximation of Fixed Points. Proceedings, 1978. Edited by H.-O. Peitgen and H.-O. Walther. XV, 503 pages. 1979.

Vol. 731: Y. Nakagami and M. Takesaki, Duality for Crossed Products of von Neumann Algebras. IX, 139 pages. 1979.

Vol. 732: Algebraic Geometry. Proceedings, 1978. Edited by K. Lønsted. IV, 658 pages. 1979.

Vol. 733: F. Bloom, Modern Differential Geometric Techniques in the Theory of Continuous Distributions of Dislocations. XII, 206 pages. 1979.

Vol. 734: Ring Theory, Waterloo, 1978. Proceedings, 1978. Edited by D. Handelman and J. Lawrence. XI, 352 pages. 1979.

Vol. 735: B. Aupetit, Propriétés Spectrales des Algèbres de Banach. XII, 192 pages. 1979.

Vol. 736: E. Behrends, M-Structure and the Banach-Stone Theorem. X, 217 pages. 1979.

Vol. 737: Volterra Equations. Proceedings 1978. Edited by S.-O. Londen and O. J. Staffans. VIII, 314 pages. 1979.

Vol. 738: P. E. Conner, Differentiable Periodic Maps. 2nd edition, IV, 181 pages. 1979.

Vol. 739: Analyse Harmonique sur les Groupes de Lie II. Proceedings, 1976–78. Edited by P. Eymard et al. VI, 646 pages. 1979.

Vol. 740: Séminaire d'Algèbre Paul Dubreil. Proceedings, 1977–78. Edited by M.-P. Malliavin. V, 456 pages. 1979.

Vol. 741: Algebraic Topology, Waterloo 1978. Proceedings. Edited by P. Hoffman and V. Snaith. XI, 655 pages. 1979.

Vol. 742: K. Clancey, Seminormal Operators. VII, 125 pages. 1979.

Vol. 743: Romanian-Finnish Seminar on Complex Analysis. Proceedings, 1976. Edited by C. Andreian Cazacu et al. XVI, 713 pages. 1979.

Vol. 744: I. Reiner and K. W. Roggenkamp, Integral Representations. VIII, 275 pages. 1979.

Vol. 745: D. K. Haley, Equational Compactness in Rings. III, 167 pages. 1979.

Vol. 746: P. Hoffman, τ-Rings and Wreath Product Representations. V, 148 pages. 1979.

Vol. 747: Complex Analysis, Joensuu 1978. Proceedings, 1978. Edited by I. Laine, O. Lehto and T. Sorvali. XV, 450 pages. 1979.

Vol. 748: Combinatorial Mathematics VI. Proceedings, 1978. Edited by A. F. Horadam and W. D. Wallis. IX, 206 pages. 1979.

Vol. 749: V. Girault and P.-A. Raviart, Finite Element Approximation of the Navier-Stokes Equations. VII, 200 pages. 1979.

Vol. 750: J. C. Jantzen, Moduln mit einem höchsten Gewicht. III, 195 Seiten. 1979.

Vol. 751: Number Theory, Carbondale 1979. Proceedings. Edited by M. B. Nathanson. V, 342 pages. 1979.

Vol. 752: M. Barr, *-Autonomous Categories. VI, 140 pages. 1979.

Vol. 753: Applications of Sheaves. Proceedings, 1977. Edited by M. Fourman, C. Mulvey and D. Scott. XIV, 779 pages. 1979.

Vol. 754: O. A. Laudal, Formal Moduli of Algebraic Structures. III, 161 pages. 1979.

Vol. 755: Global Analysis. Proceedings, 1978. Edited by M. Grmela and J. E. Marsden. VII, 377 pages. 1979.

Vol. 756: H. O. Cordes, Elliptic Pseudo-Differential Operators – An Abstract Theory. IX, 331 pages. 1979.

Vol. 757: Smoothing Techniques for Curve Estimation. Proceedings, 1979. Edited by Th. Gasser and M. Rosenblatt. V, 245 pages. 1979.

Vol. 758: C. Năstăsescu and F. Van Oystaeyen; Graded and Filtered Rings and Modules. X, 148 pages. 1979.

Vol. 759: R. L. Epstein, Degrees of Unsolvability: Structure and Theory. XIV, 216 pages. 1979.

Vol. 760: H.-O. Georgii, Canonical Gibbs Measures. VIII, 190 pages. 1979.

Vol. 761: K. Johannson, Homotopy Equivalences of 3-Manifolds with Boundaries. 2, 303 pages. 1979.

Vol. 762: D. H. Sattinger, Group Theoretic Methods in Bifurcation Theory. V, 241 pages. 1979.

Vol. 763: Algebraic Topology, Aarhus 1978. Proceedings, 1978. Edited by J. L. Dupont and H. Madsen. VI, 695 pages. 1979.

Vol. 764: B. Srinivasan, Representations of Finite Chevalley Groups. XI, 177 pages. 1979.

Vol. 765: Padé Approximation and its Applications. Proceedings, 1979. Edited by L. Wuytack. VI, 392 pages. 1979.

Vol. 766: T. tom Dieck, Transformation Groups and Representation Theory. VIII, 309 pages. 1979.

Vol. 767: M. Namba, Families of Meromorphic Functions on Compact Riemann Surfaces. XII, 284 pages. 1979.

Vol. 768: R. S. Doran and J. Wichmann, Approximate Identities and Factorization in Banach Modules. X, 305 pages. 1979.

Vol. 769: J. Flum, M. Ziegler, Topological Model Theory. X, 151 pages. 1980.

Vol. 770: Séminaire Bourbaki vol. 1978/79 Exposés 525–542. IV, 341 pages. 1980.

Vol. 771: Approximation Methods for Navier-Stokes Problems. Proceedings, 1979. Edited by R. Rautmann. XVI, 581 pages. 1980.

Vol. 772: J. P. Levine, Algebraic Structure of Knot Modules. XI, 104 pages. 1980.

Vol. 773: Numerical Analysis. Proceedings, 1979. Edited by G. A. Watson. X, 184 pages. 1980.

Vol. 774: R. Azencott, Y. Guivarc'h, R. F. Gundy, Ecole d'Eté de Probabilités de Saint-Flour VIII-1978. Edited by P. L. Hennequin. XIII, 334 pages. 1980.

Vol. 775: Geometric Methods in Mathematical Physics. Proceedings, 1979. Edited by G. Kaiser and J. E. Marsden. VII, 257 pages. 1980.

Vol. 776: B. Gross, Arithmetic on Elliptic Curves with Complex Multiplication. V, 95 pages. 1980.

Vol. 777: Séminaire sur les Singularités des Surfaces. Proceedings, 1976-1977. Edited by M. Demazure, H. Pinkham and B. Teissier. IX, 339 pages. 1980.

Vol. 778: SK1 von Schiefkörpern. Proceedings, 1976. Edited by P. Draxl and M. Kneser. II, 124 pages. 1980.

Vol. 779: Euclidean Harmonic Analysis. Proceedings, 1979. Edited by J. J. Benedetto. III, 177 pages. 1980.

Vol. 780: L. Schwartz, Semi-Martingales sur des Variétés, et Martingales Conformes sur des Variétés Analytiques Complexes. XV, 132 pages. 1980.

Vol. 781: Harmonic Analysis Iraklion 1978. Proceedings 1978. Edited by N. Petridis, S. K. Pichorides and N. Varopoulos. V, 213 pages. 1980.

Vol. 782: Bifurcation and Nonlinear Eigenvalue Problems. Proceedings, 1978. Edited by C. Bardos, J. M. Lasry and M. Schatzman. VIII, 296 pages. 1980.

Vol. 783: A. Dinghas, Wertverteilung meromorpher Funktionen in ein- und mehrfach zusammenhängenden Gebieten. Edited by R. Nevanlinna and C. Andreian Cazacu. XIII, 145 pages. 1980.

Vol. 784: Séminaire de Probabilités XIV. Proceedings, 1978/79. Edited by J. Azéma and M. Yor. VIII, 546 pages. 1980.

Vol. 785: W. M. Schmidt, Diophantine Approximation. X, 299 pages. 1980.

Vol. 786: I. J. Maddox, Infinite Matrices of Operators. V, 122 pages. 1980.